RUDIMENTS OF μ-CALCULUS

STUDIES IN LOGIC

AND

THE FOUNDATIONS OF MATHEMATICS

VOLUME 146

ELSEVIER
AMSTERDAM • LONDON • NEW YORK • OXFORD • PARIS • SHANNON • TOKYO

RUDIMENTS OF
μ-CALCULUS

A. ARNOLD
c/o LaBRI
Université Bordeaux I
351, cours de la Libération
33405 Talence, France

D. NIWIŃSKI
Institute of Informatics
University of Warsaw
ul. Banacha 2
02-097 Warsaw, Poland

2001

ELSEVIER
AMSTERDAM • LONDON • NEW YORK • OXFORD • PARIS • SHANNON • TOKYO

ELSEVIER SCIENCE B.V.
Sara Burgerhartstraat 25
P.O. Box 211, 1000 AE Amsterdam, The Netherlands

First edition 2001

Library of Congress Cataloging in Publication Data
A catalog record from the Library of Congress has been applied for.

ISBN: 0 444 50620 9
ISSN: 0049-237X

♾ The paper used in this publication meets the requirements of ANSI/NISO Z39.48-1992 (Permanence of Paper).
Transferred to digital printing 2006

To the memory of Helena Rasiowa (1917–1994)

from whom we learned the algebraic approach to logic.

Preface

The μ-calculus is basically an algebra of monotonic functions over a complete lattice, whose basic constructors are functional composition and least and greatest fixed point operators. In some sense, the μ-calculus naturally extends the concept of an inductive definition, very common in mathematical practice. An object defined by induction, typically a set, is obtained as a least fixed point of some monotonic operator, usually over a powerset lattice. For example, the set of theorems of a formal theory is the least fixed point of the consequence operator. For some concepts, however, the use of co–induction, i.e. of greatest fixed points is more appropriate. For example, a maximal dense–in–itself subspace of a topological space can be defined as the greatest fixed point of the derivative operator. In the μ-calculus, both least and greatest fixed points are considered, but, more importantly, their occurrences can be nested and mutually dependent. The alternation between the least and greatest fixed-point operators is a source of a sharp expressive power of the μ-calculus and gives rise to a proper hierarchy, just as the alternation of quantifiers is the basis of strength of first-order logic.

The μ-calculus emerged from numerous works of logicians and computer scientists and its use has become common in the works about verification of computer programs because it provides a simple way of expressing and checking their behavioural properties. In the literature, the most popular reference is perhaps the modal μ-calculus introduced by Kozen [55]; let us also mention prior works by Scott and de Bakker [88], Moschovakis [64], Emerson and Clarke [36], Park [80], and Pratt [82]. Indeed, there is a wide variety of phenomena that can be modeled in the μ-calculus, from finite automata and regular expressions, to alternating automata on infinite trees or, even more generally, infinite games with finitely presentable winning conditions.

From a point of view of computer science, a virtue of the μ-calculus is that it allows static characterization of dynamic concepts. Computation, by its nature, refers to time, and its properties are naturally expressed in terms of histories. It is possible, for instance, to model nondeterministic computations by (possibly infinite) trees, and to express computation properties using temporal operators and quantification over paths. In contrast to that approach, known as temporal logic, in a fixed-point definition of a computational property, an explicit reference to computation paths is no longer needed, since a fixed point contains information about the computation paths converging to it. In this context, the least and the greatest fixed point operators usually correspond to the references to finite (e.g., reachability) or potentially infinite (e.g., safety) periods of time, respectively. To give a simple illustration, the set of origins of all infinite paths in a graph $\langle V, E\rangle$ can be presented as the greatest fixed point of an equation $X = E^{-1}X$ (where the graph is given by a relation $E \subseteq V \times V$, $E^{-1}Y = \{x \in V : \exists y \in Y, \langle x, y\rangle \in E\}$, and the term "greatest" refers to the powerset lattice $\wp(V)$).

Another interesting and important feature of the μ-calculus is the similarity between its semantic aspects and two-player games with perfect information. Indeed, such games (more specifically, infinite parity games) turn out to be inherent to the semantics of the μ-calculus. In some sense, the converse is also true, i.e., the μ-calculus constitutes a useful framework for discussing games. In particular, one can give a μ-calculus explanation of the determinacy of certain infinite games. More precisely, in Chapter 4, we derive the Memoryless Determinacy Theorem (which says that in an infinite parity game, starting in an arbitrary position, one of the players has a winning strategy which depends only on the actual position, and not on the history of the play), from the Selection Property, where the latter is a kind of normal form result of the Boolean μ-calculus.

The μ-calculus can also be considered as a natural extension of the notion of an automaton to structures more complex than words and trees. Automata are usually well tractable algorithmically due to a straightforward rather than inductive semantics. One may even consider automata themselves as a kind of a specification language, since, for words and trees, they achieve the expressive power of the monadic second-order logic (by the fundamental results of Büchi [21] and Rabin [83]). However, automata in general lack the compositionality of logical formulas, and so do not reflect the complexity of the properties specified. The μ-calculus combines the good points of both logic and automata. It offers an elegant and well–structured mathematical notation inducing nice semantical hierarchies. On the other hand, solutions to the related algorithmic problems are already implicitly present in the structure of the fixed-point expressions, as computing a least (or, dually, greatest) fixed point is one of general paradigms of algorithms.

The aims of the book. This book presents what in our opinion constitutes the basis of the theory of the μ-calculus, considered as an algebraic system rather than a logic. We have wished to present the subject in a unified way, and in a form as general as possible. Therefore, our emphasis is on the generality of the fixed-point notation, and on the connections between μ-calculus, games, and automata, which we also explain in an algebraic way.

This book should be accessible for graduate or advanced undergraduate students both in mathematics and computer science. We have designed this book especially for researchers and students interested in logic in computer science, computer aided verification, and general aspects of automata theory. We have aimed at gathering in a single place the fundamental results of the theory, that are currently very scattered in the literature, and often hardly accessible for interested readers.

The presentation is self–contained, except for the proof of the McNaughton's Determinization Theorem (see, e.g., [97]). However, we suppose that the reader is already familiar with some basic automata theory and universal algebra. The references, credits, and suggestions for further reading are given at the end of each chapter.

We wish to stress that our presentation is far from being complete. One important omission is the issue of proof systems of the μ-calculus. For this matter, we refer the reader to the original paper by Walukiewicz [103] who established the completeness of an axiomatization proposed by Kozen [55]. Another topic not considered here is a first–order version of the μ-calculus, i.e., the fixed–point extension of first-order logic. We refer the reader to the monograph by Moschovakis [64] for general considerations, and to monographs by Ebbinghaus and Flum [29], and by Immerman [43] for fixed-point logic over finite models.

More generally, the connections between the μ-calculus and related logics of programs (see, e.g., [41]) are not considered, although they are a great motivation for the development of the μ-calculus.

Contents. In Chapter 1, we provide a background on fixed points of monotonic mappings over complete lattices. In addition to the basic definitions, we present a number of properties (typically inequalities), which can be viewed as fundamental laws (or tautologies) of the μ-calculus. In particular, the Bekič principle, and Gauss elimination principle allow us to move between scalar and vector fixed points.

A formalized language for the μ-calculus, based on the concept of fixed-point terms, is introduced in Chapter 2, together with the notion of a μ-interpretation. To stress the algebraic character of the theory, we introduce a general concept of an abstract μ-calculus, so that fixed-point terms themselves can be organized into a μ-calculus, somehow analogous to an algebra of (ordinary) terms. Then the meaning of fixed-point terms under a particular μ-interpretation is obtained by a homomorphism of μ-calculi. We will meet other examples of abstract μ-calculi later in Chapters 5 and 7, in particular the μ-calculus of automata.

Still in Chapter 2, we also show a special role played by μ-interpretations over powerset lattices, which in some sense are representative of all μ-interpretations. This leads to the Boolean μ-calculus, i.e., the calculus of monotonic mappings over the Boolean algebra $\{0, 1\}$, which we study in detail in Chapter 3. The prevalent place occupied by this calculus is somehow analogous to that of the Boolean algebra in first order logic. In particular, as we will show later in Chapters 10 and 11, most of the algorithmic problems originating from the μ-calculus (including the well–known model–checking problem) reduce to evaluation of Boolean vector fixed–point terms. Finally, we go beyond the standard Boolean μ-calculus by considering infinite powers of $\{0, 1\}$, in order to show the aforementioned Selection Property in its full generality.

The next chapter is devoted to the correspondence between the μ-calculus and games. We show that the winning sets in parity games on graphs can be defined by fixed-point terms, and conversely, the value of any fixed-point term under a powerset interpretation coincides with a winning set in some parity game induced by the term and interpretation. As we have mentioned above, the Memoryless Determinacy Theorem follows from the Selection Property of the Boolean μ-calculus.

Chapter 5 studies the connection between the μ-calculus and automata over finite and infinite words. We show that both formalisms define the same class of languages. A reader interested in this topic can read Chapter 5 without knowledge of Chapters 3 and 4.

Chapter 6 introduces the concept of a powerset algebra, the idea of which can be traced back to the work of Jónnson and Tarski [47, 48]. In this frame we present the modal μ-calculus of Kozen [55]. We also note a connection between preservation of fixed-point terms and bisimulation.

In Chapter 7, we establish an equivalence between the μ-calculus and automata which generalizes the correspondence already shown for automata on words in Chapter 5. To this end, we consider a very general concept of automaton, whose semantics can be given in an arbitrary powerset algebra and is defined in terms of parity games. Our automata generalize in particular nondeterministic and alternating automata on infinite trees. Again, we stress the algebraic character of the theory, by organizing the automata into an abstract μ-calculus. Then, the transformation from fixed-point terms to automata is presented as a homomorphism of μ-calculi which, while failing to be surjective, captures all automata up to semantic equivalence. Reading of Chapter 7 requires the knowledge of Chapters 4 and 6, but not necessarily 3 and 5.

Chapter 8 studies the problem of the hierarchy induced by the alternation of least and greatest fixed-point operators. We show that this hierarchy is indeed proper in the powerset algebra of trees.

Chapter 9 is motivated by the celebrated Rabin Complementation Lemma which, in a strengthening due to Muller and Schupp, takes form of a Simulation Theorem: An alternating automaton on trees can be simulated by a nondeterministic one. Simplification of the Rabin's original proof has constituted a longstanding challenge, pursued by many authors. (A proof based on the μ-calculus was given by Emerson and Jutla [33].) In this chapter, we explain the Simulation Theorem in the framework of the μ-calculus, as a conditional elimination of the (lattice) intersection operator.

Chapter 10 shows the decidability of the basic decision problems related to fixed-point terms: nonemptiness of an interpretation of a term in a fixed μ-interpretation, satisfiability, and semantic equivalence of fixed-point terms. As we have already remarked, most of the algorithmic problems of the μ-calculus can be reduced to the evaluation of vector Boolean fixed-point terms. This leads us to the last chapter, where we analyze various algorithms that have been proposed for this problem, for which no polynomial–time algorithm is known at the time we close this book. An interested reader can read Chapter 11 directly after Chapter 3.

André Arnold, Damian Niwiński [1]

[1] Damian Niwiński was supported by Polish KBN grants no. 8 T11C 002 11 and 8 T11C 027 16.
Both authors were supported by a "Polonium" French-Polish grant in 1998–1999.

Table of Contents

1. Complete lattices and fixed-point theorems

The μ-calculus is based on the celebrated Knaster–Tarski fixed-point theorem which states that a monotone function over a complete lattice has a least fixed point. In this chapter we review basic properties of complete lattices and show the fixed-point theorem and its variants. By duality, the Knaster–Tarski theorem also assures the existence of a greatest fixed point of a monotone function, which gives rise to definitions combining both extremal fixed points. We discuss general properties of such fixed-point definitions in Section 1.3. We then extend our concepts to vectors of functions and show the basic properties of vectorial fixed points. This concept is very useful throughout the book, although it turns out to be redundant. Indeed, we will see in Section 1.4 that the vectorial fixed points can be reduced to scalar ones by Bekič principle and Gauss elimination method.

1.1 Complete lattices

1.1.1 Least upper bounds and greatest lower bounds

Let $\langle E, \leq \rangle$ be an ordered set. That is, E is a set equipped with a (partial) order relation $\leq$, which is reflexive, antisymmetric, and transitive. As usual, we read '$x \leq y$' as "x is *less* than or equal to y"; and we say that "x is *less* than y" if $x \leq y$ and $x \neq y$.

Definition 1.1.1. Let X be a subset of E. An element $e \in E$ is an *upper bound* of X in E if $x \leq e$ holds for all x in X. Similarly, an element $e \in E$ is a *lower bound* of X if $e \leq x$ holds for all $x \in X$.

It is easy to see that if an upper bound of a set X belongs itself to X, then it is unique and is the *greatest* element of X. Similarly, if there is a lower bound of X in X, then it is unique and is the *least* element of X.

Definition 1.1.2. An element e of E is said to be the *least upper bound* of a set X if it is the least element in the set of upper bounds of X, i.e., if the following two conditions hold:

– $\forall x \in X, x \leq e$,
– if e' is such that $\forall x \in X, x \leq e'$, then $e \leq e'$.

Clearly, if a subset X of E has a least upper bound, then this least upper bound is unique. We shall denote it by $\bigvee X$.

However a least upper bound of a set needs not always exist. For instance consider the set $\{a, b, c, d\}$ ordered by $a \leq c, a \leq d, b \leq c, b \leq d$ (see Figure 1.1). The subset $\{c, d\}$ has no upper bound and the subset $\{a, b\}$ has two upper bounds, namely c and d, but none of them is a least upper bound.

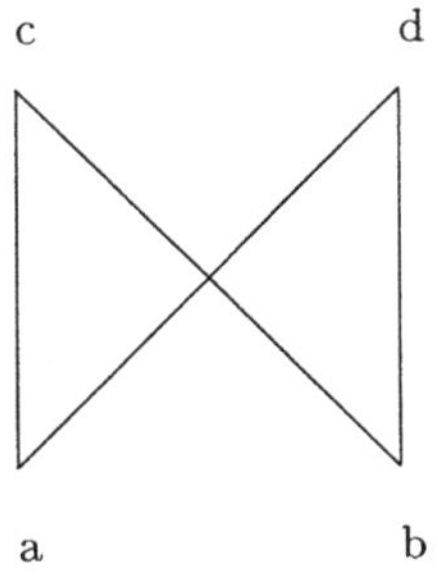

Fig. 1.1. An ordered set, not a lattice

Symmetrically, if it exists, the *greatest lower bound* of a subset X of E is the (unique) element e of E satisfying:

– $\forall x \in X, e \leq x$,
– if e' is such that $\forall x \in X, e' \leq x$, then $e' \leq e$.

The greatest lower bound of a set X will be denoted by $\bigwedge X$.

Notice that, by remark above, $\bigvee X \in X$ if and only if it is the greatest element of X; the similar claim holds for $\bigwedge X$.

1.1.2 Complete lattices

Definition 1.1.3. A *lattice* is an ordered set $\langle E, \leq \rangle$ such that for any two elements x and y of E, the set $\{x, y\}$ has a least upper bound and a greatest lower bound. In this case, we denote $\bigvee\{x, y\}$ by $x \vee y$ and $\bigwedge\{x, y\}$ by $x \wedge y$.

A *complete lattice* is an ordered set $\langle E, \leq \rangle$ such that any subset X of E has a least upper bound and a greatest lower bound.

Indeed, the above definition may be simplified by noting that the existence of $\bigvee X$ for *any* subset $X \subseteq E$ implies the existence of $\bigwedge X$ for any $X \subseteq E$, and *vice versa*.

This is due to the following remark.

Proposition 1.1.4. *Given a subset X of any ordered set E, if the set of lower bounds of X has a least upper bound then it is also the greatest lower bound of X.*

Symmetrically, if the set of upper bounds of X has a greatest lower bound, this greatest lower bound is the least upper bound of X.

Proof. Let $LB(X)$ denote the set of the lower bounds of X. Firstly, $\bigvee LB(X)$ is a lower bound of X. For, let x be any element of X. We have, $\forall y \in LB(X), y \leq x$. Therefore x is an upper bound of $LB(X)$ and $\bigvee LB(X) \leq x$.

It follows that $\bigvee LB(X)$ belongs to $LB(X)$, and thus it is the greatest element of $LB(X)$. □

Notice that, if $\langle E, \leq \rangle$ is a complete lattice then the least upper bound of the set E exists in E, and hence is the greatest element of E. We shall denote $\bigvee E$ by $\top$ and $\bigwedge E$ by $\bot$.

What are $\bigvee \emptyset$ and $\bigwedge \emptyset$? Since *any* element of E is an upper bound, and also a lower bound, of the empty subset of E, we have clearly $\bigvee \emptyset = \bot$ and $\bigwedge \emptyset = \top$.

The following result is an immediate consequence of the definition.

Proposition 1.1.5. *Let E be a complete lattice. If $X \subseteq X' \subseteq E$ then*

$$\bigwedge X' \leq \bigwedge X$$
$$\bigvee X \leq \bigvee X'$$

Observe that the inequality $\bigwedge X \leq \bigvee X$ holds for any *nonempty* set X, but not for $\emptyset$, for which $\bigwedge \emptyset = \top$ and $\bigvee \emptyset = \bot$.

Example 1.1.6. Let E be any set, and let $\mathcal{P}(E)$ be its powerset, ordered by inclusion. Then $\langle \mathcal{P}(E), \subseteq \rangle$ is a complete lattice: for any subset $\mathcal{E}$ of $\mathcal{P}(E)$,

$$\bigvee \mathcal{E} = \bigcup_{X \in \mathcal{E}} X = \{x \in E \mid \exists X \in \mathcal{E} : x \in X\}$$

and

$$\bigwedge \mathcal{E} = \bigcap_{X \in \mathcal{E}} X = \{x \in E \mid \forall X \in \mathcal{E}, x \in X\}.$$

In particular $X \vee Y = X \cup Y$ and $X \wedge Y = X \cap Y$, $\bot = \emptyset$ and $\top = E$. □

1.1.3 Some algebraic properties of lattices

It is easy to see that in a lattice, the two binary operations $\vee$ and $\wedge$ are

– idempotent: $x \vee x = x \wedge x = x$,

– commutative: $x \vee y = y \vee x$, $x \wedge y = y \wedge x$,
– associative: $x \vee (y \vee z) = (x \vee y) \vee z$, $x \wedge (y \wedge z) = (x \wedge y) \wedge z$,

and satisfy the *absorption law:*

– $x \vee (x \wedge y) = x = x \wedge (x \vee y)$.

It follows that in a lattice every nonempty finite set $\{x_1, \dots, x_n\}$ has a least upper bound $x_1 \vee \dots \vee x_n$ and a greatest lower bound $x_1 \wedge \dots \wedge x_n$.

Also, $x \leq y$ if and only if $x \vee y = y$ if and only if $x \wedge y = x$.

In a complete lattice $\langle E, \leq \rangle$, the two mappings $\bigvee$ and $\bigwedge$ from $\mathcal{P}(E)$ into E are *associative* in the following sense: for any family $(X_i)_{i \in I}$ of subsets of E,

$$\bigvee\{\bigvee X_i \mid i \in I\} = \bigvee \bigcup_{i \in I} X_i$$

and

$$\bigwedge\{\bigwedge X_i \mid i \in I\} = \bigwedge \bigcup_{i \in I} X_i.$$

1.1.4 Symmetry in lattices

If $\langle E, \leq \rangle$ is an ordered set then $\langle E, \leq_* \rangle$ with $\leq_*$ defined by

$$x \leq_* y \iff y \leq x$$

is also an ordered set. Moreover, it is easy to see that, if the least upper bound of a set $X \subseteq E$ exists in $\langle E, \leq \rangle$ then it equals to the greatest lower bound of the same set in $\langle E, \leq_* \rangle$, and *vice versa*. Hence, $\langle E, \leq \rangle$ is a (complete) lattice if and only if $\langle E, \leq_* \rangle$ is a (complete) lattice, and we have

$$\begin{aligned} x \vee y &= x \wedge_* y \\ x \wedge y &= x \vee_* y \\ \bigvee X &= \bigwedge\nolimits_* X \\ \bigwedge X &= \bigvee\nolimits_* X \end{aligned}$$

where $\bigvee_*$, $\bigwedge_*$ etc. denote the bounds with respect to the ordering $\leq_*$.

This fact explains why any property true in all lattices, has its dual, obtained by replacing $\leq$ by $\geq$ (i.e., $\leq_*$) and thus exchanging $\vee$ and $\wedge$. Therefore, in most of the proofs concerning lattices, it is enough to give a "half" of the proof, and the other "half" follows by a symmetric argument. We shall refer to this informal statement as to *the principle of symmetry*.

1.1.5 Sublattices

Let $\langle E, \leq \rangle$ be a complete lattice and let E' be a nonempty subset of E. This subset E' is naturally ordered by the restriction of $\leq$ to E', and it may be or may not be the case that $\langle E', \leq \rangle$ is a complete lattice.

But, even if $\langle E', \leq \rangle$ is a complete lattice, it is not necessary that the least upper bounds (or greatest lower bounds) of a subset X of E' are the same in E and E'.

Example 1.1.7. Let $\{1, 2, 3, 6, 12\}$ be ordered by the divisibility ordering ($n \leq m \Leftrightarrow n$ divides m). It is a complete lattice where $\bigvee\{2, 3\} = 6$. Its subset $\{1, 2, 3, 12\}$ is also a complete lattice, but in this lattice, $\bigvee\{2, 3\} = 12$. □

Definition 1.1.8. Let $\langle E, \leq \rangle$ be a complete lattice and E' be a nonempty subset of E. The ordered set $\langle E', \leq \rangle$ is said to be a *complete sublattice* of $\langle E, \leq \rangle$ if for any nonempty subset X of E, if $X \subseteq E'$ then $\bigvee X$ and $\bigwedge X$ are in E'. If it is the case $\langle E', \leq \rangle$ is obviously a complete lattice whose minimal and maximal elements are $\bigwedge E'$ and $\bigvee E'$.

Example 1.1.9. Let $\langle E, \leq \rangle$ be a complete lattice, let $d_1 < d_2$ be two elements of E, and let $E' = \{e \in E \mid d_1 \leq e \leq d_2\}$. Then $\langle E', \leq \rangle$ is a complete sublattice of $\langle E, \leq \rangle$. This is because d_1 is a lower bound and d_2 is an upper bound of any subset X of E'. □

Definition 1.1.10. Let E and E' be two complete lattices. A mapping $f : E \to E'$ is said to be *additive* if for any *nonempty* subset X of E, $f(\bigvee X) = \bigvee f(X)$. It is *multiplicative* if for any *nonempty* subset X of E, $f(\bigwedge X) = \bigwedge f(X)$. Obviously, a mapping that is either additive or multiplicative is monotonic.

Proposition 1.1.11. *Let $\langle E, \leq \rangle$ and $\langle E', \leq' \rangle$ be complete lattices and let $f : E \to E'$. If f is additive and multiplicative then $\langle f(E), \leq' \rangle$ is a complete sublattice of $\langle E', \leq' \rangle$.*

Proof. Let Y be a nonempty subset of $f(E)$ and let $X = \{e \in E \mid f(e) \in Y\}$. Clearly, $f(X) = Y$. Hence, $\bigvee Y = \bigvee f(X) = f(\bigvee X)$ and $\bigwedge Y = \bigwedge f(X) = f(\bigwedge X)$ which are both in $f(E)$. □

1.1.6 Boolean algebras

Definition 1.1.12. A lattice $\langle E, \leq \rangle$ is *distributive* if it satisfies the two conditions

- $\forall x, y, z \in E,\ x \vee (y \wedge z) = (x \vee y) \wedge (x \vee z)$,
- $\forall x, y, z \in E,\ x \wedge (y \vee z) = (x \wedge y) \vee (x \wedge z)$.

Indeed, it is easy to see that these two conditions are equivalent.

Definition 1.1.13. A lattice $\langle E, \leq \rangle$ is *complemented* if

- it has a least element $\bot$ and a greatest element $\top$, with $\bot \neq \top$,
- there exists a mapping $\gamma : E \to E$ such that
  - $\forall x \in E, x \vee \gamma(x) = \top$,
  - $\forall x \in E, x \wedge \gamma(x) = \bot$.

Example 1.1.14. For any set E, the complete lattice $\langle \mathcal{P}(E), \subseteq \rangle$ is distributive and complemented, with γ the set-theoretical complement operator. □

In a complemented lattice, it is always true that $\gamma(\bot) = \top$, since $\top = \gamma(\bot) \vee \bot = \gamma(\bot)$. For similar reasons, $\gamma(\top) = \bot$.

To get some other properties of γ, we need to assume that the lattice is distributive.

Proposition 1.1.15. *If a lattice $\langle E, \leq \rangle$ is distributive and complemented, then there exists a* unique *mapping $\gamma : E \to E$ such that*

- $\forall x \in E, x \vee \gamma(x) = \top$,
- $\forall x \in E, x \wedge \gamma(x) = \bot$.

Moreover, this mapping satisfies

- *involution property:* $\forall x \in E,\ \gamma(\gamma(x)) = x$,
- *De Morgan's law:* $\forall x, y \in E,\ \gamma(x \vee y) = \gamma(x) \wedge \gamma(y)$ *and* $\gamma(x \wedge y) = \gamma(x) \vee \gamma(y)$,
- *antimonotonicity:* $\forall x, y \in E,\ x \leq y \Leftrightarrow \gamma(y) \leq \gamma(x)$.

Proof. We first note the following property: In a distributive lattice with least element $\bot$ and greatest element $\top$, the following holds:

$$\forall x, y, z \in E,\ x \wedge y = \bot \text{ and } x \vee z = \top \Rightarrow y \leq z.$$

This is true because $z = z \vee \bot = z \vee (x \wedge y) = (z \vee x) \wedge (z \vee y) = \top \wedge (z \vee y) = z \vee y$, hence $y \leq z$.

From the above property, it follows that, in a distributive and complemented lattice, for every x, there exists a *unique* element z such that $x \wedge z = \bot$ and $x \vee z = \top$.

The involution property and De Morgan's law follow easily from this fact. To see the antimonotonicity property, recall that $x \leq y$ is equivalent to $x \vee y = y$. Hence $\gamma(x \vee y) = \gamma(y)$ and, by De Morgan's law, $\gamma(x) \wedge \gamma(y) = \gamma(y)$ which in turn is equivalent to $\gamma(y) \leq \gamma(x)$. □

Definition 1.1.16. A *Boolean algebra* is a distributive and complemented lattice. If this lattice is complete, the Boolean algebra is said to be complete.

From now on, we use the standard notation $\overline{x}$ to denote the unique complement $\gamma(x)$ of an element x in a Boolean algebra.

In a complete Boolean algebra, De Morgan's law generalizes to the following property.

Proposition 1.1.17. *For any subset X of E,*

$$\overline{\bigvee X} = \bigwedge\{\overline{x} \mid x \in X\} \text{ and } \overline{\bigwedge X} = \bigvee\{\overline{x} \mid x \in X\}.$$

1.1.7 Products of lattices

Let $\langle E_1, \leq_1\rangle, \ldots, \langle E_n, \leq_n\rangle$ be ordered sets. We shall consider the Cartesian product $E_1 \times \cdots \times E_n$ with the *product ordering:*

$$\langle x_1, \ldots, x_n\rangle \leq \langle y_1, \ldots, y_n\rangle \Leftrightarrow \forall i, x_i \leq_i y_i.$$

For $i = 1, \ldots, n$, the ith projection of $E_1 \times \cdots \times E_n$ is the mapping $\pi_i : E_1 \times \cdots \times E_n \longrightarrow E_i$ defined by $\pi_i(\langle x_1, \ldots, x_n\rangle) = x_i$.

As usual, we extend this mapping to subsets X of $E_1 \times \cdots \times E_n$ by

$$\pi_i(X) = \{\pi_i(\langle x_1, \ldots, x_n\rangle) \mid \langle x_1, \ldots, x_n\rangle \in X\}$$

Let us abbreviate $E_1 \times \cdots \times E_n$ by $\boldsymbol{E}$. Clearly, an element $\langle x_1, \ldots, x_n\rangle$ of $\boldsymbol{E}$ is an upper bound of a set $X \subseteq \boldsymbol{E}$ if and only if x_i is an upper bound of $\pi_i(X)$, for $i = 1, \ldots, n$. Hence, it is easy to see that X has a least upper bound in $\boldsymbol{E}$ if and only if each $\pi_i(X)$ has a least upper bound in E_i, and if this is the case then

$$\bigvee X = \langle\bigvee \pi_1(X), \ldots, \bigvee \pi_n(X)\rangle$$

A similar property holds for the greatest lower bound.

Therefore, if each $\langle E_i, \leq_i\rangle$ is a (complete) lattice then $E_1 \times \cdots \times E_n$ with the product ordering is also a (complete) lattice, and

$$\begin{aligned}\langle x_1, \ldots, x_n\rangle \vee \langle y_1, \ldots, y_n\rangle &= \langle x_1 \vee y_1, \ldots, x_n \vee y_n\rangle\\ \langle x_1, \ldots, x_n\rangle \wedge \langle y_1, \ldots, y_n\rangle &= \langle x_1 \wedge y_1, \ldots, x_n \wedge y_n\rangle\end{aligned}$$

It is well known [39] and can be easily proved that if an equation (on $\wedge$ and $\vee$) is true in each E_i then it is also true in the product. So, if each $\langle E_i, \leq_i\rangle$ is distributive then so is $\langle E, \leq\rangle$. If each $\langle E_i, \leq_i\rangle$ is complemented then so is $\langle E, \leq\rangle$, setting $\gamma(\langle x_1, \ldots, x_n\rangle) = \langle\gamma_1(x_1), \ldots, \gamma_n(x_n)\rangle$.

1.1.8 Functional lattices

The definitions and observations of the previous subsection can be extended without difficulty to products of infinitely many ordered sets. Recall that, for a family of ordered sets, say $\{E_i | i \in I\}$, the product is the set of functions f from I to $\bigcup_{i \in I} E_i$, such that, for all i, $f(i) \in E_i$. If I is finite and so can be identified with $\{1, \ldots, n\}$ for some n, this product can be identified with $E_1 \times \cdots \times E_n$. We shall not consider the products of infinitely many lattices except for the case when all the sets E_i are equal. In this case, the product is just the set of functions from some set into a lattice.

So, if E is an ordered set, and F is an arbitrary set, we consider the set E^F of mappings from F to E with the componentwise ordering

$$f \leq g \quad \Leftrightarrow \quad \forall x \in F, \, f(x) \leq g(x)$$

Then, if E is a lattice, so is E^F, and, for $f, g \in E^F$, the mappings $f \vee g$ and $f \wedge g$ satisfy

$$\begin{aligned} (f \vee g)(x) &= f(x) \vee g(x) \\ (f \wedge g)(x) &= f(x) \wedge g(x) \end{aligned}$$

for all $x \in F$. If E is a complete lattice, so is E^F, and, for any set $G \subseteq E^F$, $x \in F$,

$$\begin{aligned} (\bigvee G)(x) &= \bigvee \{f(x) | f \in G\} \\ (\bigwedge G)(x) &= \bigwedge \{f(x) | f \in G\} \end{aligned}$$

1.2 Fixed-point theorems

1.2.1 Monotonic and continuous mappings

Let $\langle E, \leq_E \rangle$ and $\langle F, \leq_F \rangle$ be two ordered sets.

Defi· tion 1.2.1. A mapping $f : E \to F$ is said to be *monotonic* if

$$\forall x, y \in E, \, x \leq_E y \Rightarrow f(x) \leq_F f(y).$$

Proposition 1.2.2. *Let $\langle E, \leq \rangle$ be a complete lattice and $\langle E', \leq' \rangle$ be an ordered set. Let $f : E \to E'$ be a monotonic mapping and let us assume that f has a monotonic inverse, i.e., there exists a monotonic mapping $g : f(E) \to E$ such that $\forall e' \in f(E), f(g(e')) = e'$. Then $\langle f(E), \leq' \rangle$ is a complete lattice.*

Proof. For any nonempty subset Y of $f(E)$, we have $f(g(Y)) = Y$. Let us show that $f(\bigvee g(Y))$ is the least upper bound of Y. Since $\bigvee g(Y)$ is an upper bound of $g(Y)$, $f(\bigvee g(Y))$ is an upper bound of $f(g(Y)) = Y$. Conversely, if $e' \in f(E)$ is an upper bound of Y, then $g(e')$ is an upper bound of $g(Y)$, hence, $\bigvee g(Y) \leq g(e')$ and $f(\bigvee g(Y)) \leq f(g(e')) = e'$.

The proof that $f(\bigwedge g(Y))$ is the greatest lower bound of Y is similar. □

Corollary 1.2.3. *Let $\langle E, \leq \rangle$ be a complete lattice and let $f : E \to E$ be monotonic and idempotent (i.e., $f(f(e)) = e$). Then $\langle (f(E), \leq \rangle$ is a complete lattice.*

Proof. The mapping f is an inverse of f on $f(E)$. □

Proposition 1.2.4. *Let E be a complete lattice and F be an ordered set. The set $M(F, E) \subseteq E^F$ of monotonic mappings from F to E is a complete lattice.*

Proof. Let $G = \{f_i \mid i \in I\}$ be a subset of $M(F, E)$. Since $M(F, E)$ is a subset of the complete lattice E^F, G has a least upper bound $\bigvee G$ and a greatest lower bound $\bigwedge G$ in E^F. It is enough to show that they are in $M(F, E)$, i.e., if $x, x' \in F$ with $x \leq x'$ then $(\bigvee G)(x) \leq (\bigvee G)(x')$ and $(\bigwedge G)(x) \leq (\bigwedge G)(x')$. By definition, of $\bigvee G$ and $\bigwedge G$,

$$(\bigvee G)(x) = \bigvee\{f_i(x) \mid i \in I\}, \qquad (\bigvee G)(x') = \bigvee\{f_i(x') \mid i \in I\},$$
$$(\bigwedge G)(x) = \bigwedge\{f_i(x) \mid i \in I\}, \qquad (\bigwedge G)(x') = \bigwedge\{f_i(x') \mid i \in I\}.$$

Since each f_i is monotonic, $f_i(x) \leq f_i(x')$, therefore every upper bound of $\{f_i(x') \mid i \in I\}$ is an upper bound of $\{f_i(x) \mid i \in I\}$ and every lower bound of $\{f_i(x) \mid i \in I\}$ is a lower bound of $\{f_i(x') \mid i \in I\}$. Hence, $\bigvee\{f_i(x) \mid i \in I\} \leq \bigvee\{f_i(x') \mid i \in I\}$ and $\bigwedge\{f_i(x) \mid i \in I\} \leq \bigwedge\{f_i(x') \mid i \in I\}$. □

Proposition 1.2.5. *Let E and F be two complete lattices, and let $f : E \to F$ be monotonic. For any subset X of E, $\bigvee_F f(X) \leq_F f(\bigvee_E X)$ and $f(\bigwedge_E X) \leq_F \bigwedge_F f(X)$.*

Proof. It is enough to prove that $f(\bigvee_E X)$ is an upper bound of $\bigvee_F f(X)$. (By symmetry, $f(\bigwedge_E X)$ is a lower bound of $\bigwedge_F f(X)$.

For any $y \in f(X)$ there is $x \in X$ such that $y = f(x)$. Since $x \leq_E \bigvee_E X$, $y = f(x) \leq_F f(\bigvee_E X)$. □

Interesting cases are when the inequalities in the above proposition become equalities. The following definition introduces some cases which we shall frequently meet in this book.

Definition 1.2.6. Let E and F be complete lattices, and let β be an ordinal number. A mapping $f : E \to F$ is said to be *β-sup-continuous* (resp. *β-inf-continuous*) if for any nondecreasing (resp. nonincreasing) sequence $(x_\alpha)_{\alpha<\beta}$ of elements of E,

$$\bigvee_F f(\{x_\alpha \mid \alpha < \beta\}) = f(\bigvee_E \{x_\alpha \mid \alpha < \beta\})$$

(resp. $f(\bigwedge_E \{x_\alpha \mid \alpha < \beta\}) = \bigwedge_F f(\{x_\alpha \mid \alpha < \beta\})$).

Here, we consider only the case of nondecreasing (or nonincreasing) sequences of elements of E.

If a mapping is β-sup- or inf-continuous, for $\beta \geq 2$, then it is monotonic: if $x \leq_E y$, then $y = \bigvee_E \{x, y, y, \dots\}$ (resp. $x = \bigwedge_E \{x, y, y, \dots\}$), $f(y) = \bigvee_F \{f(x), f(y), f(y), \dots\}$ (resp. $f(x) = \bigwedge_F \{f(x), f(y), f(y), \dots\}$) thus $f(x) \leq_F f(y)$.

1.2.2 Fixed points of a function

Let f be a mapping from a set E into itself.

Definition 1.2.7. A *fixed point* of f is an element x of E such that $f(x) = x$. Let $\text{Fix}(f)$ be the set of fixed points of f.

If E is an ordered set, $\text{Fix}(f)$ is an ordered subset of E, possibly empty.

The Knaster-Tarski fixed point theorem asserts that if E is a complete lattice and f is monotonic, then $\bigvee \text{Fix}(f)$ and $\bigwedge \text{Fix}(f)$ belong to $\text{Fix}(f)$ so that $\text{Fix}(f)$ is not empty and has a least and a greatest element.

Theorem 1.2.8 (Knaster–Tarski). *Let $\langle E, \leq \rangle$ be a complete lattice and f be a monotonic mapping from E into E. Then $\bigwedge \text{Fix}(f)$ and $\bigvee \text{Fix}(f)$ belong to* $\text{Fix}(f)$.

Moreover,

$$\bigwedge \text{Fix}(f) = \bigwedge \{x \in E \mid f(x) \leq x\}$$

and

$$\bigvee \text{Fix}(f) = \bigvee \{x \in E \mid x \leq f(x)\}.$$

Proof. Let $X = \{x \in E \mid f(x) \leq x\}$. We show that $\bigwedge \text{Fix}(f) = \bigwedge X \in \text{Fix}(f)$.

Let us remark that, by definition of X, $\bigwedge f(X) \leq \bigwedge X$. By monotonicity of f, $f(X) \subseteq X$, and by Proposition 1.1.5 (page 3), $\bigwedge X \leq \bigwedge f(X)$ hence $\bigwedge X = \bigwedge f(X)$. By Proposition 1.2.5, $f(\bigwedge X) \leq \bigwedge f(X)$ and thus $\bigwedge X \in X$. It follows that $f(\bigwedge X) \in X$, hence $\bigwedge X \leq f(\bigwedge X) \leq \bigwedge f(X) = \bigwedge X$. Therefore $f(\bigwedge X) = \bigwedge X$, i.e. $\bigwedge X \in \text{Fix}(f)$.

Finally, $\text{Fix}(f)$ is obviously a subset of X. Thus, $\bigwedge X \leq \bigwedge \text{Fix}(f)$. Since $\bigwedge X \in \text{Fix}(f)$ we get $\bigwedge X = \bigwedge \text{Fix}(f)$.

The claim for $\bigvee \text{Fix}(f)$ follows from the claim for $\bigwedge \text{Fix}(f)$ and the principle of symmetry. □

Example 1.2.9. Let $\langle E, \leq \rangle$ be a complete distributive lattice, let a and b be two elements of E, and let $f : E \to E$ be defined by $f(x) = a \vee (b \wedge x)$. The mapping f is monotonic. Let us show that $\mathrm{Fix}(f) = \{x \in E \mid a \leq x \leq a \vee b\}$.

Obviously, $a \leq a \vee (b \wedge x) \leq a \vee b$, therefore $x = f(x) \Rightarrow a \leq x \leq a \vee b$. Conversely, $x \leq a \vee b \Rightarrow x = x \wedge (a \vee b) = (x \wedge a) \vee (x \wedge b) \leq f(x)$ and $a \leq x \Rightarrow f(x) = a \vee (b \wedge x) \leq x \vee (b \wedge x) = x$.

Therefore the least fixed point of f is a and its greatest fixed point is $a \vee b$.

□

The Knaster–Tarski theorem induces a method by which one can prove that an element is a least (greatest) fixed point of a monotonic mapping.

Corollary 1.2.10. *Let $\langle E, \leq \rangle$ be a complete lattice and $f : E \to E$ be monotonic. Then $e \in E$ is the least fixed point of f if and only if it satisfies the following clauses :*

(i) for each $e' \in E$, $f(e') \leq e'$ implies $e \leq e'$;

(ii) $f(e) \leq e$.

Similarly, $e \in E$ is the greatest fixed point of f if and only if it satisfies the following clauses :

(i′) for each $e' \in E$, $e' \leq f(e')$ implies $e' \leq e$;

(ii′) $e \leq f(e)$.

Proof. If e is the least fixed point of f then (i) follows immediately from Theorem 1.2.8 and (ii) is trivial. Conversely, if e satisfies (i) and (ii) then e is a lower bound of the set $X = \{x \in E \mid f(x) \leq x\}$ and $e \in X$. Consequently, $e = \bigwedge X$ and hence, again by Theorem 1.2.8, e is the least fixed point of f.

The claim for a greatest fixed point follows from the principle of symmetry.

□

There is another way to characterize the least and greatest elements of Fix (f).

With each ordinal number α we associate the two elements x_α and y_α of E, inductively defined by

- $x_0 = \bot$, $y_0 = \top$,
- $x_{\alpha+1} = f(x_\alpha)$, $y_{\alpha+1} = f(y_\alpha)$,
- $x_\beta = \bigvee\{x_\alpha \mid \alpha < \beta\}$, $y_\beta = \bigwedge\{y_\alpha \mid \alpha < \beta\}$, whenever β is a limit ordinal.

Theorem 1.2.11. *Let $\langle E, \leq \rangle$ be a complete lattice and f be a monotonic mapping from E into E. Then*

- *for any ordinal β, $x_\beta = \bigvee\{f(x_\alpha) \mid \alpha < \beta\}$ and $y_\beta = \bigwedge\{f(y_\alpha) \mid \alpha < \beta\}$,*
- *the ordinal sequence (x_α) is nondecreasing and bounded from above by the least fixed point of f,*
- *the ordinal sequence (y_α) is nonincreasing and bounded from below by the greatest fixed point of f,*
- *there exists an ordinal α such that $\bigwedge \mathrm{Fix}(f) = x_\alpha$ and $\bigvee \mathrm{Fix}(f) = y_\alpha$.*

Proof. Let us prove by complete induction on β the property $P(\beta)$: $x_\beta = \bigvee\{f(x_\alpha) \mid \alpha < \beta\}$. It is trivially true for $\beta = 0$. Let us assume that $P(\alpha)$ holds for any $\alpha < \beta$, which implies that for $\alpha \leq \alpha' < \beta$ we have $x_\alpha \leq x_{\alpha'}$, and, by monotonicity of f, $x_{\alpha+1} \leq x_{\alpha'+1}$.

If $\beta = \beta' + 1$ is a successor ordinal then $\alpha < \beta \Leftrightarrow \alpha \leq \beta' < \beta$. It follows that $\bigvee\{f(x_\alpha) \mid \alpha < \beta\} = \bigvee\{x_{\alpha+1} \mid \alpha \leq \beta'\} = x_{\beta'+1} = x_\beta$.

If β is a limit ordinal then $x_\beta = \bigvee\{x_\alpha \mid \alpha < \beta\}$, and by the induction hypothesis, $x_\alpha = \bigvee\{f(x_{\alpha'}) \mid \alpha' < \alpha\}$. Hence,

$$x_\beta = \bigvee\{\bigvee\{f(x_{\alpha'}) \mid \alpha' < \alpha\} \mid \alpha < \beta\}$$

which is also equal, by associativity of least upper bounds (see page 4), to $\bigvee\{f(x_{\alpha'}) \mid \exists\alpha : \alpha' < \alpha < \beta\}$. But, since β is a limit ordinal, $\alpha' < \beta$ if and only if there exists α such that $\alpha' < \alpha < \beta\}$ (for instance, take $\alpha = \alpha' + 1$). Hence, $x_\beta = \bigvee\{f(x_{\alpha'}) \mid \alpha' < \beta\}$.

As a direct consequence of this alternative definition of x_β, the sequence (x_β) is obviously nondecreasing. Moreover, if a is the least fixed point of f, it is easy to prove by complete induction that $x_\beta \leq a$: This is true for $\beta = 0$, and if $x_\alpha \leq a$ for all $\alpha < \beta$, then $x_\beta = \bigvee\{f(x_\alpha) \mid \alpha < \beta\} \leq f(a) = a$.

Since $E' = \{x_\alpha | \alpha \text{ is an ordinal}\}$ is a well-founded ordered subset of E, there exist an ordinal β and a monotonic bijection r from E' onto the set $\{\alpha \mid \alpha < \beta\}$ Let $r' : \{\alpha \leq \beta\} \to E'$ be the monotonic mapping defined by $r'(\alpha) = x_\alpha$. If it were injective then the monotonic mapping $r \circ r' : \{\alpha \leq \beta\} \to \{\alpha < \beta\}$ would be injective too, which is clearly impossible. Thus there exist γ and γ' with $\gamma < \gamma + 1 \leq \gamma' \leq \beta$ and $r'(\gamma) = r'(\gamma')$, which implies $x_\gamma = x_{\gamma'}$, hence $x_\gamma \leq x_{\gamma+1} \leq x_{\gamma'} = x_\gamma$. It follows that $x_\gamma = x_{\gamma+1} = f(x_\gamma)$ and x_γ is a fixed point of f. Thus $x \leq x_\gamma$, But we have shown that $x_\gamma \leq x$.

The claim for the sequence (y_α) follows by the principle of symmetry. □

Example 1.2.12. We have seen in Example 1.2.9 that the least and greatest fixed point of $f(x) = a \vee (b \wedge x)$ are a and $a \vee b$. Indeed, $f(\bot) = a \vee (b \wedge \bot) = a \vee \bot = a$ and $f(\top) = a \vee (b \wedge \top) = a \vee b$. □

In some cases, the ordinal sequence which converges to an extremal fixed point need not start at $\bot$ or $\top$. Obviously it can start at $x_0 = f(\bot)$ (resp. $y_0 = f(\top)$) or at any x_α (resp. y_α). But it can start also at many other values, as shown by the next proposition.

Proposition 1.2.13. *Let b and c be the least and greatest fixed points of f.*

If $a \leq f(a)$ and $a \leq b$ then the ordinal sequence defined by $x'_0 = a$, $x'_{\alpha+1} = f(x'_\alpha)$, and $x'_\beta = \bigvee_{\alpha<\beta} x'_\alpha$ is nondecreasing and converges to b.

Similarly, if $a \geq f(a)$ *and* $a \geq c$ *then the ordinal sequence defined by* $y'_0 = a$, $y'_{\alpha+1} = f(y'_\alpha)$, *and* $y'_\beta = \bigwedge_{\alpha<\beta} y'_\alpha$ *is nonincreasing and converges to* c.

Proof. We consider the case of the least fixed point, the other case being symmetric.

Let $g(x) = a \vee f(a \vee x)$, which is a monotonic function, and let b' be its least fixed point. Since $f \leq g$, we have $b \leq b'$. Since $g(b) = a \vee f(a \vee b) = a \vee f(b) = a \vee b = b$, we have $b' \leq b$. Hence $b = b'$.

By Theorem 1.2.11, $b = b' = x_\gamma$ where $x_0 = \bot$, $x_1 = g(\bot) = a \vee f(a) = f(a) = x'_1$, and it remains to prove that for any $\alpha \geq 1$, $x_\alpha = x'_\alpha$. Indeed it is obviously enough to prove that $x_\alpha = x'_\alpha \Rightarrow x_{\alpha+1} = x'_{\alpha+1}$. Since $a \leq f(a) \leq x'_\alpha = x_\alpha$, we have $x_{\alpha+1} = a \vee f(a \vee x'_\alpha) = a \vee f(x'_\alpha) = a \vee x'_{\alpha+1} = x'_{\alpha+1}$. □

Theorem 1.2.14. *If f is ω-sup-continuous (resp. ω-inf-continuous) then* $\bigwedge \mathrm{Fix}(f) = x_\omega$ *(resp.* $\bigvee \mathrm{Fix}(f) = y_\omega$*).*

Proof. Obviously, $x_\omega \leq \bigwedge \mathrm{Fix}(f)$. On the other hand, $f(x_\omega) = f(\bigvee\{x_i \mid i \in \mathbb{N}\})$ and, because f is sup-continuous and the sequence $(x_i)_{i\in\mathbb{N}}$ is nondecreasing, we have $f(\bigvee\{x_i \mid i \in \mathbb{N}\} = \bigvee\{f(x_i) \mid i \in \mathbb{N}\} = \bigvee\{x_{i+1} \mid i \in \mathbb{N}\} = x_\omega$, and $x_\omega \in \mathrm{Fix}(f)$, thus $\bigwedge \mathrm{Fix}(f) \leq x_\omega$.

The proof is similar for y_ω. □

The following lemma shows that fixed points can be transferred from a complete lattice to another. It will be used many times in this book.

Lemma 1.2.15 (Transfer lemma). *Let E and F be two complete lattices. Let $f : E \to F$ be a mapping that is β-inf-continuous and β-sup-continuous for any ordinal β and such that* $f(\bot_E) = \bot_F$ *and* $f(\top_E) = \top_F$.

Let $g : E \to E$ and $h : F \to F$ be two monotonic mappings such that $f \circ g = h \circ f$, i,e,, the following diagram commutes.

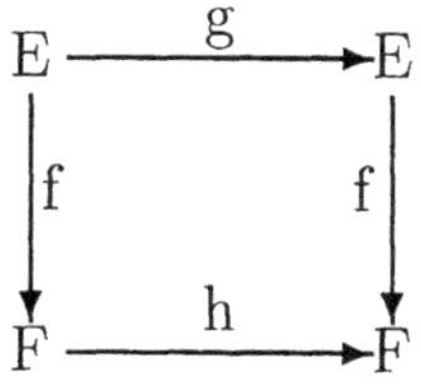

Let a and b be the least and the greatest fixed points of g. Let a' and b' be the least and the greatest fixed points of h.

Then $a' = f(a)$ *and* $b' = f(b)$.

Proof. Let $a_0 = \bot_E$, $a'_0 = \bot_F$, $a_{\alpha+1} = g(a_\alpha)$, $a'_{\alpha+1} = h(a'_\alpha)$, $a_\beta = \bigvee_{\alpha<\beta} a_\alpha$, $a'_\beta = \bigvee_{\alpha<\beta} a'_\alpha$.

We show, by induction, that for any ordinal α, $a'_\alpha = f(a_\alpha)$. For $\alpha = 0$, it is because $f(\bot_E) = \bot_F$. Next, $f(a_{\alpha+1}) = f(g(a_\alpha)) = h(f(a_\alpha)) = h(a'_\alpha) = a'_{\alpha+1}$. Finally, because of the β-sup-continuity of f, $f(a_\beta) = \bigvee_{\alpha<\beta} f(a_\alpha) = \bigvee_{\alpha<\beta} a'_\alpha = a'_\beta$.

By Theorem 1.2.11 (page 11) there exists γ such that $a = a_\gamma$ and $a' = a'_\gamma$, and the result follows. The proof for the greatest fixed point is similar. □

One can remark that a requirement that a function f be monotonic, instead of being β–continuous, is not sufficient for the above lemma to hold.

Example 1.2.16. Consider $E = F$ being the ordinal $\omega + 2 = \omega \cup \{\omega, \omega+1\}$ with the obvious ordering. Let functions g and h be defined by $g(n) = h(n) = n+1$, $g(\omega+1) = g(\omega+1) = \omega+1$, $g(\omega) = \omega+1$, and $h(\omega) = \omega$. Then the least fixed points of g and h are $\omega + 1$ and ω, respectively. Now let the function f be defined by $f : n \mapsto n$, $f : \omega \mapsto \omega + 1$, and $f : \omega + 1 \mapsto \omega + 1$. Then the diagram assumed in the lemma commutes, but f does not send the least fixed point of g onto the least fixed point of h. □

1.2.3 Nested fixed points

Throughout the book, we often refer to an *extremal fixed point* of a function to mean least or greatest, as appropriate. Before going further, we now establish a notation that we usually use for extremal fixed points.

Typically in mathematics, one defines a function in the following terms.

Let $f : E \to E$ be defined by $\forall x \in E$, $f(x) = expr$.

Here *expr* is some expression which may involve x. If it is not too complex, we can write more shortly

Let $expr : E \to E$.

assuming that the argument of the function can be identified in *expr* without ambiguity. Otherwise, lambda notation can be used

Let $f = \lambda x.expr$.

so that the role of $\forall x$ from the first definition is taken by λx. Now, a least fixed point of the mapping defined by $\lambda x.expr$ is unambiguously identified by the notation $\mu x.expr$, and similarly the greatest fixed point by $\nu x.expr$. Here, the prefixes μx and νx bind the variable x as did λx in the definition of f although with different meaning.

We will study the notation for fixed points in greater detail in Chapter 2, where we develop the μ-calculus as a formal system, somehow analogous to the λ-calculus. Here, we use the fixed-point notation introduced above just as a mathematical notation.

In particular, if $f : E \to E$ is an arbitrary mapping with no specific definition, we shall write $\mu x.f(x)$ and $\nu x.f(x)$ to denote its extremal fixed points. As usual, the symbol chosen for the bound variable does not matter, so that $\mu x.f(x) = \mu y.f(y)$ and $\nu x.f(x) = \nu y.f(y)$. In particular, if $f(x) = expr$ then $\mu x.f(x) = \mu x.expr$ and $\nu x.f(x) = \nu x.expr$.

Example 1.2.17. The extremal fixed points of the identity mapping $\lambda x.x$ over E are $\mu x.x = \bot$ and $\nu x.x = \top$. □

We start with the following observation.

Let E be a complete lattice and let $f, g : E \to E$. We shall write $f \leq g$ to mean that f is less than or equal to g in the componentwise ordering on E^E, that is, $f \leq g$ if and only if $\forall x \in E, f(x) \leq g(x)$ (see Section 1.1.8, page 8).

Proposition 1.2.18. *If $f \leq g$ then $\mu x.f(x) \leq \mu x.g(x)$ and $\nu x.f(x) \leq \nu x.g(x)$.*

Proof. Let $e = \mu x.g(x)$. Then $f(e) \leq g(e) = e$, hence, by Corollary 1.2.10, $\mu x.f(x) \leq e$.

The claim for the greatest fixed point follows by the principle of symmetry. □

Now, let E be a complete lattice and D be an ordered set. Let $M(D, E)$ be the complete lattice of all the monotonic mappings in E^D (see Proposition 1.2.4, page 9). We say that a functional mapping $h : E^D \to E^D$ *preserves monotonicity* if the image $h(g)$ of a monotonic mapping $g \in E^D$ is monotonic, in other words, the range of the restriction h' of h to $M(D, E)$ is included in $M(D, E)$.

Proposition 1.2.19. *If $h : E^D \to E^D$ is a monotonic mapping which preserves monotonicity then the least and greatest fixed points of h are in $M(D, E)$.*

Proof. If $h : E^D \to E^D$ is monotonic and preserves monotonicity, its restriction $h' : M(D, E) \to M(D, E)$ is monotonic, and since $M(D, E)$ is a complete lattice, h' has least and greatest fixed points in $M(D, E)$.

Now we prove that the least (resp. greatest) fixed points of h and h' are equal.

Let us compute the least fixed point of h and h' as explained in Theorem 1.2.11 (page 11). It is easy to prove by induction on the ordinal α that the approximations $h_\alpha, h'_\alpha \in E^D$ are always equal since $h_0 = h'_0$ is equal to the constant function $\bot$ which is monotonic. The proof is similar for the greatest fixed point by the principle of symmetry. □

In particular, let $f : E \times D \to E$ be monotonic in its first argument. Let $\hat{f} : E^D \to E^D$ be defined as follows. If g is a mapping from D to E, $\hat{f}(g)$ is the mapping from D to E such that for any $d \in D$, $\hat{f}(g)(d) = f(g(d), d)$.

Proposition 1.2.20. *If $f : E \times D \to E$, where E is a complete lattice and D is an ordered set, is monotonic in its two arguments, then the monotonic mapping $\hat{f} : E^D \to E^D$ preserves monotonicity.*

Proof. Let $g : D \to E$ be monotonic, and let $d \leq d'$, hence $g(d) \leq g(d')$. We have $\hat{f}(g)(d) = f(g(d), d) \leq f(g(d'), d') = \hat{f}(g)(d')$. □

Therefore, by Proposition 1.2.19, $\mu g.\hat{f}(g)$ and $\nu g.\hat{f}(g)$ are two monotonic mappings from D to E.

On the other hand, for any $d \in D$, the mapping $f_d : E \to E$ defined by $f_d(x) = f(x, d)$ is monotonic so that $\mu x.f_d(x)$ and $\nu x.f_d(x)$ exist. It seems natural to denote by $\mu x.f(x, y)$ (resp. $\nu x.f(x, y)$) the mapping from D to E defined by $\mu x.f(x, d) = \mu x.f_d(x)$ (resp. $\nu x.f(x, d) = \nu x.f_d(x)$).

Example 1.2.21. Let b be an element of a complete distributive lattice E and let $f : E \times E \to E$ be defined by $f(x, y) = y \vee (b \wedge x)$. For $a \in E$, $f_a(x) = a \vee (b \wedge x)$. As shown in Example 1.2.9 (page 11), $\mu x.f(x, a) = a$ and $\nu x.f(x, a) = a \vee b$, hence $\mu x.f(x, y) = y$ and $\nu x.f(x, y) = y \vee b$. □

The following proposition shows that $\mu x.f(x, y)$ and $\nu x.f(x, y)$ are also extremal fixed points of $\hat{f}$ in the sense of Proposition 1.2.20.

Proposition 1.2.22. $\mu g.\hat{f}(g) = \mu x.f(x, y)$, $\nu g.\hat{f}(g) = \nu x.f(x, y)$.

Proof. Let $g' = \mu g.\hat{f}(g)$ and $g''(y) = \mu x.f(x, y)$. We have to show that for any $d \in D$, $g'(d) = g''(d)$.

By definition of g', $g'(d) = f(g'(d), d) = f_d(g'(d))$, and by Corollary 1.2.10 (page 11), $g'(d) \geq \mu x.f_d(x) = \mu x.f(x, d) = g''(d)$.

On the other hand, for any $d \in D$, $g''(d) = f(g''(d), d) = \hat{f}(g'')(d)$, hence $g'' = \hat{f}(g'')$, which implies $g'' \geq g'$, i.e., $g''(d) \geq g'(d)$ for any d.

The proof for ν is similar by the principle of symmetry. □

Note that, with a slight abuse of notation, the equality of the previous proposition can also be presented as $\mu x.f(x, y) = \mu x(y).f(x(y), y)$, and similarly for ν, as suggested by Robert Maron.

As a consequence of the previous propositions, we get the following property.

Proposition 1.2.23. *Let E be a complete lattice and D be an ordered set. If $f : E \times D \to E$ is monotonic in its two arguments, then $\mu x.f(x, y)$ and $\nu x.f(x, y)$ are monotonic mappings from D to E.*

In particular, if D is equal to E, then $\mu x.f(x,y)$ and $\nu x.f(x,y)$ are two monotonic mappings from the complete lattice E into E, which have least and greatest fixed points. It is natural to denote the extremal fixed points $\mu x.f(x,y)$ (resp. $\nu x.f(x,y)$) by $\mu y.\mu x.f(x,y)$ and $\nu y.\mu x.f(x,y)$ (resp. $\mu y.\nu x.f(x,y)$ and $\nu y.\nu x.f(x,y)$).

Another important case is when D is a product of complete lattices. In a general setting, let $E_1, \ldots, E_n$ be complete lattices and let

$$f(x_1, \ldots, x_n) : E_1 \times \cdots \times E_{i-1} \times E_i \times E_{i+1} \times \cdots \times E_n \to E_i$$

be monotonic in all its arguments. By Proposition 1.2.23, the mappings $\mu x_i.f(x_1, \ldots, x_n)$ and $\nu x_i.f(x_1, \ldots, x_n)$ are also monotonic in all their arguments.

If E_j is equal to E_i, we can also consider the mappings

$$\mu x_j.\mu x_i.f(x_1, \ldots, x_n),\ \mu x_j.\nu x_i.f(x_1, \ldots, x_n), \text{ etc.,}$$

as well as the mappings

$$\mu x_i.\mu x_j.f(x_1, \ldots, x_n),\ \mu x_i.\nu x_j.f(x_1, \ldots, x_n), \text{ etc.}$$

If, moreover, for some k not equal to i or j, $E_k = E_i = E_j$, then the mappings $\theta x_j.\theta' x_i.f(x_1, \ldots, x_n)$ are monotonic and range over E_k. Hence, the mappings denoted by

$$\mu x_k.\theta x_j.\theta' x_i.f(x_1, \ldots, x_n) \quad \text{and} \quad \nu x_k.\theta x_j.\theta' x_i.f(x_1, \ldots, x_n)$$

are also well defined. We shall consider such mappings in Section 1.3 (page 19), usually when all the E_i's are equal.

1.2.4 Duality

Definition 1.2.24. If E is a Boolean algebra, for any mapping $f : E \to E$, we define the *dual* mapping $\widetilde{f} : E \to E$ by $\widetilde{f}(x) = \overline{f(\overline{x})}$. It is obvious that $\widetilde{f} = g$ implies $\widetilde{g} = f$. If f is monotonic, $\widetilde{f}$ is monotonic too.

The extremal fixed points $\mu x.\widetilde{f}(x)$ and $\nu x.\widetilde{f}(x)$ are related to $\mu x.f(x)$ and $\nu x.f(x)$ by the following proposition.

Proposition 1.2.25.

$$\begin{aligned} \mu x.\widetilde{f}(x) &= \overline{\nu x.f(x)} \\ \nu x.\widetilde{f}(x) &= \overline{\mu x.f(x)} \end{aligned}$$

Proof. Observe first that if a is a fixed point of f, i.e., $a = f(a)$, then $\overline{a} = \overline{f(a)} = \widetilde{f}(\overline{a})$, i.e., $\overline{a}$ is a fixed point of $\widetilde{f}$. Next, in any Boolean algebra, we have $a \leq b$ if and only if $\overline{b} \leq \overline{a}$. Then clearly a is a maximal fixed point of f if and only if $\overline{a}$ is a minimal fixed point of $\widetilde{f}$ and *vice versa.* □

Example 1.2.26. Assume that the lattice E of the Example 1.2.9 is a Boolean algebra. The dual $\widetilde{f}$ of the function $f(x) = a \vee (b \wedge x)$ is defined by $\widetilde{f}(x) = \overline{a \vee (b \wedge \overline{x})} = \overline{a} \wedge (\overline{b} \vee x)$ and one can show that Fix $(\widetilde{f}) = \{x \in E \mid \overline{a} \wedge \overline{b} \leq x \leq \overline{a}\}$. Thus, $\mu x.\widetilde{f}(x) = \overline{a} \wedge \overline{b} = \overline{a \vee b} = \overline{\nu x.f(x)}$ and $\nu x.\widetilde{f}(x) = \overline{a} = \overline{\mu x.f(x)}$ □

This result can be generalized to functional fixed points of the previous section.

Definition 1.2.27. Let $f : E_1 \times \cdots \times E_n \to E$ where E and all the E_j are Boolean algebras. Then the *dual* of f is the mapping $\widetilde{f} : E_1 \times \cdots \times E_n \to E$ defined by $f(x_1, \dots, x_n) = \overline{f(\overline{x_1}, \dots, \overline{x_n})}$.

Proposition 1.2.28. *Let $f : E_1 \times \cdots \times E_n \to E_i$ where all the E_j are Boolean algebras, and moreover, E_i is complete. If*

$$\begin{aligned} g_1(x_1, \dots, x_{i-1}, x_{i+1}, \dots, x_n) &= \mu x_i.f(x_1, \dots, x_n), \\ g_2(x_1, \dots, x_{i-1}, x_{i+1}, \dots, x_n) &= \nu x_i.f(x_1, \dots, x_n), \end{aligned}$$

then

$$\begin{aligned} \widetilde{g_1}(x_1, \dots, x_{i-1}, x_{i+1}, \dots, x_n) &= \nu x_i.\widetilde{f}(x_1, \dots, x_n), \\ \widetilde{g_2}(x_1, \dots, x_{i-1}, x_{i+1}, \dots, x_n) &= \mu x_i.\widetilde{f}(x_1, \dots, x_n). \end{aligned}$$

Proof. By definition of the dual mapping, $\widetilde{g_1}(x_1, \dots, x_{i-1}, x_{i+1}, \dots, x_n) = \overline{\mu x_i.f(\overline{x_1}, \dots, \overline{x_{i-1}}, x_i, \overline{x_{i+1}}, \dots, \overline{x_n})}$. Let $e_j \in E_j$ for $j = 1, \dots, i-1, i+1, \dots, n$, and let us compute

$$\widetilde{g_1}(e_1, \dots, e_{i-1}, e_{i+1}, \dots, e_n) = \overline{\mu x_i.f(\overline{e_1}, \dots, \overline{e_{i-1}}, x_i, \overline{e_{i+1}}, \dots, \overline{e_n})}.$$

By Theorem 1.2.8 (page 10), $\mu x_i.f(\overline{e_1}, \dots, \overline{e_{i-1}}, x_i, \overline{e_{i+1}}, \dots, \overline{e_n}) = \bigwedge\{e_i \in E_i \mid f(\overline{e_1}, \dots, \overline{e_{i-1}}, e_i, \overline{e_{i+1}}, \dots, \overline{e_n}) \leq e_i\}$. But

$$\begin{aligned} \overline{\bigwedge\{e_i \in E_i \mid f(\overline{e_1}, \dots, \overline{e_{i-1}}, e_i, \overline{e_{i+1}}, \dots, \overline{e_n}) \leq e_i\}} &= \\ \bigvee\{\overline{e_i} \in E_i \mid f(\overline{e_1}, \dots, \overline{e_{i-1}}, e_i, \overline{e_{i+1}}, \dots, \overline{e_n}) \leq e_i\} &= \\ \bigvee\{e_i \in E_i \mid f(\overline{e_1}, \dots, \overline{e_{i-1}}, \overline{e_i}, \overline{e_{i+1}}, \dots, \overline{e_n}) \leq \overline{e_i}\} &= \\ \bigvee\{e_i \in E_i \mid e_i \leq \overline{f(\overline{e_1}, \dots, \overline{e_{i-1}}, \overline{e_i}, \overline{e_{i+1}}, \dots, \overline{e_n})}\} &= \\ \bigvee\{e_i \in E_i \mid e_i \leq \widetilde{f}(e_1, \dots, e_{i-1}, e_i, e_{i+1}, \dots, e_n)\} &= \\ \nu x_i.\widetilde{f}(e_1, \dots, e_{i-1} x_i, e_{i+1}, \dots, e_n). & \end{aligned}$$

The other equality is proved similarly, by the principle of symmetry. □

Remark Note that the only properties we have used to prove the previous proposition are $a \leq b \Rightarrow \bar{b} \leq \bar{a}$ and $\bar{\bar{a}} = a$. Therefore, if each E_i is equipped with an antimonotonic involution $\gamma_i : E_i \to E_i$, we can define the dual $\tilde{f}$ of $f : E_1 \times \cdots \times E_n \to E_i$ with respect to $(\gamma_1, \dots \gamma_n)$ by $\tilde{f}(x_1, \dots, x_n) = \gamma_i(f(\gamma_1(x_1), \dots, \gamma_n(x_n))$ and the previous proposition still holds.

1.3 Some properties of fixed points

This section is devoted to some properties of fixed points that are true in any complete lattices and can therefore be considered as tautologies.

Notation. In this section and in all the book, we use θ, sometimes with subscripts and superscripts to denote either of the symbols μ or ν.

In all this section, E is a given arbitrary complete lattice, all the mappings we consider are monotonic with respect to all their arguments and the number of their arguments should be clear from the context. We also use the same symbol $\leq$ to denote the ordering on E as well as the orderings on the lattices E^{E^n}.

We shall also make large use of Corollary 1.2.10 (page 11) without explicitly mentioning where it is used.

From the very definition of $\theta x.f(x, y)$, we have:

Proposition 1.3.1. *If* $g(y) = \theta x.f(x, y)$ *then* $g(h(z)) = \theta x.f(x, h(z))$.

The following property is so useful that it deserves to be named the "golden lemma" of the μ-calculus.

Proposition 1.3.2. $\theta x.\theta y.h(x, y) = \theta x.h(x, x) = \theta y.\theta x.h(x, y)$.

Proof. We give the proof for $\theta = \mu$. The proof for $\theta = \nu$ is similar, by the principle of symmetry.

Let $h'(x) = \mu y.h(x, y) = h(x, h'(x))$, $a = \mu x.h'(x) = \mu x.\mu y.h(x, y)$ and $b = \mu x.h(x, x)$. We have $a = h'(a) = h(a, h'(a)) = h(a, a)$. Thus $b = \mu x.h(x, x) \leq a$. On the other hand, $b = h(b, b)$, hence, $b \geq \mu y.h(b, y)$ and $b \geq \mu x.\mu y.h(x, y) = a$. □

Proposition 1.3.3. *If* $a = \theta x.\theta' y.h(x, y)$ *then* $a = \theta x.h(x, a)$.

Proof. We give the proof for $\theta = \mu$. The case $\theta = \nu$ is similar, by the principle of symmetry.

Let $h'(x) = \theta' y.h(x,y)$, so that $a = \mu x.h'(x)$ and let $b = \mu x.h(x,a) = h(b,a)$. We have to prove $a = b$. Since $a = h'(a)$ and $h'(x) = h(x, h'(x))$, we have $a = h(a,a)$, thus $b \leq a$.

By monotonicity of h' we deduce $h'(b) \leq h'(a) = a$, thus $h'(b) = h(b, h'(b)) \leq h(b,a) = b$. Since $h'(b) \leq b$ we get $a = \mu x.h'(x) \leq b$. □

Proposition 1.3.4. $\mu x.\nu y.h(x,y) \leq \nu y.\mu x.h(x,y)$

Proof. Let $a = \mu x.\nu y.h(x,y)$ and let $h'(y) = \mu x.h(x,y)$. By Proposition 1.3.3, $a = \mu x.h(x,a) = h'(a)$, thus $a \leq \nu y.h'(y) = \nu y.\mu x.h(x,y)$. □

In general the inequality is strict as shown by the following example.

Example 1.3.5. Let $[0,1]$ be the segment of the real line, ordered by the natural ordering, which is a complete lattice. Let $h(x,y) = \frac{1}{2}(x+y) : [0,1]^2 \to [0,1]$ which is monotonic in both its arguments.

It is obvious that x (resp. y) is the unique z such that $z = \frac{1}{2}(x+z)$ (resp. $z = \frac{1}{2}(z+y)$), hence $\mu x.h(x,y) = y$ and $\nu y.h(x,y) = x$. It follows that $\mu x.\nu y.h(x,y) = \mu x.x = 0$ and $\nu y.\mu x.h(x,y) = \nu y.y = 1$. □

The two monotonic binary operators $\vee$ and $\wedge$ of the lattice E can also be considered as functions that are written infixed. Then, when we write $\theta x.expr$, we assume that the scope of θx extends to the end of *expr*, e.g., $\theta x.x \vee y$ should be read $\theta x.(x \vee y)$ and not $(\theta x.x) \vee y$.

Proposition 1.3.6. *Let D be any set. Let $f, g : D \to E$ and $h : D \times E \to E$ be such that $f(y) \leq \theta x.h(y,x) \leq g(y)$. Then*

$$\theta x.h(y,x) = \theta x.h(y, f(y) \vee (g(y) \wedge x)).$$

Proof. Let $h'(y,x) = h(y, f(y) \vee (g(y) \wedge x))$. Let $h_1(y) = \theta x.h(y,x)$ and $h'_1(y) = \theta x.h'(y,x)$. We have $h_1(y) = h(y, h_1(y))$. Since $f(y) \leq h_1(y) \leq g(y)$ we have $f(y) \vee (g(y) \wedge h_1(y)) = h_1(y)$, thus, (1) $h'(y, h_1(y)) = h(y, h_1(y)) = h_1(y)$.

Case $\theta = \mu$. From (1) we get $h'_1(y) \leq h_1(y)$, and thus $h'_1(y) \leq g(y)$. It follows that $h'_1(y) = g(y) \wedge h'_1(y) \leq f(y) \vee (g(y) \wedge h'_1(y))$ and, by monotonicity of $h(y,x)$, $h(y, h'_1(y)) \leq h(y, f(y) \vee (g(y) \wedge h'_1(y))) = h'(y, h'_1(y)) = h'_1(y)$. Thus $h_1(y) = \mu x.h(y,x) \leq h'_1(y)$.

Case $\theta = \nu$. From (1) we get $h_1(y) \leq h'_1(y)$, and thus $f(y) \leq h'_1(y)$. It follows that $f(y) \vee (g(y) \wedge h'_1(y)) \leq h'_1(y) \vee (g(y) \wedge h'_1(y)) = h'_1(y)$ and, by monotonicity of $h(y,x)$, $h'_1(y) = h'(y, h'_1(y)) = h(y, f(y) \vee (g(y) \wedge h'_1(y))) \leq h(y, h'_1(y))$. Thus $h'_1(y) \leq \nu x.h(y,x) = h_1(y)$. □

Proposition 1.3.7. *Let D be any set and let $f, g : D \to E$, $h : D \times E^n \to E$ be such that $f(y) \leq \theta_1 x_1. \cdots .\theta_n x_n.h(y, x_1, \ldots, x_n) \leq g(y)$. Then*
$$\theta_1 x_1. \cdots .\theta_n x_n.h(y, x_1, \ldots, x_n) = \theta_1 x_1. \cdots .\theta_n x_n.h(y, x_1, \ldots, x_{n-1}, f(y) \vee (g(y) \wedge x_n)).$$

Proof. The proof is by induction on n. For $n = 1$ the result has been proved in the previous proposition.

Let $h'(y, x_1) = \theta_2 x_2. \cdots .\theta_n x_n.h(y, x_1, \ldots, x_n)$ and $a(y) = \theta_1 x_1.h'(y, x_1) = h'(y, a(y))$. By hypothesis, $f(y) \leq a(y) \leq g(y)$.
Let $f'(y, x_1) = f(y) \wedge h'(y, x_1)$ and $g'(y, x_1) = g(y) \vee h'(y, x_1)$. Obviously, $f'(y, x_1) \leq h'(y, x_1) \leq g'(y, x_1)$, so that, by the induction hypothesis where $D \times E$ plays the role of D,

$$h'(y, x_1) = \theta_2 x_2. \cdots .\theta_n x_n.h(y, x_1, \ldots, x_{n-1}, f'(y, x_1) \vee (g'(y, x_1) \wedge x_n)).$$

Thus

$$a(y) = \theta_1 x_1.\theta_2 x_2. \cdots .\theta_n x_n.h(y, x_1, \ldots, x_{n-1}, f'(y, x_1) \vee (g'(y, x_1) \wedge x_n)).$$

By Proposition 1.3.2 (page 19)

$$a(y) = \theta x.\theta_1 x_1.\theta_2 x_2. \cdots .\theta_n x_n.h(y, x_1, \ldots, x_{n-1}, f'(y, x) \vee (g'(y, x) \wedge x_n)).$$

Thus,

$$a(y) = \theta_1 x_1.\theta_2 x_2. \cdots .\theta_n x_n.h(y, x_1, \ldots, x_{n-1}, f'(y, a(y)) \vee (g'(y, a(y)) \wedge x_n)).$$

But $f'(y, a(y)) = f(y) \wedge h'(y, a(y)) = f(y) \wedge a(y) = f(y)$ and $g'(y, a(y)) = g(y) \vee h'(y, a(y)) = g(y) \vee a(y) = g(y)$, that completes the proof. □

In case D is a set with only one element, E^D can be identified with E and $D \times E^n$ with E^n, and the previous property becomes

Corollary 1.3.8. *If $a \leq \theta_1 x_1. \cdots \theta_n x_n.h(x_1, \ldots, x_n) \leq b$, then*

$$\theta_1 x_1. \cdots .\theta_n x_n.h(x_1, \ldots, x_n) = \theta_1 x_1. \cdots .\theta_n x_n.h(x_1, \ldots, x_{n-1}, a \vee (b \wedge x_n)).$$

Another consequence is a generalization of Proposition 1.3.3 (page 19).

Proposition 1.3.9. *Let $a = \theta_1 x_1. \cdots .\theta_n x_n.f(x_1, \ldots, x_n)$. Then*

$$a = \theta_1 x_1. \cdots .\theta_{i-1} x_{i-1}.\theta_{i+1} x_{i+1}. \cdots .\theta_n x_n.f(x_1, \ldots, x_{i-1}, a, x_{i+1}, \ldots, x_n).$$

Proof. We apply the previous corollary with $y = x_i$, $h(x_1, \dots, x_{i-1}, y) = \theta_{i+1}x_{i+1}.\cdots.\theta_n x_n.f(x_1, \dots, x_{i-1}, y, x_{i+1}, \dots, x_n)$ and $b = a$, and we get

$$a = \theta_1 x_1.\cdots.\theta_n x_n.\theta y.h(x_1, \dots, x_n, a \vee (a \wedge y)).$$

Thus the result follows by the remark that

$$\theta y.h(x_1, \dots, x_n, a \vee (a \wedge y)) = h(x_1, \dots, x_n, a).$$

This is because if

$$g(x_1, \dots, x_n) = \theta y.h(x_1, \dots, x_n, a \vee (a \wedge y))$$

then $g(x_1, \dots, x_n) = h(x_1, \dots, x_n, a \vee (a \wedge g(x_1, \dots, x_n)))$, but $a \vee (a \wedge g(x_1, \dots, x_n)) = a$. □

Proposition 1.3.10. *Let* $f(z, z_1, \dots, z_n) : E^{n+1} \to E$. *Then*
$\mu x.\theta_1 z_1.\cdots.\theta_n z_n.\theta y.f(x \wedge y, z_1, \dots, z_n) = \mu x.\theta_1 z_1.\cdots.\theta_n z_n.f(x, z_1, \dots, z_n)$,
$\nu x.\theta_1 z_1.\cdots.\theta_n z_n.\theta y.f(x \wedge y, z_1, \dots, z_n) = \theta_1 z_1.\cdots.\theta_n z_n.\theta y.f(y, z_1, \dots, z_n)$.

Proof. (First equality) Let

$$\begin{aligned} g(x) &= \theta_1 z_1.\cdots.\theta_n z_n.\theta y.f(x \wedge y, z_1, \dots, z_n), \\ h(x) &= \theta_1 z_1.\cdots.\theta_n z_n.f(x, z_1, \dots, z_n). \end{aligned}$$

Let $a = \mu x.g(x)$ and $b = \mu x.h(x)$; we have to prove $a = b$. By Proposition 1.3.9 (page 21), we have, for any $e \in E$,

$$g(e) = \theta_1 z_1.\cdots.\theta_n z_n.f(e \wedge g(e), z_1, \dots, z_n) = h(e \wedge g(e)).$$

From this equality we get (1) $a = g(a) = h(a \wedge g(a)) = h(a)$ and (2) $g(b) = h(b \wedge g(b)) \leq h(b) = b$ From (1), it follows that $b = \mu x.h(x) \leq a$, and from (2) it follows $a = \mu x.g(x) \leq b$.

(Second equality) Let

$$\begin{aligned} a &= \nu x.\theta_1 z_1.\cdots.\theta_n z_n.\theta y.f(x \wedge y, z_1, \dots, z_n), \\ b &= \theta_1 z_1.\cdots.\theta_n z_n.\theta y.f(y, z_1, \dots, z_n). \end{aligned}$$

Obviously,

$$\begin{aligned} a &= \theta_1 z_1.\cdots.\theta_n z_n.\theta y.f(a \wedge y, z_1, \dots, z_n) \\ &\leq \theta_1 z_1.\cdots.\theta_n z_n.\theta y.f(y, z_1, \dots, z_n) = b. \end{aligned}$$

By Proposition 1.3.2 (page 19) and by Proposition 1.3.9 (page 21),

$$\begin{aligned} b &= \theta_1 z_1 \cdots .\theta_n z_n.\theta y.f(y, z_1, \ldots, z_n) \\ &= \theta_1 z_1. \cdots .\theta_n z_n.\theta y.\theta y'.f(y' \wedge y, z_1, \ldots, z_n) \\ &= \theta_1 z_1. \cdots .\theta_n z_n.\theta y.f(b \wedge y, z_1, \ldots, z_n) \end{aligned}$$

from which it follows that $b \leq \nu x.\theta_1 z_1. \cdots \theta_n z_n.\theta y.f(x \wedge y, z_1, \ldots, z_n) = a$. □

Corollary 1.3.11.
$\mu x.\theta_1 z_1. \cdots .\theta_n z_n.\theta y.f(x, z_1, \ldots, z_n, y)$
$= \mu x.\theta_1 z_1. \cdots .\theta_n z_n.\theta y.f(x \wedge y, z_1, \ldots, z_n, y)$,

$\nu x.\theta_1 z_1. \cdots .\theta_n z_n.\theta y.f(x, z_1, \ldots, z_n, y)$
$= \nu x.\theta_1 z_1. \cdots .\theta_n z_n.\theta y.f(x, z_1, \ldots, z_n, x \wedge y)$.

Proof. By the previous proposition,
$\mu x.\theta_1 z_1. \cdots .\theta_n z_n.\theta y.f(x, z_1, \ldots, z_n, y) =$
$\mu x.\theta_1 z_1. \cdots .\theta_n z_n.\theta y.\theta y'.f(x \wedge y', z_1, \ldots, z_n, y)$ and
$\nu x.\theta_1 z_1. \cdots .\theta_n z_n.\theta y.f(x, z_1, \ldots, z_n, y) =$
$\nu x'.\nu x.\theta_1 z_1. \cdots .\theta_n z_n.\theta y.f(x, z_1, \ldots, z_n, x' \wedge y)$.
By Proposition 1.3.2 (page 19),
$\mu x.\theta_1 z_1. \cdots .\theta_n z_n.\theta y.\theta y'.f(x \wedge y', z_1, \ldots, z_n, y) =$
$\mu x.\theta_1 z_1. \cdots .\theta_n z_n.\theta y.f(x \wedge y, z_1, \ldots, z_n, y)$ and
$\nu x'.\nu x.\theta_1 z_1. \cdots .\theta_n z_n.\theta y.f(x, z_1, \ldots, z_n, x' \wedge y) =$
$\nu x.\theta_1 z_1. \cdots .\theta_n z_n.\theta y.f(x, z_1, \ldots, z_n, x \wedge y)$. □

Proposition 1.3.12. *For $h : E \to E'$ and $h' : E' \to E$,*

$$\theta x.h(h'(x)) = h(\theta x.h'(h(x))).$$

Proof. We give the proof for $\theta = \mu$, the case $\theta = \nu$ is symmetrical.

Let $a = \mu x.h(h'(x))$ and $b = \mu x.h'(h(x))$. We have to prove that $a = h(b)$.

Since $b = h'(h(b))$, we have $h(b) = h(h'(h(b)))$. Therefore, $h(b)$ is a fixed point of $h(h'(x))$ and $a \leq h(b)$. Since $a = h(h'(a))$, we have $h'(a) = h'(h(h'(a)))$ thus $b \leq h'(a)$ and $h(b) \leq h(h'(a)) = a$. □

Example 1.3.13. Let $h : [0,1] \to [0,1]$ be defined by $h(x) = \frac{x+1}{3}$ and $h' : [0,1] \to [0,1]$ be defined by $h'(x) = \frac{x+5}{7}$.

Then $h(h'(x)) = \frac{x+12}{21}$ and $h'(h(x)) = \frac{x+16}{21}$. It follows that $\mu x.h(h'(x)) = \frac{12}{20} = \frac{3}{5}$ and that $\mu x.h'(h(x)) = \frac{16}{20} = \frac{4}{5}$. Indeed, $h(\frac{4}{5}) = (\frac{4}{5} + 1)/3 = \frac{9}{15} = \frac{3}{5}$.

Corollary 1.3.14. *For $h : E \to E$,*

$$\theta x.h(x) = \theta x.h(h(x)).$$

Proof. Let $a = \mu x.h(x)$ and $a' = \mu x.h(h(x))$. We have $a = h(a) = h(h(a))$, hence, $a' \leq a$. By the previous proposition, $h(a') = a'$, hence, $a \leq a'$.

The proof for ν is similar, by the principle of symmetry. □

1.4 Fixed points on product lattices

An extremal fixed point of a function f can be viewed as an extremal solution of an equation $x = f(x)$. However, we are usually interested in solving systems of equations rather than single equations. This amounts to computing fixed points in product lattices.

We have remarked already in Section 1.1.7 that a product of complete lattices is itself a complete lattice. In general, it is common to identify a function f of an argument x ranging over a product, say $f : E_1 \times \cdots \times E_m \to E$, with a function say f' of m arguments $x_1, \ldots, x_m$, where x_i ranges over E_i respectively, such that

$$f(x) \quad = \quad f'(\pi_1(x), \ldots, \pi_m(x))$$

or, equivalently,

$$f'(x_1, \ldots, x_m) \quad = \quad f(\langle x_1, \ldots, x_m \rangle)$$

(Here π_i is the projection of $E_1 \times \cdots \times E_m$ onto E_i.) Now, if E and $E_1, \ldots, E_m$ are ordered then the monotonicity of f with respect to the product ordering on $E_1 \times \cdots \times E_m$ amounts to the monotonicity of f' with respect to all arguments, i.e. to the condition $a_1 \leq_1 b_1 \wedge \ldots \wedge a_m \leq_m b_m \Rightarrow f'(a_1, \ldots, a_m) \leq f'(b_1, \ldots, b_m)$. In the sequel we will not make any notational distinction between f and f'.

On the other hand, if the range E of a function f is a product, say $E = E'_1 \times \cdots \times E'_k$, f can be identified with a tuple of functions $\langle f_1, \ldots, f_k \rangle$, where f_i is a function of range E'_i given by $f_i = \pi_i \circ f$ (here π_i stands for the projection of E on E'_i). Then monotonicity of f is equivalent to monotonicity of $f_1, \ldots, f_k$.

In this section we examine the extremal fixed points of a monotonic mapping of a form $f : E_1 \times \cdots \times E_n \times E \to E_1 \times \cdots \times E_n$ and look at how these fixed points are related to those of the components $f_i : E_1 \times \cdots \times E_n \times E \to E_i$, $i = 1, \ldots, n$. By the above, we can present f by $f(\boldsymbol{x}, x)$, where $\boldsymbol{x}$ ranges over $E_1 \times \cdots \times E_n$ and x over E, but we can also, equivalently, present it by $\langle f_1(x_1, \ldots, x_n, x), \ldots, f_n(x_1, \ldots, x_n, x) \rangle$, where x_i ranges over E_i, for $i = 1, \ldots, n$. We will therefore denote an extremal fixed point $\theta \boldsymbol{x}.f(\boldsymbol{x}, x)$ of f, which is a mapping from $E \to E_1 \times \cdots \times E_n$ by

$$\theta \langle x_1, \ldots, x_n \rangle . \langle f_1(x_1, \ldots, x_n, x), \ldots, f_n(x_1, \ldots, x_n, x) \rangle$$

or, more pictorially, by

$$\theta \begin{bmatrix} x_1 \\ \vdots \\ x_n \end{bmatrix} . \begin{bmatrix} f_1(x_1, \ldots, x_n, x) \\ \vdots \\ f_n(x_1, \ldots, x_n, x) \end{bmatrix} .$$

The latter notation is motivated by the fact that the extremal fixed point in consideration amounts to an extremal solution of a system of equations

$$\begin{aligned} x_1 &= f_1(x_1, \ldots, x_n, x) \\ &\ldots \\ x_n &= f_n(x_1, \ldots, x_n, x) \end{aligned}$$

Note that the vector $\langle x_1, \ldots, x_n \rangle$ (or its vertical counterpart) is a notation which formally stands for a single argument ranging over $E_1 \times \cdots \times E_n$, and the above fixed-point notation is equivalent to

$$\theta \boldsymbol{x} . \langle f_1(\pi_1(\boldsymbol{x}), \ldots, \pi_n(\boldsymbol{x})), \ldots, f_n(\pi_1(\boldsymbol{x}), \ldots, \pi_n(\boldsymbol{x})) \rangle .$$

More generally, if $\boldsymbol{f}$ is a monotonic mapping from $(E_1 \times \cdots \times E_n)^k \times E$ into $E_1 \times \cdots \times E_n$ and if $\boldsymbol{x}_1, \ldots \boldsymbol{x}_k$ are variable ranging over $E_1 \times \cdots \times E_n$, we can write $\theta_1 \boldsymbol{x}_1 . \cdots \theta . \boldsymbol{x}_k . \boldsymbol{f}(\boldsymbol{x}_1, \ldots, \boldsymbol{x}_k, x)$ in the form

$$\theta_1 \begin{bmatrix} x_1^{(1)} \\ \vdots \\ x_n^{(1)} \end{bmatrix} . \cdots . \theta_k \begin{bmatrix} x_1^{(k)} \\ \vdots \\ x_n^{(k)} \end{bmatrix} . \begin{bmatrix} f_1(x_1^{(1)}, \ldots, x_n^{(1)}, \ldots, x_1^{(k)}, \ldots, x_n^{(k)}, x) \\ \vdots \\ f_n(x_1^{(1)}, \ldots, x_n^{(1)}, \ldots, x_1^{(k)}, \ldots, x_n^{(k)}, x) \end{bmatrix}$$

where the $x_i^{(j)}$ are pairwise distinct variables.

1.4.1 The replacement lemma

In case of fixed points on product lattices, we can generalize Proposition 1.3.9 (page 21) as follows.

Let $\boldsymbol{g}(x)$ be equal to

$$\theta_1 \begin{bmatrix} x_1^{(1)} \\ \vdots \\ x_n^{(1)} \end{bmatrix} . \cdots . \theta_k \begin{bmatrix} x_1^{(k)} \\ \vdots \\ x_n^{(k)} \end{bmatrix} . \begin{bmatrix} f_1(x_1^{(1)}, \ldots, x_n^{(1)}, \ldots, x_1^{(k)}, \ldots, x_n^{(k)}, x) \\ \vdots \\ f_n(x_1^{(1)}, \ldots, x_n^{(1)}, \ldots, x_1^{(k)}, \ldots, x_n^{(k)}, x) \end{bmatrix} ,$$

which we can also write $\langle g_1(x), \ldots, g_n(x) \rangle$.

Let I be a subset of $\{1,\dots,n\} \times \{1,\dots,n\} \times \{1,\dots,k\}$.

For $i \in \{1,\dots,n\}$, let $f_i' : (E_1 \times \cdots \times E_n)^k \times E$ into E_i be defined by $f_i'(e_1^{(1)},\dots,e_n^{(1)},\dots,e_1^{(k)},\dots,e_n^{(k)},e) = f_i(d_1^{(1)},\dots,d_n^{(1)},\dots,d_1^{(k)},\dots,d_n^{(k)},e)$ where $d_p^{(j)}$ is equal to $e_p^{(j)}$ if $(i,p,j) \in I$ and to $g_p(e)$ otherwise.

It follows that

$$\theta_1 \begin{bmatrix} x_1^{(1)} \\ \vdots \\ x_n^{(1)} \end{bmatrix} \cdots\cdots .\theta_k \begin{bmatrix} x_1^{(k)} \\ \vdots \\ x_n^{(k)} \end{bmatrix} . \begin{bmatrix} f_1'(x_1^{(1)},\dots,x_n^{(1)},\dots,x_1^{(k)},\dots,x_n^{(k)},x) \\ \vdots \\ f_n'(x_1^{(1)},\dots,x_n^{(1)},\dots,x_1^{(k)},\dots,x_n^{(k)},x) \end{bmatrix}$$

is equal to

$$\theta_1 \begin{bmatrix} x_1^{(1)} \\ \vdots \\ x_n^{(1)} \end{bmatrix} \cdots\cdots .\theta_k \begin{bmatrix} x_1^{(k)} \\ \vdots \\ x_n^{(k)} \end{bmatrix} . \begin{bmatrix} f_1(y_1^{(1)},\dots,y_n^{(1)},\dots,y_1^{(k)},\dots,y_n^{(k)},x) \\ \vdots \\ f_n(y_1^{(1)},\dots,y_n^{(1)},\dots,y_1^{(k)},\dots,y_n^{(k)},x) \end{bmatrix}$$

where $y_p^{(j)}$ is equal to $x_p^{(j)}$ if $(i,p,j) \in I$ and to $g_p(x)$ otherwise.

In other words, we substitute $g_p(x)$ for some occurrences of variables indexed by p. The following proposition shows that this transformation does not alter the result.

Proposition 1.4.1. $\boldsymbol{g}(x) =$

$$\theta_1 \begin{bmatrix} x_1^{(1)} \\ \vdots \\ x_n^{(1)} \end{bmatrix} \cdots\cdots .\theta_k \begin{bmatrix} x_1^{(k)} \\ \vdots \\ x_n^{(k)} \end{bmatrix} . \begin{bmatrix} f_1'(x_1^{(1)},\dots,x_n^{(1)},\dots,x_1^{(k)},\dots,x_n^{(k)},x) \\ \vdots \\ f_n'(x_1^{(1)},\dots,x_n^{(1)},\dots,x_1^{(k)},\dots,x_n^{(k)},x) \end{bmatrix}.$$

Proof. For $j = 1,\dots,k$, let $\boldsymbol{y}_j = \langle y_1^{(j)},\dots,y_n^{(j)}\rangle$.

Let $\boldsymbol{h}(\boldsymbol{x}_1,\boldsymbol{y}_1,\dots,\boldsymbol{x}_k,\boldsymbol{y}_k,x)$ be the vector

$$\begin{bmatrix} f_1(z_{1,1}^{(1)},\dots,z_{n,1}^{(1)},\dots,z_{1,1}^{(k)},\dots,z_{n,1}^{(k)},x) \\ \vdots \\ f_n(z_{1,n}^{(1)},\dots,z_{n,n}^{(1)},\dots,z_{1,n}^{(k)},\dots,z_{n,n}^{(k)},x) \end{bmatrix}$$

where $z_{i,p}^{(j)}$ is equal to $x_p^{(j)}$ if $(i,p,j) \in I$ and to $y_p^{(j)}$ otherwise.

It follows that $\boldsymbol{h}(\boldsymbol{x}_1,\boldsymbol{g}(x)\dots,\boldsymbol{x}_k,\boldsymbol{g}(x),x)$ is equal to

$$\begin{bmatrix} f_1'(x_1^{(1)},\dots,x_n^{(1)},\dots,x_1^{(k)},\dots,x_n^{(k)},x) \\ \vdots \\ f_n'(x_1^{(1)},\dots,x_n^{(1)},\dots,x_1^{(k)},\dots,x_n^{(k)},x) \end{bmatrix}$$

and that $\boldsymbol{h}(\boldsymbol{x}_1, \boldsymbol{x}_1, \ldots, \boldsymbol{x}_k, \boldsymbol{x}_k, x)$ is equal to

$$\begin{bmatrix} f_1(x_1^{(1)}, \ldots, x_n^{(1)}, \ldots, x_1^{(k)}, \ldots, x_n^{(k)}, x) \\ \vdots \\ f_n(x_1^{(1)}, \ldots, x_n^{(1)}, \ldots, x_1^{(k)}, \ldots, x_n^{(k)}, x) \end{bmatrix}.$$

Therefore $\boldsymbol{g}(x) = \theta_1 \boldsymbol{x}_1. \cdots .\theta_k \boldsymbol{x}_k.\boldsymbol{h}(\boldsymbol{x}_1, \boldsymbol{x}_1, \ldots, \boldsymbol{x}_k, \boldsymbol{x}_k, x)$.
By Proposition 1.3.2 (page 19), this is equal to

$$\theta_1 \boldsymbol{x}_1.\theta_1 \boldsymbol{y}_1. \cdots .\theta_k \boldsymbol{x}_k.\theta_k \boldsymbol{y}_k.\boldsymbol{h}(\boldsymbol{x}_1, \boldsymbol{y}_1, \ldots, \boldsymbol{x}_k, \boldsymbol{y}_k, x),$$

and by Proposition 1.3.9 (page 21), to

$$\theta_1 \boldsymbol{x}_1. \cdots .\theta_k \boldsymbol{x}_k.\boldsymbol{h}(\boldsymbol{x}_1, \boldsymbol{g}(x), \ldots, \boldsymbol{g}(x), \boldsymbol{x}_k, x),$$

that completes the proof. □

1.4.2 Bekič principle

We start this section with an important property which is called the *Bekič principle* [13, 80]. Let $\langle E_1, \leq_1 \rangle$ and $\langle E_2, \leq_2 \rangle$ be complete lattices. Let $f_1 : E_1 \times E_2 \to E_1$ and $f_2 : E_1 \times E_2 \to E_2$ be mappings monotonic with respect to their two arguments. The mapping

$$\begin{array}{rccc} \boldsymbol{f}: & E_1 \times E_2 & \to & E_1 \times E_2 \\ & \langle x, y \rangle & \mapsto & \langle f_1(x, y), f_2(x, y) \rangle \end{array}$$

is monotonic and so has extremal fixed points in $E_1 \times E_2$

$$\theta \langle x, y \rangle . \langle f_1(x, y), f_2(x, y) \rangle .$$

Lemma 1.4.2 (Bekič principle).

$$\theta \begin{bmatrix} x \\ y \end{bmatrix} . \begin{bmatrix} f_1(x, y) \\ f_2(x, y) \end{bmatrix} = \begin{bmatrix} \theta x.f_1(x, \theta y.f_2(x, y)) \\ \theta y.f_2(\theta x.f_1(x, y), y) \end{bmatrix}.$$

Proof. Let $\langle a, b \rangle = \theta \langle x, y \rangle . \langle f_1(x, y), f_2(x, y) \rangle$, let $a' = \theta x.f_1(x, \theta y.f_2(x, y))$ and $b' = \theta y.f_2(\theta x.f_1(x, y), y)$.

We give the proof for $\theta = \mu$, the case $\theta = \nu$ is similar by the principle of symmetry.

We have $a = f_1(a, b)$ and $b = f_2(a, b)$, since $\langle a, b \rangle$ is a fixed point of $\boldsymbol{f}$. Hence, $\mu x.f_1(x, b) \leq_1 a$ and $\mu y.f_2(a, y) \leq_2 b$. By monotonicity of f_1, $f_1(a, \mu y.f_2(a, y)) \leq_1 f_1(a, b) = a$, thus $a' \leq_1 a$, and, by monotonicity of f_2, $f_2(\mu x.f_1(x, b), b) \leq_2 f_2(a, b) = b$, thus $b' \leq_2 b$.

On the other hand, let $b'' = \mu y.f_2(a', y) = f_2(a', b'')$ and $a'' = \mu x.f_1(x, b') = f_1(a'', b')$. We have $a' = f_1(a', \mu y.f_2(a', y)) = f_1(a', b'')$ and $b' = f_2(\mu x.f_1(x, b'), b') = f_2(a'', b')$. It follows that $\langle a', b''\rangle$ and $\langle a'', b'\rangle$ are fixed points of $\boldsymbol{f}$, thus both greater than the least fixed point $\langle a, b\rangle$. Then, $a \leq_1 a'$ and $b \leq_2 b'$. □

It may happen that f_i does not depend on all its arguments, for instance $f_i(x, y) = g(x)$. Here, we state the Bekič principle for some such degenerate cases.

Corollary 1.4.3. *Let $f : E_1 \to E_1$, $g : E_2 \to E_2$, $h : E_1 \times E_2 \to E_2$. Then*

$$\begin{aligned}\theta\langle x, y\rangle.\langle f(x), g(y)\rangle &= \langle \theta x.f(x), \theta y.g(y)\rangle,\\ \theta\langle x, y\rangle.\langle f(x), h(x, y)\rangle &= \langle \theta x.f(x), \theta y.h(\theta x.f(x), y)\rangle.\end{aligned}$$

We mention another consequence of the Bekič principle which will be useful later on.

Proposition 1.4.4. *Let i be a fixed integer in $\{0, 1, \ldots, n\}$. Let $f_j : E_1^i \times E_2^{n-i} \to E_j$ for $j = 1, 2$.*
Then $\theta_1 \begin{bmatrix} x_1 \\ y_1 \end{bmatrix}.\cdots.\theta_n \begin{bmatrix} x_n \\ y_n \end{bmatrix}.\begin{bmatrix} f_1(x_1, \ldots, x_i, y_{i+1}, \ldots, y_n) \\ f_2(x_1, \ldots, x_i, y_{i+1}, \ldots, y_n) \end{bmatrix}$ *is equal to* $\begin{bmatrix} h_1' \\ h_2(h_1', \ldots, h_1') \end{bmatrix}$ *where*
$h_2(x_1, \ldots, x_i) = \theta_{i+1}y_{i+1}.\cdots.\theta_n y_n.f_2(x_1, \ldots, x_i, y_{i+1}, \ldots, y_n)$ *and*
$h_1' = \theta_1 x_1.\cdots.\theta_i x_i.f_1(x_1, \ldots, x_i, h_2(x_1, \ldots, x_i), \ldots, h_2(x_1, \ldots, x_i))$.

Proof. Let $\begin{bmatrix} h_1(x_1, \ldots, x_i) \\ h_2(x_1, \ldots, x_i) \end{bmatrix} =$

$$\theta_{i+1} \begin{bmatrix} x_{i+1} \\ y_{i+1} \end{bmatrix}.\cdots.\theta_n \begin{bmatrix} x_n \\ y_n \end{bmatrix}.\begin{bmatrix} f_1(x_1, \ldots, x_i, y_{i+1}, \ldots, y_n) \\ f_2(x_1, \ldots, x_i, y_{i+1}, \ldots, y_n) \end{bmatrix}.$$

By Proposition 1.4.1 (page 26) we get $\begin{bmatrix} h_1(x_1, \ldots, x_i) \\ h_2(x_1, \ldots, x_i) \end{bmatrix} =$

$$\theta_{i+1} \begin{bmatrix} x_{i+1} \\ y_{i+1} \end{bmatrix}.\cdots.\theta_n \begin{bmatrix} x_n \\ y_n \end{bmatrix}. \begin{bmatrix} f_1(x_1, \ldots, x_n, h_2(x_1, \ldots, x_i), \ldots, h_2(x_1, \ldots, x_i)) \\ f_2(x_1, \ldots, x_i, y_{i+1}, \ldots, y_n) \end{bmatrix}$$

Using the Bekič principle, it is easy to prove by induction that for $i + 1 \leq j \leq n$,

$$\theta_j \begin{bmatrix} x_j \\ y_j \end{bmatrix}.\cdots.\theta_n \begin{bmatrix} x_n \\ y_n \end{bmatrix}. \begin{bmatrix} f_1(x_1, \ldots, x_n, h_2(x_1, \ldots, x_i), \ldots, h_2(x_1, \ldots, x_i)) \\ f_2(x_1, \ldots, x_i, y_{i+1}, \ldots, y_n) \end{bmatrix}$$

is equal to $\begin{bmatrix} f_1(x_1,\dots,x_n,h_2(x_1,\dots,x_i),\dots,h_2(x_1,\dots,x_i)) \\ \theta_j y_j.\cdots.\theta_n y_n.f_2(x_1,\dots,x_i,y_{i+1},\dots,y_n) \end{bmatrix}$, hence $h_2(x_1,\dots,x_i) = \theta_{i+1}y_{i+1}.\cdots.\theta_n y_n.f_2(x_1,\dots,x_i,y_{i+1},\dots,y_n)$. Now, let

$$\begin{bmatrix} a_1 \\ a_2 \end{bmatrix} = \theta_1 \begin{bmatrix} x_1 \\ y_1 \end{bmatrix}.\cdots.\theta_n \begin{bmatrix} x_n \\ y_n \end{bmatrix}.\begin{bmatrix} f_1(x_1,\dots,x_i,y_{i+1},\dots,y_n) \\ f_2(x_1,\dots,x_i,y_{i+1},\dots,y_n) \end{bmatrix} =$$

$$\theta_1 \begin{bmatrix} x_1 \\ y_1 \end{bmatrix}.\cdots.\theta_i \begin{bmatrix} x_i \\ y_i \end{bmatrix}.$$
$$\begin{bmatrix} f_1(x_1,\dots,x_n,h_2(x_1,\dots,x_i),\dots,h_2(x_1,\dots,x_i)) \\ h_2(x_1,\dots,x_n) \end{bmatrix}.$$

By the same technique as above, we get $a_1 = \theta_1 x_1.\cdots.\theta_i x_i.f_1(x_1,\dots,x_n,h_2(x_1,\dots,x_i),\dots,h_2(x_1,\dots,x_i)) = h'_1$ and $a_2 = h_2(a_1,\dots,a_1) = h_2(h'_1,\dots,h'_1)$. □

The following result, appeared in [104], can be seen as a generalization of the Bekič principle, as it will be shown in Example 1.4.6, page 30.

Proposition 1.4.5. *For $i = 1,\dots,n$, let $f_i : E_1 \times \cdots \times E_n \to E_i$, and let*

$$\boldsymbol{f} : E_1 \times \cdots \times E_n \to E_1 \times \cdots \times E_n$$

be defined by $\boldsymbol{f}(x_1,\dots,x_n) = \langle f_1(x_1,\dots,x_n),\dots,f_n(x_1,\dots,x_n)\rangle$.

Now let $f'_i : E_1 \times \cdots \times E_n \to E_i$ be defined by

$$f'_i(x_1,\dots,x_n) = f_i(g_1^{(i)}(x_1,\dots,x_n),\dots,g_n^{(i)}(x_1,\dots,x_n))$$

where $g_j^{(i)}(x_1,\dots,x_n)$ is equal either to x_j or to $f_j(x_1,\dots,x_n)$ or to $\theta x_j.f_j(x_1,\dots,x_n)$.

Let $\boldsymbol{f}' : E_1 \times \cdots \times E_n \to E_1 \times \cdots \times E_n$ be defined by $\boldsymbol{f}'(x_1,\dots,x_n) = \langle f'_1(x_1,\dots,x_n),\dots,f'_n(x_1,\dots,x_n)\rangle$.

Then $\theta\langle x_1,\dots,x_n\rangle.\boldsymbol{f}(x_1,\dots,x_n) = \theta\langle x_1,\dots,x_n\rangle.\boldsymbol{f}'(x_1,\dots,x_n)$.

Proof. Let $\langle a_1,\dots,a_n\rangle = \theta\langle x_1,\dots,x_n\rangle.\boldsymbol{f}(x_1,\dots,x_n)$
and $\langle a'_1,\dots,a'_n\rangle = \theta\langle x_1,\dots,x_n\rangle.\boldsymbol{f}'(x_1,\dots,x_n)$.

We consider first the case $\theta = \mu$. Obviously, $a_i = f_i(a_1,\dots,a_n)$, thus

$$\theta x_i.f_i(a_1,\dots,a_{i-1},x_i,a_{i+1},\dots,a_n) \le a_i.$$

It follows that for any i and j, $g_j^{(i)}(a_1,\dots,a_n) \le a_j$, thus

$$f'_i(a_1,\dots,a_n) \le f_i(a_1,\dots,a_n) = a_i,$$

and $\boldsymbol{f}'(a_1,\dots,a_n) \le \langle a_1,\dots,a_n\rangle$ hence $\langle a'_1,\dots,a'_n\rangle \le \langle a_1,\dots,a_n\rangle$.

Conversely, we know that $\langle a_1, \dots, a_n\rangle$ is the last element of an increasing sequence of elements of $E_1 \times \dots \times E_n$, indexed by $\alpha \leq \gamma$. We show that, for each $\langle b_1, \dots, b_n\rangle$ in this sequence, we have, for any i,

$$b_i \leq a'_i, \quad b_i \leq f_i(a'_1, \dots, a'_n), \quad b_i \leq \theta x_i.f_i(a'_1, \dots, a'_{i-1}, x_i, a'_{i+1}, \dots, a'_n).$$

Indeed, it is enough to show that if $\langle b_1, \dots, b_n\rangle$ has this property, then $\langle b'_1, \dots, b'_n\rangle = \boldsymbol{f}(b_1, \dots, b_n)$ also has this property.

From the definition of the $g_j^{(i)}$ and the induction hypothesis, we have $b_j \leq g_j^{(i)}(a'_1, \dots, a'_n)$ hence, $b'_i = f_i(b_1, \dots, b_n) \leq f'_i(a'_1, \dots, a'_n) = a'_i$.

We also have, by induction hypothesis,

$$b'_i = f_i(b_1, \dots, b_n) \leq f_i(a'_1, \dots, a'_n).$$

Finally, since $b_i \leq \theta x_i.f_i(a'_1, \dots, a'_{i-1}, x_i, a'_{i+1}, \dots, a'_n)$, we have

$$\begin{aligned} b'_i &\leq f_i(a'_1, \dots a'_{i-1}, b_i, a'_{i+1}, \dots, a'_n) \\ &\leq f_i(a'_1, \dots a'_{i-1}, \theta x_i.f_i(a'_1, \dots, a'_{i-1}, x_i, a'_{i+1}, \dots, a'_n), a'_{i+1}, \dots, a'_n) \\ &= \theta x_i.f_i(a'_1, \dots, a'_{i-1}, x_i, a'_{i+1}, \dots, a'_n). \end{aligned}$$

The proof for the case $\theta = \nu$ is symmetrical. □

Example 1.4.6. , Let $\langle a, b\rangle = \theta\langle x, y\rangle.\langle f_1(x, y), f_2(x, y)\rangle$. Let us substitute $\theta x.f_1(x, y)$ for x in $f_2(x, y)$, and $\theta y.f_2(x, y)$ for y in $f_1(x, y)$. We get

$$\langle a, b\rangle = \theta\langle x, y\rangle.\langle f_1(x, \theta y.f_2(x, y)), f_2(\theta x.f_1(x, y), y)\rangle$$

which is obviously equal to $\langle \theta x.f_1(x, \theta y.f_2(x, y)), \theta y.f_2(\theta x.f_1(x, y), y)\rangle$. □

1.4.3 Gauss elimination

An important consequence of the generalized Bekič principle is the following result.

Proposition 1.4.7 (Gauss elimination principle). *Let $f_1(x, y)$ and $f_2(x, y)$ as above, let $g_1(y) = \theta x.f_1(x, y)$, and let*

$$\langle a, b\rangle = \theta\langle x, y\rangle.\langle f_1(x, y), f_2(x, y)\rangle.$$

Then $b = \theta y.f_2(g_1(y), y)$ and $a = g_1(b)$.

Proof. Let $g_1(y) = \theta x.f_1(x,y)$. By the Bekič principle, $b = \theta y.f_2(g_1(y),y)$. By Proposition 1.4.1 (page 26) we have

$$\begin{aligned}\theta\langle x,y\rangle.\langle f_1(x,y), f_2(x,y)\rangle &= \theta\langle x,y\rangle.\langle f_1(x,b), f_2(a,b)\rangle \\ &= \langle \theta x.f_1(x,b), f_2(a,b)\rangle.\end{aligned}$$

It follows that $a = \theta x.f_1(x,b) = g_1(b)$. □

This proposition is called the *Gauss elimination principle* because it allows us to compute

$$\langle a_1,\dots,a_n\rangle = \theta\langle x_1,\dots,x_n\rangle.\langle f_1(x_1,\dots,x_n),\dots,f_n(x_1,\dots,x_n)\rangle$$

by a recursive method similar to the Gauss elimination method in linear algebra:
1. compute $g_1(x_2,\dots,x_n) = \theta x_1.f_1(x_1,x_2,\dots,x_n)$
2. compute

$$\begin{bmatrix} a_2 \\ \vdots \\ a_n \end{bmatrix} = \theta \begin{bmatrix} x_2, \\ \vdots \\ x_n \end{bmatrix} . \begin{bmatrix} f_2(g_1(x_2,\dots,x_n),x_2,\dots,x_n) \\ \vdots \\ f_n(g_1(x_2,\dots,x_n),x_2,\dots,x_n) \end{bmatrix}$$

by recursively using the same method,
3. compute $a_1 = g_1(a_2,\dots,a_n)$.

Example 1.4.8. Let us compute $\mu\langle x,y,z\rangle.\langle x \vee z \vee a, x \wedge z, y \vee b\rangle$. We first compute $g_1(y,z) = \mu x.(x \vee z \vee a) = z \vee a$. Then we compute

$$\mu\langle y,z\rangle.\langle (z \vee a) \wedge z, y \vee b\rangle.$$

To do that, we compute $g_2(z) = \mu y.(z \vee a) \wedge z = z$, next $\mu z.g_2(z) \vee b = \mu z.(z \vee b) = b$, so that $\mu\langle y,z\rangle.\langle (z \vee a) \wedge z, y \vee b\rangle = \langle b,b\rangle$. It follows that $\mu\langle x,y,z\rangle.\langle x \vee z \vee a, x \wedge z, y \vee b\rangle = \langle g_1(b,b),b,b\rangle = \langle a \vee b, b, b\rangle$. □

1.4.4 Systems of equations

We generalize this elimination principle by introducing the notion of a system of equations, which can be seen as an alternative presentation of the vectorial μ-calculus. One can find this alternative viewpoint used in [1], [59] and [89], for instance.

Definition 1.4.9. For $i = 1, \dots, n$, let $f_i : E_1 \times \cdots \times E_n \times E \to E_i$, and let x_i be a variable ranging over E_i. A system of equations is a sequence

$$x_1 \stackrel{\theta_1}{=} f_1(x_1, \dots, x_n, x), \dots, x_i \stackrel{\theta_i}{=} f_i(x_1, \dots, x_n, x), \dots, x_n \stackrel{\theta_n}{=} f_n(x_1, \dots, x_n, x)$$

where θ_i is either μ or ν.

The *solution* of this system is the element

$$\langle h_1(x), \dots, h_n(x) \rangle : E \to E_1 \times \cdots \times E_n$$

obtained by the following sequence of transformations. First we compute the solution of the first equation $x_1 \stackrel{\theta_1}{=} f_1(x_1, \dots, x_n, x)$ that is $g_1(x_2, \dots, x_n, x) = \theta_1 x_1.f_1(x_1, \dots, x_n, x)$.

We substitute this solution for x_1 in all the remaining equations, and we get the system

$x_2 \stackrel{\theta_2}{=} f_1(g_1(x_2, \dots, x_n, x), x_2, \dots, x_n, x), \dots$

$x_n \stackrel{\theta_n}{=} f_n(g_1(x_2, \dots, x_n, x), x_2, \dots, x_n, x).$

We solve the next equation and we get $x_2 = g_2(x_3, \dots, x_n, x)$. We again substitute this solution for x_2 in the remaining equations, and we solve again the next equation getting $x_3 = g_3(x_4, \dots, x_n, x)$. By iterating this process, we compute successively $x_i = g_i(x_{i+1}, \dots, x_n, x), \dots, x_n = g_n(x)$.

Now, we set $h_n(x) = g_n(x)$, $h_{n-1}(x) = g_{n-1}(h_n(x), x), \dots,$ $h_i(x) = g_i(h_{i+1}(x), \dots, h_n(x), x), \dots,$ $h_1(x) = g_1(h_2(x), \dots, h_n(x), x)$.

More formally, the solution of a system of equations can be defined by induction on n.

Definition 1.4.10. If $n = 1$ the solution of $x_1 \stackrel{\theta_1}{=} f_1(x_1, x)$ is $\theta_1 x_1.f_1(x_1, x)$. Otherwise, the solution of

$$x_1 \stackrel{\theta_1}{=} f_1(x_1, \dots, x_n, x), \dots, x_i \stackrel{\theta_i}{=} f_i(x_1, \dots, x_n, x), \dots, x_n \stackrel{\theta_n}{=} f_n(x_1, \dots, x_n, x)$$

is $\langle g_1(h_2(x), \dots, h_n(x), x), h_2(x), \dots, h_n(x) \rangle$ where $g_1(x_2, \dots, x_n, x) = \theta_1 x_1.f_1(x_1, \dots, x_n, x)$ and $\langle h_2(x), \dots, h_n(x) \rangle$ is the solution of

$x_2 \stackrel{\theta_2}{=} f_1(g_1(x_2, \dots, x_n, x), x_2, \dots, x_n, x), \dots$

$x_n \stackrel{\theta_n}{=} f_n(g_1(x_2, \dots, x_n, x), x_2, \dots, x_n, x).$

Proposition 1.4.11. *The solution of the system*

$$x_1 \stackrel{\theta_1}{=} f_1(x_1, \dots, x_n, x), \dots, x_i \stackrel{\theta_i}{=} f_i(x_1, \dots, x_n, x), \dots, x_n \stackrel{\theta_n}{=} f_n(x_1, \dots, x_n, x)$$

is

$$\theta_n \begin{bmatrix} z_1^{(n)} \\ \vdots \\ z_n^{(n)} \end{bmatrix} \cdots .\theta_1 \begin{bmatrix} z_1^{(1)} \\ \vdots \\ z_n^{(1)} \end{bmatrix} . \begin{bmatrix} f_1(z_1^{(1)}, \dots, z_i^{(i)}, \dots, z_n^{(n)}, x) \\ \vdots \\ f_n(z_1^{(1)}, \dots, z_i^{(i)}, \dots, z_n^{(n)}, x) \end{bmatrix}.$$

Note that in this notation we consider that $f_j(z_1^{(1)}, \dots, z_i^{(i)}, \dots, z_n^{(n)}, x)$ is indeed a mapping from $(E_1 \times \cdots \times E_n)^n \times E$ to E_j which does not depend on arguments $z_\ell^{(i)}$ with $i \neq \ell$.

Proof. Without loss of generality, we may assume that x has a fixed value, thus, we do not have to deal with the extra argument x so that we have to prove

$$\begin{bmatrix} h_1 \\ \vdots \\ h_n \end{bmatrix} = \theta_n \begin{bmatrix} z_1^{(n)} \\ \vdots \\ z_n^{(n)} \end{bmatrix} \cdots .\theta_1 \begin{bmatrix} z_1^{(1)} \\ \vdots \\ z_n^{(1)} \end{bmatrix} . \begin{bmatrix} f_1(z_1^{(1)}, \dots, z_i^{(i)}, \dots, z_n^{(n)}) \\ \vdots \\ f_n(z_1^{(1)}, \dots, z_i^{(i)}, \dots, z_n^{(n)}) \end{bmatrix}.$$

The proof is by induction on n. For $n = 1$, the solution of $x_1 \stackrel{\theta_1}{=} f_1(x_1)$ is $\theta_1 x_1.f_1(x_1)$.

Otherwise the solution of the system

$$x_1 \stackrel{\theta_1}{=} f_1(x_1, \dots, x_n), \dots, x_i \stackrel{\theta_i}{=} f_i(x_1, \dots, x_n), \dots, x_n \stackrel{\theta_n}{=} f_n(x_1, \dots, x_n).$$

is $\langle g(h_2, \dots, h_n), h_2, \dots, h_n\rangle$ where $g(x_2, \dots, x_n) = \theta_1 x_1.f_1(x_1, \dots, x_n)$ and $\langle h_2, \dots, h_n\rangle$ is the solution of the system

$$x_2 \stackrel{\theta_1}{=} f_1(g(x_2, \dots, x_n), x_2, \dots, x_n), \dots x_n \stackrel{\theta_n}{=} f_n(g(x_2, \dots, x_n), x_2, \dots, x_n).$$

and, by the induction hypothesis,

$$\begin{bmatrix} h_2 \\ \vdots \\ h_n \end{bmatrix} = \theta_n \begin{bmatrix} z_2^{(n)} \\ \vdots \\ z_n^{(n)} \end{bmatrix} \cdots \theta_2 \begin{bmatrix} z_2^{(2)} \\ \vdots \\ z_n^{(2)} \end{bmatrix} . \begin{bmatrix} f_2(g(z_2^{(2)}, \dots, z_n^{(n)}), \dots, z_n^{(n)}) \\ \vdots \\ f_n(g(z_2^{(2)}, \dots, z_n^{(n)}), \dots, z_n^{(n)}) \end{bmatrix}.$$

Now, let $\langle b_1, \dots, b_n\rangle =$

$$\theta_n \begin{bmatrix} z_1^{(n)} \\ \vdots \\ z_n^{(n)} \end{bmatrix} \cdots .\theta_1 \begin{bmatrix} z_1^{(1)} \\ \vdots \\ z_n^{(1)} \end{bmatrix} . \begin{bmatrix} f_1(z_1^{(1)}, \dots, z_i^{(i)}, \dots, z_n^{(n)}) \\ \vdots \\ f_n(z_1^{(1)}, \dots, z_i^{(i)}, \dots, z_n^{(n)}) \end{bmatrix}.$$

By Proposition 1.4.5 (page 29),
$\theta_1\langle z_1^{(1)}, z_2^{(1)}, \ldots, z_n^{(1)}\rangle.$
$\langle f_1(z_1^{(1)}, \ldots, z_n^{(n)}), f_2(z_1^{(1)}, \ldots, z_n^{(n)}), \ldots, f_n(z_1^{(1)}, \ldots, z_n^{(n)})\rangle =$

$$\theta_1 \begin{bmatrix} z_1^{(1)} \\ z_2^{(1)} \\ \vdots \\ z_n^{(1)} \end{bmatrix} . \begin{bmatrix} g(z_2^{(2)}, \ldots, z_n^{(n)}) \\ f_2(g(z_2^{(2)}, \ldots, z_n^{(n)}), \ldots, z_n^{(n)}) \\ \vdots \\ f_n(g(z_2^{(2)}, \ldots, z_n^{(n)}), \ldots, z_n^{(n)}) \end{bmatrix}.$$

Since none of these components depends on $z_1^{(1)}, z_2^{(1)}, \ldots, z_n^{(1)}$, this is equal to $\langle g(z_2^{(2)}, \ldots, z_n^{(n)}), f_2(g(z_2^{(2)}, \ldots, z_n^{(n)}), \ldots, z_n^{(n)}), \ldots,$
$f_n(g(z_2^{(2)}, \ldots, z_n^{(n)}), \ldots, z_n^{(n)})\rangle$ so that

$$\begin{bmatrix} b_1 \\ b_2 \\ \vdots \\ b_n \end{bmatrix} = \theta_n \begin{bmatrix} z_1^{(n)} \\ z_2^{(n)} \\ \vdots \\ z_n^{(n)} \end{bmatrix} \cdots .\theta_2 \begin{bmatrix} z_1^{(2)} \\ z_2^{(2)} \\ \vdots \\ z_n^{(2)} \end{bmatrix} . \begin{bmatrix} g(z_2^{(2)}, \ldots, z_n^{(n)}) \\ f_2(g(z_2^{(2)}, \ldots, z_n^{(n)}), \ldots, z_n^{(n)}) \\ \vdots \\ f_n(g(z_2^{(2)}, \ldots, z_n^{(n)}), \ldots, z_n^{(n)}) \end{bmatrix}.$$

By Proposition 1.4.1 (page 26) we get $\langle b_1, b_2, \ldots, b_n\rangle =$

$$\theta_n \begin{bmatrix} z_1^{(n)} \\ z_2^{(n)} \\ \vdots \\ z_n^{(n)} \end{bmatrix} \cdots .\theta_2 \begin{bmatrix} z_1^{(2)} \\ z_2^{(2)} \\ \vdots \\ z_n^{(2)} \end{bmatrix} . \begin{bmatrix} g(b_2, \ldots, b_n) \\ f_2(g(z_2^{(2)}, \ldots, z_n^{(n)}), \ldots, z_n^{(n)}) \\ \vdots \\ f_n(g(z_2^{(2)}, \ldots, z_n^{(n)}), \ldots, z_n^{(n)}) \end{bmatrix}.$$

Now, using the Bekič principle, it is easy to prove by induction on i, that

$$\theta_i \begin{bmatrix} z_1^{(i)} \\ z_2^{(i)} \\ \vdots \\ z_n^{(i)} \end{bmatrix} \cdots .\theta_2 \begin{bmatrix} z_1^{(2)} \\ z_2^{(2)} \\ \vdots \\ z_n^{(2)} \end{bmatrix} . \begin{bmatrix} g(b_2, \ldots, b_n) \\ f_2(g(z_2^{(2)}, \ldots, z_n^{(n)}), \ldots, z_n^{(n)}) \\ \vdots \\ f_n(g(z_2^{(2)}, \ldots, z_n^{(n)}), \ldots, z_n^{(n)}) \end{bmatrix} =$$

$$\begin{bmatrix} g(b_2, \ldots, b_n) \\ \theta_i \begin{bmatrix} z_2^{(i)} \\ \vdots \\ z_n^{(i)} \end{bmatrix} \cdots .\theta_2 \begin{bmatrix} z_2^{(2)} \\ \vdots \\ z_n^{(2)} \end{bmatrix} . \begin{bmatrix} f_2(g(z_2^{(2)}, \ldots, z_n^{(n)}), \ldots, z_n^{(n)}) \\ \vdots \\ f_n(g(z_2^{(2)}, \ldots, z_n^{(n)}), \ldots, z_n^{(n)}) \end{bmatrix} \end{bmatrix}.$$

It follows that

$$\langle b_1, b_2, \ldots, b_n\rangle = \langle g(b_2, \ldots, b_n), h_2, \ldots, h_n\rangle = \langle g(h_2, \ldots, h_n), h_2, \ldots, h_n\rangle.$$

□

In case $\theta_i = \theta_{i+1}$ we can use Proposition 1.3.2 (page 19) to reduce the number of fixed point operators, until two of them that are consecutive are distinct.

Corollary 1.4.12. *The solution of the system*

$$x_1 \stackrel{\theta}{=} f_1(x_1, \dots, x_n), \dots, x_i \stackrel{\theta}{=} f_i(x_1, \dots, x_n), \dots, x_n \stackrel{\theta}{=} f_n(x_1, \dots, x_n)$$

is $\theta\langle x_1, \dots, x_n\rangle.\langle f_1(x_1, \dots, x_n), \dots, f_n(x_1, \dots, x_n)\rangle$.

Note that this corollary gives an alternative proof of the Gauss elimination principle.

As a consequence of Proposition 1.4.11 we have another characterization of the solution of a system of equations.

Proposition 1.4.13. *The solution of the system of equations*

$$x_1 \stackrel{\theta_1}{=} f_1(x_1, \dots, x_n), \dots, x_i \stackrel{\theta_i}{=} f_i(x_1, \dots, x_n), \dots, x_n \stackrel{\theta_n}{=} f_n(x_1, \dots, x_n)$$

is equal to

$$\langle h_1(h'_{i+1}, \dots, h'_n), \dots, h_i(h'_{i+1}, \dots, h'_n), h'_{i+1}, \dots, h'_n\rangle$$

where $\langle h_1(x_{i+1}, \dots, x_n), \dots, h_i(x_{i+1}, \dots, x_n)\rangle$ *is the solution of*

$$x_1 \stackrel{\theta_1}{=} f_1(x_1, \dots, x_n), \dots, x_i \stackrel{\theta_i}{=} f_i(x_1, \dots, x_n)$$

and $\langle h'_{i+1}, \dots h'_n\rangle$ *is the solution of*

$$\begin{aligned} x_{i+1} &\stackrel{\theta_{i+1}}{=} f_{i+1}(h_1(x_{i+1}, \dots, x_n), \dots, h_i(x_{i+1}, \dots, x_n), x_{i+1} \dots, x_n), \\ \dots, & \\ x_n &\stackrel{\theta_n}{=} f_n(h_1(x_{i+1}, \dots, x_n), \dots, h_i(x_{i+1}, \dots, x_n), x_{i+1} \dots, x_n). \end{aligned}$$

Proof. By the previous proposition, we know that the solution $\langle a_1, \dots, a_n\rangle$ of the system is equal to

$$\theta_n \begin{bmatrix} z_1^{(n)} \\ \vdots \\ z_n^{(n)} \end{bmatrix} \cdots . \theta_1 \begin{bmatrix} z_1^{(1)} \\ \vdots \\ z_n^{(1)} \end{bmatrix} . \begin{bmatrix} f_1(z_1^{(1)}, \dots, z_n^{(n)}) \\ \vdots \\ f_n(z_1^{(1)}, \dots, z_n^{(n)}) \end{bmatrix}$$

which has the form

$$\theta_n \begin{bmatrix} x_n \\ y_n \end{bmatrix} \cdots . \theta_1 \begin{bmatrix} x_1 \\ y_1 \end{bmatrix} . \begin{bmatrix} f_1(x_1, \dots, x_i, y_{i+1}, \dots y_n) \\ f_2(x_1, \dots, x_i, y_{i+1}, \dots y_n) \end{bmatrix}$$

if we set $\boldsymbol{x}_j = \langle z_1^{(j)}, \ldots, z_i^{(j)} \rangle$, $\boldsymbol{y}_j = \langle z_{i+1}^{(j)}, \ldots, z_n^{(j)} \rangle$,
$\boldsymbol{f}_1(\boldsymbol{x}_1, \ldots, \boldsymbol{x}_i, \boldsymbol{y}_{i+1}, \ldots \boldsymbol{y}_n) = \langle f_1(z_1^{(1)}, \ldots, z_n^{(n)}), \ldots, f_i(z_1^{(1)}, \ldots, z_n^{(n)}) \rangle$,
$\boldsymbol{f}_2(\boldsymbol{x}_1, \ldots, \boldsymbol{x}_i, \boldsymbol{y}_{i+1}, \ldots \boldsymbol{y}_n) = \langle f_{i+1}(z_1^{(1)}, \ldots, z_n^{(n)}) \ldots, f_n(z_1^{(1)}, \ldots, z_n^{(n)}) \rangle$,
which makes sense since $z_j^{(j)}$ is a component of $\boldsymbol{x}_j$ if $j \leq i$ and of $\boldsymbol{y}_j$ otherwise.

Thus, by Proposition 1.4.4 (page 28),
$\langle a_1, \ldots, a_n \rangle = \langle h_1(h'_{i+1}, \ldots, h'_n), \ldots, h_i(h'_{i+1}, \ldots, h'_n), \ldots, h'_{i+1}, \ldots, h'_n \rangle$
where $\langle h_1(x_{i+1}, \ldots, x_n), \ldots, h_i(x_{i+1}, \ldots, x_n) \rangle$ and $\langle h'_{i+1}, \ldots, h'_n \rangle$ have the required property, by Proposition 1.4.11 (page 32). □

The following proposition shows that we do not change the solution of a system of equations

$$x_1 \stackrel{\theta_1}{=} f_1(x_1, \ldots, x_n), \ldots, x_i \stackrel{\theta_i}{=} f_i(x_1, \ldots, x_n), \ldots, x_n \stackrel{\theta_n}{=} f_n(x_1, \ldots, x_n)$$

if we substitute $f_i(x_1, \ldots, x_n)$ or $\theta x_i.f_i(x_1, \ldots, x_n)$ for x_i in $f_j(x_1, \ldots, x_n)$, provided $i \leq j$. This result generalizes Proposition 1.4.5 (page 29).

Proposition 1.4.14. *The three systems*

$$x_1 \stackrel{\theta_1}{=} f_1(x_1, \ldots, x_n), \ldots, x_j \stackrel{\theta_j}{=} f_j(x_1, \ldots, x_n), \ldots, x_n \stackrel{\theta_n}{=} f_n(x_1, \ldots, x_n),$$

$$x_1 \stackrel{\theta_1}{=} f_1(x_1, \ldots, x_n), \ldots, x_j \stackrel{\theta_j}{=} f_j(x_1, \ldots, x_{i-1}, f_i(x_1, \ldots, x_n), x_{i+1} \ldots, x_n), \ldots, x_n \stackrel{\theta_n}{=} f_n(x_1, \ldots, x_n),$$

$$x_1 \stackrel{\theta_1}{=} f_1(x_1, \ldots, x_n), \ldots, x_j \stackrel{\theta_j}{=} f_j(x_1, \ldots, x_{i-1}, \theta_i x_i.f_i(x_1, \ldots, x_n), x_{i+1}, \ldots, x_n), \ldots, x_n \stackrel{\theta_n}{=} f_n(x_1, \ldots, x_n)$$

have the same solution whenever $i \leq j$.

Proof. Because of the previous characterization of the solution of a system of equations, and by fixing the values of $x_{j+1}, \ldots, x_n$, it is enough to show that the three systems

$$x_1 \stackrel{\theta_1}{=} f_1(x_1, \ldots, x_j), \ldots, x_j \stackrel{\theta_j}{=} f_j(x_1, \ldots, x_j),$$

$$x_1 \stackrel{\theta_1}{=} f_1(x_1, \ldots, x_j), \ldots, x_j \stackrel{\theta_j}{=} f_j(x_1, \ldots, x_{i-1}, f_i(x_1, \ldots, x_j), x_{i+1} \ldots, x_j),$$

$$x_1 \stackrel{\theta_1}{=} f_1(x_1, \ldots, x_j), \ldots, x_j \stackrel{\theta_j}{=} f_j(x_1, \ldots, x_{i-1}, \theta_i x_i.f_i(x_1, \ldots, x_j), x_{i+1}, \ldots, x_j)$$

have the same solution. We prove it by induction on j. If $j = 1$, the result is a consequence of Proposition 1.4.5 (page 29). Otherwise, by the previous proposition, the three solutions of these equations are respectively equal to $\langle h_1(h), \ldots, h_{j-1}(h), h\rangle$, $\langle h_1(h'), \ldots, h_{j-1}(h'), h'\rangle$, $\langle h_1(h''), \ldots, h_{j-1}(h''), h''\rangle$, where $\langle h_1(x_j), \ldots, h_{j-1}(x_j)\rangle$ is the solution of

$$x_1 \overset{\theta_1}{=} f_1(x_1, \ldots, x_j), \ldots, x_{j-1} \overset{\theta_{j-1}}{=} f_{j-1}(x_1, \ldots, x_j),$$

and where $h = \theta_j x_j.f_j(h_1(x_j), \ldots, h_i(x_j), \ldots, h_{j-1}(x_j), x_j)$,
$h' = \theta_j x_j.f_j(h_1(x_j), \ldots, f_i(h_1(x_j), \ldots, h_{j-1}(x_j), x_j), \ldots, h_{j-1}(x_j), x_j)$, and
$h'' = \theta_j x_j.f_j(h_1(x_j), \ldots, g(x_j), \ldots, h_{j-1}(x_j), x_j)$, with
$g(x_j) = \theta_i x_i.f_i(h_1(x_j), \ldots, x_i, \ldots, h_{j-1}(x_j), x_j)$. Therefore, the result holds if $h_i(x) = f_i(h_1(x), \ldots, h_{j-1}(x), x) = \theta_i x_i.f_i(h_1(x), \ldots, x_i, \ldots, h_{j-1}(x), x)$.

By Proposition 1.4.11 (page 32) $\langle h_1(z_j^{(j)}), \ldots, h_{j-1}(z_j^{(j)})\rangle =$

$$\theta_{j-1}\begin{bmatrix} z_1^{(j-1)} \\ \vdots \\ z_{j-1}^{(j-1)} \end{bmatrix} \cdots .\theta_1 \begin{bmatrix} z_1^{(1)} \\ \vdots \\ z_{j-1}^{(1)} \end{bmatrix} . \begin{bmatrix} f_1(z_1^{(1)}, \ldots, z_j^{(j)}) \\ \vdots \\ f_{j-1}(z_1^{(1)}, \ldots, z_j^{(j)}) \end{bmatrix}.$$

It follows that $\langle h_1(z_j^{(j)}), \ldots, h_{j-1}(z_j^{(j)})\rangle =$
$\langle f_1(h_1(z_j^{(j)}), \ldots, h_{j-1}(z_j^{(j)}), z_j^{(j)}), \ldots, f_{j-1}(h_1(z_j^{(j)}), \ldots, h_{j-1}(z_j^{(j)}), z_j^{(j)})\rangle$, and, in particular, $h_i(x) = f_i(h_1(x), \ldots, h_{j-1}(x), x)$.

We also have $\langle h_1(z_j^{(j)}), \ldots, h_{j-1}(z_j^{(j)})\rangle =$

$$\begin{aligned} \theta_i\langle z(i)_1, \ldots, z_{j-1}^{(i)}\rangle \quad . \quad & \langle f_1(h_1(z_j^{(j)}), \ldots, z_i^{(i)}, \ldots, h_{j-1}(z_j^{(j)}), z_j^{(j)}), \\ & \ldots, \\ & f_{j-1}(h_1(z_j^{(j)}), \ldots, z_i^{(i)}, \ldots, h_{j-1}(z_j^{(j)}), z_j^{(j)})\rangle, \end{aligned}$$

from which we get, by the Bekič principle,
$h_i(x) = \theta_i z_i^{(i)}.f_i(h_1(x), \ldots, z_i^{(i)}, \ldots, h_{j-1}(x), x)$. □

Note that in this lemma the condition $i \leq j$ is necessary, as shown by the following example.

Example 1.4.15. Let the system $x \overset{\mu}{=} y, y \overset{\nu}{=} x$. The solution of the first equation is $\mu x.y = y$. Substituting y for x in the second one yields $y \overset{\nu}{=} y$ whose solution is $y = \nu y.y = \top$. Thus, the solution of this system is $\langle \top, \top\rangle$.

Now, let us substitute x, the value of y defined by the second equation, for y in the first equation, in violation of the hypothesis. We get the system $x \overset{\mu}{=} x, y \overset{\nu}{=} x$. The solution of the first equation is $x = \mu x.x = \bot$. The second equation becomes $y \overset{\nu}{=} \bot$ whose solution is $y = \nu y.\bot = \bot$. And the solution of the system is $\langle \bot, \bot\rangle$. □

We end this section by giving a kind of converse of Proposition 1.4.11 (page 32).

Proposition 1.4.16. *Let*

$$\begin{bmatrix} a_1 \\ \vdots \\ a_k \end{bmatrix} = \theta_n \begin{bmatrix} z_1^{(n)} \\ \vdots \\ z_k^{(n)} \end{bmatrix} \cdots . \theta_1 \begin{bmatrix} z_1^{(1)} \\ \vdots \\ z_k^{(1)} \end{bmatrix} . \begin{bmatrix} f_1(z_1^{(1)}, \ldots, z_j^{(i)}, \ldots, z_k^{(n)}) \\ \vdots \\ f_k(z_1^{(1)}, \ldots, z_j^{(i)}, \ldots, z_k^{(n)}) \end{bmatrix}.$$

Then $\langle a_1, \ldots, a_k \rangle^n$, the concatenation of n times the vector $\langle a_1, \ldots, a_k \rangle$ is the solution of the system of equations

$$\begin{array}{ll}
z_1^{(1)} \stackrel{\theta_1}{=} f_1(z_1^{(1)}, \ldots, z_j^{(i)}, \ldots, z_k^{(n)}), \ldots, & z_k^{(1)} \stackrel{\theta_1}{=} f_k(z_1^{(1)}, \ldots, z_j^{(i)}, \ldots, z_k^{(n)}), \\
z_1^{(2)} \stackrel{\theta_2}{=} z_1^{(1)} \qquad , \ldots, & z_k^{(2)} \stackrel{\theta_2}{=} z_k^{(1)}, \\
\ldots & \\
z_1^{(i+1)} \stackrel{\theta_{i+1}}{=} z_1^{(i)} \qquad , \ldots, & z_k^{(i+1)} \stackrel{\theta_{i+1}}{=} z_k^{(i)}, \\
\ldots & \\
z_1^{(n)} \stackrel{\theta_n}{=} z_1^{(n-1)} \qquad , \ldots, & z_k^{(n)} \stackrel{\theta_n}{=} z_k^{(n-1)}.
\end{array}$$

Proof. The proof is by induction on n. For $n = 1$, the result is just Corollary 1.4.12 (page 35). Let us substitute z_i for $z_i^{(n+1)}$, and let $\langle h_1(z_1, \ldots, z_k), \ldots, h_k(z_1, \ldots, z_k) \rangle =$

$$\theta_n \begin{bmatrix} z_1^{(n)} \\ \vdots \\ z_k^{(n)} \end{bmatrix} \cdots . \theta_1 \begin{bmatrix} z_1^{(1)} \\ \vdots \\ z_k^{(1)} \end{bmatrix} . \begin{bmatrix} f_1(z_1^{(1)}, \ldots, z_j^{(i)}, \ldots, z_k^{(n)}, z_1, \ldots, z_k) \\ \vdots \\ f_k(z_1^{(1)}, \ldots, z_j^{(i)}, \ldots, z_k^{(n)}, z_1, \ldots, z_k) \end{bmatrix}.$$

Then, by the induction hypothesis, $\langle h_1(z_1, \ldots, z_k), \ldots, h_k(z_1, \ldots, z_k) \rangle^n$ is the solution of the associated system Σ, and let us compute the solution of

$$\Sigma' = \Sigma, z_1 \stackrel{\theta_{n+1}}{=} z_1^{(n)}, \ldots z_k \stackrel{\theta_{n+1}}{=} z_k^{(n)}.$$

By Proposition 1.4.13 (page 35) this solution is

$$\langle h_1(b_1, \ldots, b_k), \ldots, h_k(b_1, \ldots, b_k) \rangle^n \langle b_1, \ldots, b_k \rangle$$

where $\langle b_1, \ldots, b_k \rangle$ is the solution of the system

$$z_1 \stackrel{\theta_{n+1}}{=} h_1(z_1, \ldots, z_k), \ldots, z_k \stackrel{\theta_{n+1}}{=} h_k(z_1, \ldots, z_k).$$

By Corollary 1.4.12 (page 35),

$$\begin{aligned}
\langle b_1, \ldots, b_k \rangle &= \theta_{n+1} \langle z_1, \ldots, z_k \rangle . \langle h_1(z_1, \ldots, z_k), \ldots, h_k(z_1, \ldots, z_k) \rangle \\
&= \langle a_1, \ldots, a_k \rangle \\
&= \langle h_1(a_1, \ldots, a_k), \ldots, h_k(a_1, \ldots, a_k) \rangle
\end{aligned}$$

□

1.5 Conway identities

In [14], the notion of a fixed point in an algebraic theory is characterized by three axioms called *Conway identities.* With the notations used in this chapter these three axioms can be stated as follows.

- Parameter identity.
  If $g(y) = \theta x.f(x,y)$ then, for $z \neq x$, $g(h(z)) = \theta x.f(x,h(z))$ This identity is Proposition 1.3.1 (page 19).
- Double dagger identity.
  $\theta x.\theta y.f(x,y,z) = \theta x.f(x,x,z)$. If we fix a value for z and set $h(x,y) = f(x,y,z)$ this identity is equivalent to Proposition 1.3.2 (page 19).
- Composition identity.
  $\theta x.f(g(x,y),y) = f(\theta z.g(f(z,y),y),y)$. Here again if we fix a value for y, this identity is equivalent to Proposition 1.3.12 (page 23).

1.6 Bibliographical notes and sources

The idea of a lattice goes back to the work of Dedekind in number theory. The concept of a complete lattice first appeared in the work of Birkhoff [15]; the book of this author [16] gives a detailed exposition of lattice theory.

The fixed-point theorem (Theorem 1.2.8, see also Theorem 1.2.11) was shown for functions over sets in 1928 by Knaster [54], and generalized to complete lattices by Tarski [93]. The Bekič principle is originally from [13]. To our knowledge, the concept of mutually dependent least and greatest fixed points appeared explicitly for the first time in the studies on program semantics by Park [80] and by Emerson and Clarke [36], although the idea of considering the greatest fixed points as well as the least ones is earlier, and can be traced back to works by D. Scott [87], Dijkstra [27], Basu and Yeh [12], Flon and Suzuki [38], Arnold and Nivat [7], and others. Inequality $\mu x.\nu y.h(x,y) \leq \nu y.\mu x.h(x,y)$ was, to our knowledge, first observed in [74].

2. The μ-calculi: Syntax and semantics

In the previous chapter, we have not focused much on the notation for fixed-point definitions. We have used standard mathematical notation for functions, together with the notation $\mu x.t$ for the least solution of equation $x = t(x)$, provided that the domain and the interpretation of t were clear from the context, similarly for $\nu x.t$ and the greatest fixed point. Thus we could form unambiguous expressions like

$$f(x, g(x, y)), \quad \mu x.f(x, g(x, y)), \quad \nu y.\mu x.f(x, g(x, y)).$$

However, when we are interested in relations between structures, or in properties of classes of structures, it is useful to specify a syntactic counterpart of the calculus. A formal syntax together with a concept of semantic interpretation allows us to move between numerous interpretations, and to distinguish properties which are common for classes of interpretations.

We will follow the approach of universal algebra, where the basic expressions are terms, formed from variables and function symbols. When interpreted in the usual fashion, a term with variables, say, x, y, z denotes a function of arity 3, or, more specifically, with arity $\{x, y, z\}$. Here, we have two more syntactic constructs: μx and νx, which will bind a variable x as do the quantifiers in first–order logic, i.e., decreasing the arity of an expression. For example, the expressions shown above have the arities $\{x, y\}$, $\{y\}$, and $\emptyset$, respectively. On the semantic side, μ and ν will of course be interpreted as the least and the greatest fixed–point operators, respectively.

Now, it is well known that a classical interpretation of terms is compositional, i.e., the syntactic construction of substitution of terms is reflected by the composition of functions. Some properties can be observed both on the syntactic and semantic levels, for example associativity of substitution (or composition). Similar analogies can be found for fixed-point operators.

Therefore we find it convenient to define an abstract concept of a μ-calculus which comprise both syntactic and semantic calculi. We will further see that this concept will be also useful for organizing other objects, specifically automata, where the role of μ and ν will be taken by some kind of iterations (see Chapter 7).

2.1 μ-calculi

We fix an infinite set *Var* of variables. Unless stated otherwise, we assume that this set is countable; usually in this book it is enough, but in some cases we will need sets of variables of any cardinality. As usual, we use syntactic variables $x, y, z, \ldots$, often with subscripts or superscripts, to range over (formal) variables in *Var*.

Any mapping ρ from *Var* to a set E is called a *substitution* (into E).

If ρ is a substitution into a set E, x a variable, and e an element of E, we denote by $\rho\{e/x\}$ the substitution ρ' defined by $\rho'(x) = e$, and $\rho'(y) = \rho(y)$ for $y \neq x$. More generally, if $x_1, \ldots, x_n$ are *distinct* variables and if $e_1, \ldots, e_n$ are elements of E (not necessarily distinct), we denote by $\rho\{e_1/x_1, \ldots, e_n/x_n\}$ the substitution ρ' defined by

$$\rho'(y) = \begin{cases} e_i & \text{if } y = x_i \in \{x_1, \ldots x_n\}, \\ \rho(y) & \text{if } y \notin \{x_1, \ldots x_n\}. \end{cases}$$

For any mapping $f : E \to E'$, and for any subset $A \subseteq E$, $f \restriction A : A \to E'$ is the restriction of f to A.

Definition 2.1.1. A *μ-calculus* is a sextuple $\mathcal{T} = \langle T, id, ar, comp, \mu, \nu \rangle$, where:

- T is any set, its elements are referred to as *objects* of the μ-calculus.
- id is a mapping from *Var* to T. We write $\hat{x}$ for the element $id(x)$ in T.
- ar is a mapping that with each $t \in T$ associates a subset of *Var* called the *arity* of t. If $x \in ar(t)$, we say that x *occurs free* in t, and we call the elements of $ar(t)$ *free variables* of t.
- $comp$ is a mapping associating a term $comp(t, \rho)$ with any term t and any substitution ρ; $comp(t, \rho)$ is also written $t[\rho]$.
- μ and ν are two mappings from $Var \times T$ into T, the value of the mapping θ on x and t is written $\theta x.t$ ($\theta = \mu, \nu$).

We additionally assume that the following axioms hold.

1. $ar(\hat{x}) = \{x\}$,
2. $ar(t[\rho]) = ar'(t, \rho)$, where $ar'(t, \rho) = \bigcup_{y \in ar(t)} ar(\rho(y))$,
3. $ar(\theta x.t) = ar(t) - \{x\}$,
4. $\hat{x}[\rho] = \rho(x)$, for $x \in Var$,
5. $t[\rho] = t[\rho']$ whenever $\rho \restriction ar(t) = \rho' \restriction ar(t)$,
6. $(t[\rho])[\pi] = t[\rho \star \pi]$, where $\rho \star \pi$ is the substitution defined by $\rho \star \pi(x) = \rho(x)[\pi]$,
7. if $ar'(\theta x.t, \rho) \neq Var$, there exists a variable $y \notin ar'(\theta x.t, \rho)$ (possibly equal to x), such that $(\theta x.t)[\rho] = \theta y.(t[\rho\{\hat{y}/x\}])$.

Comments on the definition. The nature of the objects is deliberately left unspecified. They can be functions as in the previous example (see also Section 2.2, page 44), or syntactic objects (terms) like those defined later on (Section 2.3, page 46), or even more complex objects like extended and constrained languages (see Definitions 5.1.7, page 110, and 5.3.4, page 128), or automata (chapter 7). What is important are the operations one can perform on these objects.

The *arity* $ar(t)$ of an object t can be thought of as the set of variables on which t "depends". So, Axiom 3 states that $\theta x.t$ depends on the same variables as t except for x.

The object $t[\rho]$ describes the effect of substituting $\rho(x)$ for x in t. Of course this substitution has no effect on the variables that are not in the arity of t, as expressed by Axiom 5.

Axioms 1 and 4 state that each single variable can be viewed as an object. Axiom 4 can be also read as $id \star \rho = \rho$.

Axiom 2 characterizes the free variables of $t[\rho]$ as those occurring free in some $\rho(x)$, where x occurs free in t.

Axiom 6 states that substitution is associative (see Lemma 2.1.2 below).

Finally, Axiom 7 explains how it is possible to perform substitution in presence of bound variables, using a suitable renaming of variables (called α-conversion in the λ-calculus). Let $t = \mu x.f(x, y)$ and let ρ be a substitution that substitutes $h(x)$ for y and leaves x unchanged. A textual substitution gives $\mu x.f(x, h(x))$ which is not the intended result because the free occurrence of x in $h(x)$ falls in the scope of the quantifier μx and becomes bound. Instead, we rewrite $\mu x.f(x, y)$ into $\mu z.f(z, y)$ and now we can safely apply the substitution, getting $\mu z.f(z, h(x))$. Obviously, we have to replace x by a variable that does not occur in what we substitute for y but, in order to satisfy Axiom 6, not any such z can be used. Then we only require that there exists at least one such z that can be substituted for x.

Lemma 2.1.2 (Associativity of composition). *Let $\langle T, id, \mathrm{ar}, \mathrm{comp}, \mu, \nu\rangle$ be a μ-calculus and let σ, ρ, π be substitutions. Then $(\sigma \star \rho) \star \pi = \sigma \star (\rho \star \pi)$.*

Proof. For any element $t \in T$, we have $t[\sigma][\rho] = t[\sigma \star \rho]$ and $t[\sigma][\rho][\pi] = t[\sigma\star\rho][\pi] = t[(\sigma\star\rho)\star\pi]$. On the other hand $t[\sigma][\rho][\pi] = t[\sigma][\rho\star\pi] = t[\sigma\star(\rho\star\pi)]$.

Applying the equality $t[(\sigma \star \rho) \star \pi] = t[\sigma \star (\rho \star \pi)]$ for $t = \hat{x}$ we get the result, since that for any variable x, $\hat{x}[\rho'] = \rho'(x)$. □

It should also be noted that, because of Axiom 4, $id \star \rho = \rho$. However we do not in general have $t[id] = t$. A counterexample is given in Example 2.3.5 (page 48).

The following lemma gives a property of the substitution of variables used in Axiom 7 in case id is right neutral.

Lemma 2.1.3. *Let $X = ar(t[\rho\{\hat{z}/x\}])$ and $X' = ar(t[\rho\{\hat{z}'/x\}])$, with $z \neq z'$. If $z, z' \notin ar'(t, \rho)$, and $\rho(y)[id] = \rho(y)$ for all $y \in ar(t)$, then $z \in X \Leftrightarrow z' \in X'$ and $X - \{z\} = X' - \{z'\}$.*

Proof. First, let us show that $t[\rho\{\hat{z}'/x\}] = t[\rho\{\hat{z}/x\}][id\{\hat{z}'/z\}]$. Let $\pi = \rho\{\hat{z}/x\} \star id\{\hat{z}'/z\}$. For $y \in ar(t)$, $y \neq x$, $\pi(y) = \rho(y)[id\{\hat{z}'/z\}]$. Since $z \notin ar(\rho(y))$, $\rho(y)[id\{\hat{z}'/z\}] = \rho(y)[id] = \rho(y)$. For $y = x$, we have $\pi(x) = \hat{z}[id\{\hat{z}'/z\}] = \hat{z}'$.

It follows that $X' = ar(t[\rho\{\hat{z}/x\}][id\{\hat{z}'/z\}] = ar'(t[\rho\{\hat{z}/x\}], id\{\hat{z}'/z\})$ which is equal to X if $z \notin X$ and to $(X - \{z\}) \cup \{z'\}$ otherwise. In both cases $X' \subseteq (X - \{z\}) \cup \{z'\}$. Since $z' \notin ar'(t, \rho)$ we have $z' \notin X$. It follows that $z' \in X' \Rightarrow z \in X$ and $X' - \{z'\} \subseteq X - \{z\}$.

By reasons of symmetry, exchanging the role of z and z', we get $z \in X \Rightarrow z' \in X'$ and $X - \{z\} \subseteq X' - \{z'\}$. □

The definition of the notion of a homomorphism of μ-calculi is straightforward. Since this definition involves two μ-calculi, we should have to denote differently the operations *id*, *ar*,... of each one. However, the context forbids any ambiguity, so that we can use the same notation.

Definition 2.1.4. A *homomorphism* of a μ-calculus T into a μ-calculus T' is a mapping $\sigma : T \to T'$ such that

- $\forall x \in Var, \quad \sigma(\hat{x}) = \hat{x}$
- $ar(\sigma(t)) = ar(t)$,
- $\sigma(t[\rho]) = \sigma(t)[\sigma \circ \rho]$, where $\sigma \circ \rho$ is defined by $(\sigma \circ \rho)(x) = \sigma(\rho(x))$,
- $\sigma(\theta x.t) = \theta x.\sigma(t)$.

It is easy to see that homomorphisms are closed under composition. It is also obvious that $\sigma \circ (\rho \star \pi) = (\sigma \circ \rho) \star (\sigma \circ \pi)$ and that $\sigma \circ (\rho\{t/x\}) = (\sigma \circ \rho)\{\sigma(t)/x\}$.

2.2 Functional μ-calculi

Let D be a complete lattice. We define a μ-calculus whose objects are the monotonic mappings of any arity on D. We denote it by $\mathcal{F}(D)$ and name it the *functional* μ-calculus over D.

This μ-calculus is defined as follows. For any subset X of *Var*, let $\mathcal{F}(D, X)$ be the set of all monotonic mapping from D^X to D. In particular, $\mathcal{F}(D, \emptyset)$ can be identified with D.

Let $\mathcal{F}(D)$ be the disjoint union of all $\mathcal{F}(D, X)$, so that $ar(f)$ is the unique X such that $f \in \mathcal{F}(D, X)$.

We make $\mathcal{F}(D)$ a μ-calculus by setting

- $\hat{x} : D^{\{x\}} \to D$ is the mapping that identifies $D^{\{x\}}$ and D (i.e., the pair (x, d) with d). By definition, the arity of $\hat{x}$ is $\{x\}$.
- Let $Y = ar'(f, \rho)$ where f belongs to $\mathcal{F}(D, X)$, (i.e., $Y = \bigcup_{x \in X} ar(\rho(x)))$ where ρ is a substitution, i.e., $\forall x \in Var,\ \rho(x) : D^{ar(\rho(x))} \to D$, and let $g \in D^Y$. For each $x \in X$, let g_x be the restriction of g to $ar(\rho(x)) \subseteq Y$, and let $g' \in D^X$ be defined by $g'(x) = \rho(x)(g_x)$. Then $f[\rho]$ is the mapping in $\mathcal{F}(D, Y)$ defined by $f[\rho](g) = f(g')$.
- Let $X = ar(f)$ and $X' = X - \{x\}$. For any $g : X' \to D$, and any d in D, let $g_d : X \to D$ be such that $g_d(y) = g(y)$ for $y \in X'$, and, if $x \in X$ (in this case $X = X' \cup \{x\}$, otherwise, $X = X'$) then $g_d(x) = d$. Then we define $(\theta x.f)(g)$ as the extremal fixed point of the mapping that sends d to $f(g_d)$. In particular, if $x \notin ar(f)$, $\theta x.f = f$.

□

Let us show that $\mathcal{F}(D)$ is a μ-calculus.

It is easy to check that the first five axioms of a μ-calculus are satisfied and we will show that Axioms 6 and 7 hold.

A substitution ρ such that $\forall x \in Var, \rho(x) \in \mathcal{F}(D, \emptyset)$ can be seen as a mapping from Var to D. Conversely, a mapping $v : Var \to D$, called a *valuation*, can be seen as a substitution.

The following property is a straightforward consequence of the definition of the composition.

Proposition 2.2.1. *If v is a valuation from Var to D, then $f[v] = f(v')$ where v' is the restriction of v to $ar(f)$.*

Since two mappings of the same arity are equal if they have the same value at every point of their domain we get the following extensionality principle.

Proposition 2.2.2. *Let f and f' be two elements of $\mathcal{F}(D)$ having the same arity. Then $f = f'$ if and only if $\forall v : Var \to D, f[v] = f'[v]$.*

Let us show that Axiom 6 holds.

Proposition 2.2.3. $f[\sigma][\rho] = f[\sigma \star \rho]$.

Proof. Let us prove this equality when $\rho \in D^{Var}$ is a valuation. Let $X = ar(f)$, $Y_x = ar(\sigma(x))$, and $Y = \bigcup_{x \in X} Y_x = ar(f[\sigma])$. We have $f[\sigma][\rho] = f[\sigma](\rho \upharpoonright Y)$, and, by definition of composition, $f[\sigma](\rho \upharpoonright Y) = f(g)$ where $g : X \to D$ is defined by $g(x) = \sigma(x)(\rho \upharpoonright Y \upharpoonright Y_x) = \sigma(x)(\rho \upharpoonright Y_x) = \sigma(x)[\rho] = \sigma \star \rho(x)$, so that $g = (\sigma \star \rho) \upharpoonright X$. Hence $f(g) = f[\sigma \star \rho]$.

Now, by Proposition 2.2.1, to prove the result, we have to prove that for any $v : Var \to D$, $f[\sigma][\rho][v] = f[\sigma \star \rho][v]$. Since $f[\sigma][\rho][v] = f[\sigma][\rho \star v] = f[\sigma \star (\rho \star v)]$ and $f[\sigma \star \rho][v] = f[(\sigma \star \rho) \star v]$, let us prove that the two substitutions $\pi_1 = \sigma \star (\rho \star v)$ and $\pi_2 = (\sigma \star \rho) \star v$ are equal: For any variable x, $\pi_1(x) = \sigma(x)[\rho \star v] = \sigma(x)[\rho][v]$ and $\pi_2(x) = (\sigma \star \rho)(x)[v] = \sigma(x)[\rho][v]$. □

Now, let us prove that a stronger version of Axiom 7 holds.

Proposition 2.2.4. *If $ar'(\theta x.f, \rho) \neq Var$, for any variable $y \notin ar'(\theta x.f, \rho)$, $(\theta x.f)[\rho] = \theta y.(t[\rho\{\hat{y}/x\}])$.*

Proof. From the definition of $\theta x.f$ it is easy to see that for any $v : Var \to D$, $(\theta x.f)[v]$ is the extremal fixed point of the mapping that sends $d \in D$ to $f[v\{d/x\}]$.

Since $(\theta x.f)[\rho][v] = (\theta x.f)[\rho \star v]$, it is the extremal fixed point of the mapping that sends d to $f[(\rho \star v)\{d/x\}]$. Similarly, $(\theta y.(f[\rho\{\hat{y}/x\}]))[v]$ is the extremal fixed point of the mapping that sends d to $f[\rho\{\hat{y}/x\}][v\{d/y\}]$.

Let us prove that $\pi_1 = (\rho \star v)\{d/x\}$ and $\pi_2 = (\rho\{\hat{y}/x\}) \star (v\{d/y\})$ are equal on $ar(f) \cup \{x\}$.

Clearly, $\pi_1(x) = d$ and $\pi_2(x) = \hat{y}[v\{d/y\}] = d$. If $z \in ar(f) - \{x\}$, then $\pi_1(z) = (\rho \star v)(z) = \rho(z)[v]$, and $\pi_2(z) = \rho\{\hat{y}/x\}(z)[v\{d/y\}] = \rho(z)[v\{d/y\}]$. Since $y \notin ar'(\theta x.f, \rho) \supseteq ar(\rho(z))$, v and $v\{d/y\}$ are equal on $ar(\rho(z))$. □

Since $f[id] = f$ we get the following result that allows us to rename a bound variable, provided we do not capture free variables.

Corollary 2.2.5. *If $y \notin ar(\theta x.f)$ then $\theta x.f = \theta y.(f[id\{\hat{y}/x\})$*

Finally, let us remark that if D is a Boolean algebra, we have $\widetilde{f[\rho]} = \tilde{f}[\tilde{\rho}]$ where $\tilde{\rho}$ is the mapping from Var to T defined by $\tilde{\rho}(x) = \widetilde{\rho(x)}$.

2.3 Fixed-point terms

In Section 2.1 (page 42) we introduced a very abstract notion of the μ-calculus, based on a set of objects of arbitrary nature. In the previous section, we have seen an example of such a calculus made out of the semantical concepts considered in Chapter 1: complete lattices, monotone functions, and extremal fixed-point operators. Now we will develop the syntactic aspect of the theory, first mentioned in Section 1.2.3, page 14. We define fixed-point terms by extending the usual concept of term, familiar from universal algebra and first-order logic, by extremal fixed-point operators (considered syntactically). We organize these fixed-point terms into a μ-calculus, syntactic in flavour, which can be viewed as an analogue of a familiar concept of a free algebra of terms. We will then see that a natural interpretation of fixed-point terms (i.e., according to the lines of Chapter 1) is a homomorphism from the μ-calculus of fixed-point terms to the functional μ-calculus of Section 2.2 (page 44).

2.3.1 Syntax

Let *Var* be a fixed infinite set of variables as defined page 42.

We also fix a family of countably infinite alphabets Fun_n, $n < \omega$, that are pairwise disjoint and disjoint from *Var*.

The symbols in Fun_n are considered as *functional letters of arity* (or *rank*) n. Let $Fun = \bigcup_{n<\omega} Fun_n$.

For any subset F of *Fun*, we define inductively three sets of terms: the base terms, the functional terms and the fixed-point terms over F.

Definition 2.3.1. The set $base\mathcal{T}(F)$ of base terms over F is defined by

- Each variable in *Var* is a base term.
- If $f \in F$ is of rank n, and if $x_1, \ldots, x_n$ are variables, not necessarily distinct, then $f(x_1, \ldots, x_n)$ is a base term.

Definition 2.3.2. The set $funct\mathcal{T}(F)$ of functional terms over F is defined inductively by

- Each variable in *Var* is a functional term.
- If $f \in F$ is of rank n, and if $t_1, \ldots, t_n$ are functional terms, then $f(t_1, \ldots, t_n)$ is a functional term.

Definition 2.3.3. The set $fix\mathcal{T}(F)$ of fixed-point terms over F is defined inductively by

- Each variable in *Var* is a fixed-point term.
- If $f \in F$ is of rank n, and if $t_1, \ldots, t_n$ are fixed-point terms, then $f(t_1, \ldots, t_n)$ is a fixed-point term.
- If $x \in Var$ and t is a fixed-point term, so are $\mu x.t$ and $\nu x.t$.

Obviously, $base\mathcal{T}(F) \subseteq funct\mathcal{T}(F) \subseteq fix\mathcal{T}(F)$.

Note that when we use the word "term" without any other mention, we mean "fixed-point term".

2.3.2 The μ-calculus of fixed-point terms

In this section, we show how the set of fixed-point terms can be viewed as a μ-calculus. (As in the previous chapter, we shall use the letter θ, sometimes with subscripts and superscripts, as appropriate) for denoting μ or ν.)

The *arity* of a fixed-point term t is the finite set $ar(t) \subseteq Var$, defined by induction on the structure of t.

- $ar(x) = \{x\}$,
- $ar(f(t_1, t_2, \ldots, t_n)) = ar(t_1) \cup ar(t_2) \cup \cdots \cup ar(t_n)$,
- $ar(\theta x.t) = ar(t) - \{x\}$.

It is obvious that Axioms 1 and 3 hold, setting $\hat{x} = x$. We shall refer to the variables in $ar(t)$ as to the *free variables* of t. If $ar(t)$ is empty, the term t is called a *closed* term.

Let us define *comp* and let us show that Axioms 2 and 4–7 hold. To define *comp* we use the well-founded total ordering of the variables.

Definition 2.3.4. Let t be a fixed-point term in $fix\mathcal{T}(F)$ and let $\rho : Var \to fix\mathcal{T}(F)$ be a substitution. We define the composition $comp(t, \rho) = t[\rho] \in fix\mathcal{T}(F)$ by induction on the structure of t as follows.

- If $t = x$ then $t[\rho] = \rho(x)$.
- If $t = f(t_1, \ldots, t_n)$ with $f \in Fun_n$, then $t[\rho] = f(t_1[\rho], \ldots, t_n[\rho])$.
- If $t = \theta x.t'$, let z be the first variable in Var that does not belong to $ar'(t, \rho)$. (Note that z may equal x if x itself is the first one with this property.) We set $(\theta x.t')[\rho] = \theta z.(t'[\rho\{\hat{z}/x\}])$.

Note that this last equality is exactly as in Axiom 7.

Example 2.3.5. Let id be the identity on Var and let us compute $(\theta y.y)[id]$, assuming that x is the unique variable before y. Since $ar(y) = \{y\}$, $ar'(y, id) = ar(id(y)) = ar(y) = \{y\}$. The variable z selected by the construction above is x, and $(\theta y.y)[id] = \theta x.(y[id\{x/y\}]) = \theta x.x$. □

This example shows that, surprisingly enough, t is not always equal to $t[id]$. However, we shall see in Section 2.4.2 (see Proposition 2.4.2, page 52) in which sense t and $t[id]$ can be considered equivalent.

Proposition 2.3.6. *The tuple* $\langle \text{fix}\mathcal{T}(F), ar, \text{comp}, \mu, \nu\rangle$ *defined above is a* μ*-calculus.*

Proof. It remains to check Axioms 2 and 4–6.

Axiom 4 holds by definition.

The proof of Axioms 2, 5, 6 is by induction on t and there is no problem when $t = x$ or $t = f(t_1, \ldots, t_n)$. Thus, let us consider only the case when $t = \theta x.t'$.

First note that if $\rho \restriction ar(t) = \rho' \restriction ar(t)$ then $ar'(t, \rho) = ar'(t, \rho')$, thus $(\theta x.t')[\rho] = \theta z.t'[\rho\{\hat{z}/x\}]$ and $(\theta x.t')[\rho'] = \theta z.t'[\rho'\{\hat{z}/x\}]$ (with the same z). By definition of $\rho\{\hat{z}/x\}$ and of $\rho'\{\hat{z}/x\}$, we have $\rho\{\hat{z}/x\} \restriction ar(t) \cup \{x\} = \rho'\{\hat{z}/x\} \restriction ar(t) \cup \{x\}$. Since $ar(t') \subseteq ar(t) \cup \{x\}$, by the induction hypothesis, $t'[\rho\{\hat{z}/x\}] = t'[\rho'\{\hat{z}/x\}]$, hence the result.

Let us prove that $ar(t[\rho]) = ar'(t, \rho)$. Since $t[\rho] = \theta z.(t'[\rho\{\hat{z}/x\}])$, and by the induction hypothesis, we have $ar(t[\rho]) = ar(t'[\rho\{\hat{z}/x\}]) - \{z\} = ar'(t', \rho\{\hat{z}/x\}) - \{z\}$. But, since $ar(t) = ar(t') - \{x\}$, $ar'(t', \rho\{\hat{z}/x\}) = ar'(t, \rho) \cup \{z \mid x \in ar(t')\}$, thus, $ar(t[\rho]) = ar'(t, \rho) - \{z\}$. Since $z \notin ar'(t, \rho)$, $ar(t[\rho]) = ar'(t, \rho)$.

Finally, let us show that for any substitutions ρ and π, $\theta x.t'[\rho][\pi] = \theta x.t'[\rho \star \pi]$. We have $(\theta x.t')[\rho] = \theta z.t'[\rho\{\hat{z}/x\}]$, where z is the first variable that does not occur in the set $ar(t[\rho])$. In turn, $t[\rho][\pi] = \theta z.t'[\rho\{\hat{z}/x\}][\pi] = \theta z'.t'[\rho\{\hat{z}/x\}][\pi\{\hat{z}'/z\}]$, where z' is the first variable that does not occur in $ar(t[\rho][\pi])$. By the induction hypothesis, $t[\rho][\pi] = \theta z'.t'[\rho\{\hat{z}/x\} \star \pi\{\hat{z}'/z\}]$. On the other hand, $t[\rho \star \pi] = \theta z''.t'[(\rho \star \pi)\{\hat{z}''/x\}]$, where z'' is the first variable not in $ar(t[\rho \star \pi])$. Let us show that $ar(t[\rho][\pi]) = ar(t[\rho \star \pi])$, from which we will get $z' = z''$. Since $ar(t[\rho]) = \bigcup_{y \in ar(t)} ar(\rho(y))$ and $ar(t[\rho][\pi]) = \bigcup_{y \in ar(t[\rho])} ar(\pi(y))$, we get

$$\begin{aligned}
ar(t[\rho][\pi]) &= \bigcup_{y \in ar(t)} \bigcup_{y' \in ar(\rho(y))} ar(\pi(y')) \\
&= \bigcup_{y \in ar(t)} ar(\rho(y)[\pi]) \\
&= \bigcup_{y \in ar(t)} ar(\rho \star \pi(y)) \\
&= ar(t[\rho \star \pi]).
\end{aligned}$$

It remains to prove that $\sigma_1 = \rho\{\hat{z}/x\} \star \pi\{\hat{z}'/z\}$ and $\sigma_2 = (\rho \star \pi)\{\hat{z}'/x\}$ have the same restriction to $ar(t') \subseteq ar(t) \cup \{x\}$. First, $\sigma_1(x) = \rho\{\hat{z}/x\} \star \pi\{\hat{z}'/z\}(x) = \rho\{\hat{z}/x\}(x)[\pi\{\hat{z}'/z\}] = \hat{z}[\pi\{\hat{z}'/z\}] = z' = \sigma_2(x)$. For $y \neq x$ and $y \in ar(t)$, $\sigma_1(y) = \rho\{\hat{z}/x\}(y)[\pi\{\hat{z}'/z\}] = \rho(y)[\pi\{\hat{z}'/z\}]$ and $\sigma_2(y) = (\rho \star \pi)\{\hat{z}'/x\}(y) = \rho \star \pi(y) = \rho(y)[\pi]$. Since $y \in ar(t)$, we have $ar(\rho(y)) \subseteq ar(t[\rho])$, and, by the choice of z, $z \notin ar(\rho(y))$. It follows that π and $\pi\{\hat{z}'/z\}$ have the same restriction to $ar(\rho(y))$, hence, $\sigma_1(y) = \sigma_2(y)$.

□

2.3.3 Semantics

Definition 2.3.7. A *μ-interpretation* for a subset F of *Fun* is a pair $\mathcal{I} = (\langle D_\mathcal{I}, \leq \rangle, I)$, where $\langle D_\mathcal{I}, \leq \rangle$ is a complete lattice that is called a *universe* or *domain* of the interpretation $\mathcal{I}$, and I is a function that with each $f \in F \cap Fun_n$ associates a monotonic mapping $f^\mathcal{I} : D_\mathcal{I}^n \to D_\mathcal{I}$.

Given an interpretation $\mathcal{I}$, we define a homomorphism from the μ-calculus $fix\mathcal{T}(F)$ of fixed-point terms over F to $\mathcal{F}(D_\mathcal{I})$, the functional μ-calculus over $D_\mathcal{I}$ (see Section 2.2, page 44). This homomorphism associates with each fixed-point term t a monotonic mapping $[t]_\mathcal{I} : D_\mathcal{I}^{ar(t)} \to D_\mathcal{I}$, called the *interpretation of t under $\mathcal{I}$*, defined by induction on t as follows.

Definition 2.3.8.

– If $t = x \in Var$ then $[x]_\mathcal{I} = \hat{x}$.

- If $t = f(t_1, \dots, t_n)$, then $[t]_{\mathcal{I}}$ is the mapping from $D_{\mathcal{I}}^X$ to $D_{\mathcal{I}}$ where $X = ar(t) = \bigcup_{i=1}^n ar(t_i)$, defined by $[t]_{\mathcal{I}}(v) = f^{\mathcal{I}}([t_1]_{\mathcal{I}}(v_1), \dots, [t_n]_{\mathcal{I}}(v_n))$, where v_i is the restriction of v to $ar(t_i)$.
- If $t = \theta x.t'$ then $[t]_{\mathcal{I}} = \theta x.[t']_{\mathcal{I}}$.

According to the usual convention in logic, we will allow the notation $[t]_{\mathcal{I}}(v)$ also if $v : Z \to D_{\mathcal{I}}$ is a valuation of some superset $Z \supseteq ar(t)$ with the meaning $[t]_{\mathcal{I}}(v \upharpoonright ar(t))$.

Taking into account Proposition 2.2.1 (page 45) we get directly the following from the definition of $[f(t_1, \dots, t_n)]_{\mathcal{I}}$.

Lemma 2.3.9. $[f(t_1, \dots, t_n)]_{\mathcal{I}}[v] = f^{\mathcal{I}}([t_1]_{\mathcal{I}}[v], \dots, [t_n]_{\mathcal{I}}[v])$.

Proposition 2.3.10. *The mapping that, with t, associates $[t]_{\mathcal{I}}$ defined above is a homomorphism from* fix$\mathcal{T}(F)$ *to* $\mathcal{F}(D_{\mathcal{I}})$. *It is the unique homomorphism σ such that $\sigma(f(x_1, \dots, x_{ar(f)}))(v) = f^{\mathcal{I}}(v(x_1), \dots, v(x_{ar(f)}))$, for each $f \in F$, and any valuation $v : \{x_1, \dots, x_{ar(f)}\} \to D_{\mathcal{I}}$.*

Proof. We need only to prove by induction on t that $[t[\rho]]_{\mathcal{I}} = [t]_{\mathcal{I}}[[\rho]_{\mathcal{I}}]$, the other points being obvious from the definition of $[t]_{\mathcal{I}}$.

Obviously, since $x[\rho] = \rho(x)$, $[x[\rho]]_{\mathcal{I}} = [\rho(x)]_{\mathcal{I}} = \hat{x}[[\rho]_{\mathcal{I}}]$.

Let us show that $[f(t_1, \dots, t_n)[\rho]]_{\mathcal{I}}[v] = [f(t_1, \dots, t_n)]_{\mathcal{I}}[[\rho]_{\mathcal{I}}][v]$ for any $v \in D_{\mathcal{I}}^{Var}$. Since $f(t_1, \dots, t_n)[\rho] = f(t_1[\rho], \dots, t_n[\rho])$, we get, by the above lemma, $[f(t_1, \dots, t_n)[\rho]]_{\mathcal{I}}[v] = f^{\mathcal{I}}([t_1[\rho]]_{\mathcal{I}}[v], \dots, [t_n[\rho]]_{\mathcal{I}}[v])$ that is equal, by the induction hypothesis, to $f^{\mathcal{I}}([t_1]_{\mathcal{I}}[[\rho]_{\mathcal{I}}][v], \dots, [t_n]_{\mathcal{I}}[[\rho]_{\mathcal{I}}][v])$. Since the composition is associative, this is equal to $f^{\mathcal{I}}([t_1]_{\mathcal{I}}[[\rho]_{\mathcal{I}} \star v], \dots, [t_n]_{\mathcal{I}}[[\rho]_{\mathcal{I}} \star v])$, and, again by the lemma, to $[f(t_1, \dots, t_n)]_{\mathcal{I}}[[\rho]_{\mathcal{I}} \star v] = [f(t_1, \dots, t_n)]_{\mathcal{I}}[[\rho]_{\mathcal{I}}][v]$.

Finally, since $(\theta x.t)[\rho] = \theta z.t[\rho\{\hat{z}/x\}]$, with $z \notin ar'(t, \rho)$, we have

$$
\begin{aligned}
[(\theta x.t)[\rho]]_{\mathcal{I}} &= [\theta z.t[\rho\{\hat{z}/x\}]]_{\mathcal{I}} \\
&= \theta z.[t[\rho\{\hat{z}/x\}]]_{\mathcal{I}} \\
&= \theta z.([t]_{\mathcal{I}}[[\rho\{\hat{z}/x\}]_{\mathcal{I}}]) \\
&= \theta z.([t]_{\mathcal{I}}[[\rho]_{\mathcal{I}}\{\hat{z}/x\}]).
\end{aligned}
$$

On the other hand since $z \notin ar'(t, \rho) = ar'([t]_{\mathcal{I}}, [\rho]_{\mathcal{I}})$, we have, by Proposition 2.2.4 (page 46), $[\theta x.t]_{\mathcal{I}}[[\rho]_{\mathcal{I}}] = (\theta x.[t]_{\mathcal{I}})[[\rho]_{\mathcal{I}}] = \theta z.([t]_{\mathcal{I}}[[\rho]_{\mathcal{I}}\{\hat{z}/x\}])$.

The uniqueness of this homomorphism is obvious, by definition of $[t]_{\mathcal{I}}$. □

2.4 Quotient μ-calculi

Let F be a set of functional symbols. An equivalence $\sim$ on *fix*$\mathcal{T}(F)$ is said to be a *congruence* if

- $t \sim t' \Rightarrow ar(t) = ar(t')$,
- $t \sim t' \Rightarrow \theta x.t \sim \theta x.t'$,
- $t \sim t' \Rightarrow t[\rho] \sim t'[\rho]$,
- $\forall x \in ar(t), \rho(x) \sim \rho'(x) \Rightarrow t[\rho] \sim t[\rho']$.

Then it is easy to see that the quotient $\mathit{fix}\mathcal{T}(F)/\sim$ is still a μ-calculus and that the mapping that sends t on its equivalence class is a homomorphism. We give two examples of such quotients.

2.4.1 Families of interpretations

Let $\mathcal{C}$ be a family of interpretations. We write $\tau =_{\mathcal{C}} \tau'$ if $[\tau]_{\mathcal{I}} = [\tau']_{\mathcal{I}}$ for any $\mathcal{I} \in \mathcal{C}$. It follows easily by the definition of a functional μ-calculus that $=_{\mathcal{C}}$ is a congruence.

2.4.2 Variants of fixed-point terms

As usual when expressions contain bound variables, we can say that two expressions that differ only by the names of their bound variables are in some sense equivalent. The transformation consisting in changing the names of bound variables is called *α-conversion* as in the λ-calculus. The α-conversion can also be applied to the μ-calculus. For instance, it is natural to consider that $\theta x.f(x)$ and $\theta y.f(y)$ are equivalent fixed-point terms with respect to α-conversion. Also, as suggested by the case of functional μ-calculi, if $f(y)$ is independent of x, we would like $\theta x.f(y)$ be equivalent to $f(y)$, and we require that this equivalence be a congruence.

Formally, we define a relation on the set $\mathit{fix}\mathcal{T}(F)$ of fixed-point terms over F, denoted by $\approx$, as the least equivalence relation such that

(v1) if $f \in F$ is of rank n and if $t_i \approx t'_i$, for $i = 1, \ldots, n$, then $f(t_1, \ldots, t_n) \approx f(t'_1, \ldots, t'_n)$,
(v2) if $t \approx t'$ then $\theta x.t \approx \theta x.t'$ for any variable x,
(v3) if $y \notin ar(t)$ then $\theta x.t \approx \theta y.(t[id\{\hat{y}/x\}])$,
(v4) if $x \notin ar(t)$ then $\theta x.t \approx t$.

We say that t' is a *variant of* t if $t \approx t'$.

Example 2.4.1. We show that the closed term $\mu x_1.\nu x_2.f(x_1, x_2)$ is a variant of $\mu y_1.\nu y_2.f(y_1, y_2)$ when $x_1 \neq x_2$ and $y_1 \neq y_2$.

By (v3), $\mu x_1.\nu x_2.f(x_1, x_2) \approx \mu y_1.((\nu x_2.f(x_1, x_2))[id\{\hat{y_1}/x_1\}])$. By definition of the μ-calculus on terms,

$$(\nu x_2.f(x_1, x_2))[id\{\hat{y_1}/x_1\}] = \nu x'_2.(f(x_1, x'_2)[id\{\hat{y_1}/x_1\}\{\hat{x'_2}/x_2\}])$$

with $x'_2 \notin \{x_1, x_2, y_1, y_2\}$. Hence, $f(x_1, x'_2)[id\{\hat{y_1}/x_1\}\{\hat{x'_2}/x_2\}] = f(y_1, x'_2)$, and $\mu x_1.\nu x_2.f(x_1, x_2) = \mu y_1.\nu x'_2.f(y_1, x'_2)$.
Again by (v3) we have $\nu x'_2.f(y_1, x'_2) = \nu y_2.f(y_1, y_2)$, and the result follows by (v2). □

We prove that $\approx$ is a congruence.

Proposition 2.4.2. *(i)* $t[id] \approx t$,
(ii) if for any $x \in ar(t)$, $\rho(x) \approx \rho'(x)$ *then* $t[\rho] \approx t[\rho']$.

Proof. The proof of (i) and (ii) is by induction on t and the only non trivial case to consider is when $t = \theta x.t'$.

By definition, $t[id] = \theta z.(t'[id\{z/x\}])$ which is a variant of $\theta x.t'$.

We have $t[\rho] = \theta z.(t'[\rho\{z/x\}])$ and $t[\rho'] = \theta z'.(t'[\rho'\{z'/x\}])$. We can choose a variable y, neither in $ar(t'[\rho\{z/x\}])$ nor in $ar(t'[\rho'\{z'/x\}])$, so that

$$\theta z.(t'[\rho\{z/x\}]) \approx \theta y.(t'[\rho\{z/x\}][id\{\hat{y}/z\}])$$

and

$$\theta z'.(t'[\rho'\{z'/x\}]) \approx \theta y.(t'[\rho'\{z'/x\}][id\{\hat{y}/z'\}]).$$

But, on $ar(t')$, $\rho\{z/x\}\star id\{\hat{y}/z\} = \rho\{\hat{y}/x\}$ and $\rho'\{z'/x\}\star id\{\hat{y}/z'\} = \rho'\{\hat{y}/x\}$. We can apply the induction hypothesis and we get $t'[\rho\{\hat{y}/x\}] \approx t'[\rho'\{\hat{y}/x\}]$, hence the result. □

Proposition 2.4.3. *if* $t \approx t'$ *then*

(i) $ar(t) = ar(t')$,
(ii) $t[\rho] \approx t'[\rho]$ *for any substitution* ρ,
(iii) $\|t\|_{\mathcal{I}} = \|t'\|_{\mathcal{I}}$ *for any interpretation* $\mathcal{I}$.

Proof. The proof is by induction on the definition of $\approx$. Let $t \approx t'$. We have to consider the following cases.

Case 1. $t = f(t_1, \ldots, t_n)$, $t' = f(t'_1, \ldots, t'_n)$, with $t_i \approx t'_i$.

(i) $ar(t) = \bigcup_{i=1}^{n} ar(t_i) = \bigcup_{i=1}^{n} ar(t'_i) = ar(t')$,
(ii) $t[\rho] = f(t_1[\rho], \ldots, t_n[\rho]) \approx f(t'_1[\rho], \ldots, t'_n[\rho]) = t'[\rho]$,
(iii) For any $v : Var \to D_{\mathcal{I}}$, we have $\|t\|_{\mathcal{I}}[v] = f^{\mathcal{I}}(\|t_1\|_{\mathcal{I}}[v], \ldots, \|t_n\|_{\mathcal{I}}[v]) = f^{\mathcal{I}}(\|t'_1\|_{\mathcal{I}}[v], \ldots, \|t'_n\|_{\mathcal{I}}[v]) = \|t'\|_{\mathcal{I}}[v]$.

Case 2. $t = \theta x.s$ and $t' = \theta x.s'$ with $s \approx s'$.

(i) $ar(t) = ar(s) - \{x\} = ar(s') - \{x\} = ar(t')$.
(ii) By definition, $t[\rho] = \theta z.(s[\rho\{\hat{z}/x\}])$ where z is the least variable not in $ar'(t, \rho) = ar(t[\rho]) = ar(t'[\rho]) = ar'(t', \rho)$. Thus, $t'[\rho] = \theta z.(s'[\rho\{\hat{z}/x\}])$. Since $s \approx s'$, we get $s[\rho\{\hat{z}/x\}] \approx s'[\rho\{\hat{z}/x\}]$, and the result follows.

(iii) $[t]_{\mathcal{I}} = \theta x.[s]_{\mathcal{I}} = \theta x.[s']_{\mathcal{I}} = [t']_{\mathcal{I}}$.

Case 3. $t = \theta x.s$ and $t' = \theta y.(s[id\{\hat{y}/x\}])$, where $y \notin ar(s)$.

(i) $ar(t) = ar(s) - \{x\}$ and $ar(t') = Y - \{y\}$ where $Y = ar'(s, id\{\hat{y}/x\})$. But $ar'(s, id\{\hat{y}/x\}) = (ar(s) - \{x\}) \cup \{y \mid x \in ar(s)\}$, hence $Y - \{y\} = (ar(s) - \{x\}) - \{y\}$. Since $y \notin ar(s)$, $Y - \{y\} = ar(s) - \{x\} = ar(t)$.
(ii) Since $ar(t) = ar(t')$, $(\theta x.s)[\rho] = \theta z.(s[\rho\{\hat{z}/x\}])$ and $(\theta y.s[id\{\hat{y}/x\}])[\rho] = \theta z.(s[id\{\hat{y}/x\}][\rho\{\hat{z}/y\}])$. Let $\rho' = id\{\hat{y}/x\} \star \rho\{\hat{z}/y\}$. For $v \in ar(s)$ and $v \neq x$, we have $\rho'(v) = \rho\{\hat{z}/y\}(v)$, and, since $y \notin ar(s)$, v is not equal to y and $\rho'(v) = \rho(v)$. It is obvious that $\rho'(x) = \hat{z}$. It follows that $\theta z.(s[id\{\hat{y}/x\}][\rho\{\hat{z}/y\}]) = \theta z.(s[\rho\{\hat{z}/x\}])$.
(iii) $[t]_{\mathcal{I}} = \theta x.[s]_{\mathcal{I}}$ and $[t']_{\mathcal{I}} = \theta y.[s[id\{\hat{y}/x\}]]_{\mathcal{I}} = \theta x.[s]_{\mathcal{I}}$, by Proposition 2.2.5 (page 46).

Case 4. Let $t = \theta x.t'$ and $x \notin ar(t')$.

(i) $ar(t) = ar(t') - \{x\} = ar(t')$.
(ii) $t[\rho] = (\theta x.t')[\rho] = \theta z.(t'[\rho\{\hat{z}/x\}])$, with $z \notin ar(t[\rho]) = ar(t'[\rho])$. Since $x \notin ar(t')$, $\rho\{\hat{z}/x\} = \rho$ on $ar(t')$. Hence, $t[\rho] = \theta z.(t'[\rho])$, and, since $z \notin ar(t'[\rho])$, $\theta z.(t'[\rho]) \approx t'[\rho]$.
(iii) Since $x \notin ar(t')$, $[t]_{\mathcal{I}} = \theta x.[t']_{\mathcal{I}} = [t']_{\mathcal{I}}$.

□

2.5 Powerset interpretations

Most often in applications, the domain of a μ-interpretation is a complete lattice $\mathcal{P}(E)$ of all subsets of some set E, ordered by the subset ordering $\subseteq$. We call such an interpretation a *powerset interpretation*. Clearly, not every μ-interpretation is of that form; for example, by a cardinality argument, no countably infinite $D_{\mathcal{I}}$ can be identified with $\mathcal{P}(E)$. However, we show in this section that any μ-interpretation naturally induces a certain powerset interpretation, and actually is embedded in it, such that the interpretation of fixed-point terms is preserved by this embedding. Therefore, we do not lose much generality by the restriction to powerset interpretations.

Let F be a set of functional symbols and let $\mathcal{I}$ be an interpretation of domain E that associates with each symbol $f \in F$ a monotonic mapping $f^{\mathcal{I}} : E^n \to E$ where n is the arity of f.

Let $E_+ = E - \{\bot\}$. Let $D : E \to \mathcal{P}(E_+)$ and $S : \mathcal{P}(E_+) \to E$ be defined by $D(e) = \{e' \in E_+ \mid e' \leq e\}$ and for $E' \subseteq E_+$, $S(E') = \bigvee E'$. Obviously these two mappings are monotonic and for any $e \in E$, $S(D(e)) = e$.

With the interpretation $\mathcal{I}$ we associate an interpretation $\mathcal{J}$ whose domain is $\mathcal{P}(E_+)$, defined as follows. For any $f \in F$, let $f^{\mathcal{J}} : \mathcal{P}(E_+)^n \to \mathcal{P}(E_+)$ be defined by

$$f^{\mathcal{J}}(E_1, \ldots, E_n) = D(f^{\mathcal{I}}(S(E_1), \ldots, S(E_n))).$$

Finally, for $v : \mathit{Var} \to \mathcal{P}(E_+)$, $S \circ v : \mathit{Var} \to E$ is the substitution defined by $(S \circ v)(x) = S(v(x))$.

Proposition 2.5.1. *For any term t and any valuation $v : \mathit{Var} \to \mathcal{P}(E_+)$, we have*

1. $S([\![t]\!]_{\mathcal{J}}[v]) = [\![t]\!]_{\mathcal{I}}[S \circ v]$,
2. *if t is not a variant of a term consisting of a single variable $x \in \mathit{Var}$ then* $[\![t]\!]_{\mathcal{J}}[v] = D([\![t]\!]_{\mathcal{I}}[S \circ v])$.

Proof. The proof is by induction on t, where we use the obvious remark that if $[\![t]\!]_{\mathcal{J}}[v] = D([\![t]\!]_{\mathcal{I}}[S \circ v])$ then $S([\![t]\!]_{\mathcal{J}}[v]) = [\![t]\!]_{\mathcal{I}}[S \circ v]$.

If $t = x$ then $S([\![t]\!]_{\mathcal{J}}[v]) = [\![t]\!]_{\mathcal{I}}[S \circ v] = S(v(x))$.

If $t = f(t_1, \ldots, t_n)$ then

$$[\![t]\!]_{\mathcal{J}}[v] = f^{\mathcal{J}}(E_1, \ldots, E_n) = D(f^{\mathcal{I}}(S(E_1), \ldots, S(E_n)))$$

with $E_i = [\![t_i]\!]_{\mathcal{J}}[v]$. By the induction hypothesis, $S(E_i) = [\![t_i]\!]_{\mathcal{I}}[S \circ v]$ hence, $[\![t]\!]_{\mathcal{J}}[v] = D([\![t]\!]_{\mathcal{I}}[S \circ v])$.

If $t = \theta x.t'$, there are two cases to consider according to whether t' is a variant of a variable or not.

If $t' \approx y \in \mathit{Var}$ we again have two cases according to whether y is equal to x or not. If $t' \approx y \neq x$ then $t = \theta x.t' \approx y$ and $S([\![t]\!]_{\mathcal{J}}[v]) = [\![t]\!]_{\mathcal{I}}[S \circ v] = S(v(y))$. If $x = y$ then $t \approx \theta x.x$. In this case we have

- If $\theta = \mu$ then $[\![t]\!]_{\mathcal{J}}[v] = [\![\mu x.x]\!]_{\mathcal{J}} = \emptyset$ and $[\![t]\!]_{\mathcal{J}}[v] = \bot$. Since $\emptyset = D(\bot)$ the result is proved.
- If $\theta = \nu$ then $[\![t]\!]_{\mathcal{J}}[v] = E_+$ and $[\![t]\!]_{\mathcal{J}}[v] = \top$ and $E_+ = D(\top)$.

In other cases, $[\![t]\!]_{\mathcal{J}}[v] = \theta x.h(x)$ where $h(x) = [\![t']\!]_{\mathcal{J}}[v\{x/x\}]$, and $[\![t]\!]_{\mathcal{I}}[S \circ v] = \theta x.g(x)$ where $g(x) = [\![t']\!]_{\mathcal{I}}[S \circ v\{x/x\}]$. By the induction hypothesis, for any $E' \subseteq E_+$, $h(E') = D(g(S(E')))$, hence $\theta x.h(x) = \theta x.D(g(S(x)))$. By Proposition 1.3.12 (page 23), $\theta x.D(g(S(x))) = D(\theta x.g(S(D(x))))$ which is also equal to $D(\theta x.g(x))$. □

An immediate consequence of this proposition is:

Theorem 2.5.2. *For any term t that is not a variant of a variable and for any valuation $v : \mathit{Var} \to E$, for any $e \neq \bot$, $e \leq [\![t]\!]_{\mathcal{I}}[v] \Leftrightarrow e \in [\![t]\!]_{\mathcal{J}}[D \circ v]$.*

Since a closed term is never a variant of a variable, we deduce from this theorem the following corollary.

Corollary 2.5.3. *Let $\mathcal{I}$ be an interpretation and let $\mathcal{J}$ be the powerset interpretation constructed above. For any closed terms t_1 and t_2, $[t_1]_{\mathcal{I}} = [t_2]_{\mathcal{I}}$ if and only if $[t_1]_{\mathcal{J}} = [t_2]_{\mathcal{J}}$.*

2.6 Alternation-depth hierarchy

2.6.1 Clones in a μ-calculus

Definition 2.6.1. Let T be a μ-calculus. A subset $C \subseteq T$ is called a *clone* if it contains *Var* and is closed under composition in the following sense: If $t \in C$ and if ρ is a substitution such that $\rho(y) \in C$ for any $y \in ar(t)$ then $t[\rho] \in C$.

and for any interpretation $\mathcal{I}$, A clone C is a *μ-clone* if it is additionally closed under the μ operator in the following sense: if t is in C, so is $\mu x.t$. Similarly, C is a *ν-clone* if it is closed under the ν operator and it is a *fixed-point clone* if it closed under both μ and ν, that is, if it is both a μ-clone and a ν-clone.

It is easy to see that the intersection of a nonempty family of clones of any type is again a clone of this type. Therefore, for any set $T' \subseteq T$, there exists a least clone containing T'; we shall denote it by $Comp(T')$.[1] Similarly, there exists a least μ-clone, a least ν-clone, and a least fixed-point clone containing T', which we shall denote respectively by $\mu(T')$, $\nu(T')$, and $fix(T')$.

Example 2.6.2. The set $functT(F)$ is a clone. It is the least clone containing the base terms, thus $functT(F) = Comp(baseT(F))$. The set $fixT(F)$ is a fixed-point clone that is equal to $fix(functT(F)) = fix(baseT(F))$. □

2.6.2 A hierarchy of clones

Given a subset T' of T we define a hierarchy of elements of T, relative to T', by

$$\Sigma_0(T') = \Pi_0(T') = Comp(T'),$$

and, for $k < \omega$,

$$\begin{aligned} \Sigma_{k+1}(T') &= \mu(\Pi_k(T')) \\ \Pi_{k+1}(T') &= \nu(\Sigma_k(T')) \end{aligned}$$

[1] Note that the existence of this clone can be also inferred from Knaster–Tarski theorem since $Comp(T')$ is the least fixed point of an obvious closure operator.

Proposition 2.6.3. *If σ is a homomorphism from T_1 to T_2, then for any $T' \subseteq T_1$, we have $\sigma(\Sigma_k(T')) \subseteq \Sigma_k(\sigma(T'))$ and $\sigma(\Pi_k(T')) \subseteq \Pi_k(\sigma(T'))$.*

Since $T_1 \subseteq T_2$ implies $\theta(T_1) \subseteq \theta(T_2)$, for $\theta \in \{\mu, \nu\}$, we have some obvious inclusions between the classes $\Sigma_k(T)$ and $\Pi_k(T)$ which can be summarized as

$$\Sigma_k(T) \cup \Pi_k(T) \subseteq Comp(\Sigma_k(T) \cup \Pi_k(T)) \subseteq \Sigma_{k+1}(T) \cap \Pi_{k+1}(T).$$

This is depicted in the diagram of Figure 2.1.

We denote by $\mathit{fix}(T')$ the set $\bigcup_{k \geq 0} \Sigma_k(T') = \bigcup_{k \geq 0} \Pi_k(T')$.

2.6.3 The syntactic hierarchy

When T is the μ-calculus over $\mathit{fix}\mathcal{T}(F)$ (see Section 2.3.1, page 47, and Proposition 2.3.6, page 48) we denote by $\Sigma_k(F)$ and $\Pi_k(F)$ the *syntactic hierarchy* of fixed point terms defined by

$$\Sigma_0(F) = \Pi_0(F) = \mathit{funct}\mathcal{T}(F),$$

and, for $k < \omega$,

$$\begin{aligned} \Sigma_{k+1}(F) &= \mu(\Pi_k(F)) \\ \Pi_{k+1}(F) &= \nu(\Sigma_k(F)) \end{aligned}$$

Moreover, it is clear that $\mathit{fix}\mathcal{T}(F) = \bigcup_{k \geq 0} \Sigma_k(F) = \bigcup_{k \geq 0} \Pi_k(F)$.

2.6.4 The Emerson-Lei hierarchy

A slightly different definition of a hierarchy for the set $\mathit{fix}\mathcal{T}(F)$ has been proposed by Emerson and Lei [35], in the context of the modal μ-calculus (see Section 6.2, page 145). Their definition is originally based on a concept of an alternation-depth of a formula which is defined "top-down". We can rephrase that definition in our setting, by inductively defining the classes Σ_k^{EL} and Π_k^{EL} of fixed-point terms as follows. Let $\mu_{EL}(T')$ be the closure of a set T' under the application of symbols in F and under the μ-operator; note that this class may be not closed under composition. Let $\nu_{EL}(T')$ be defined similarly. Let $\Sigma_0^{EL}(F) = \Pi_0^{EL}(F) = \mathit{funct}\mathcal{T}(F)$ and let $\Sigma_{k+1}^{EL}(F) = Comp(\mu_{EL}(\Pi_k^{EL}))$, $\Pi_{k+1}^{EL}(F) = Comp(\nu_{EL}(\Sigma_k^{EL}))$.

Of course, $\bigcup_{k \geq 0} \Sigma_k^{EL}(F) = \bigcup_{k \geq 0} \Pi_k^{EL}(F) = \mathit{fix}\mathcal{T}(F)$ and it is easy to see that $\Sigma_k^{EL}(F) \subseteq \Sigma_k(F)$ and $\Pi_k^{EL}(F) \subseteq \Pi_k(F)$, but these inclusions are strict.

For example, the term $\mu x.\nu y.f(x, y, \mu z.\nu w.f(x, z, w))$, where $f \in F$, is in $\Sigma_2(F)$ but not in $\Sigma_2^{EL}(F)$. To see that it is in $\Sigma_2(F)$, note that so are

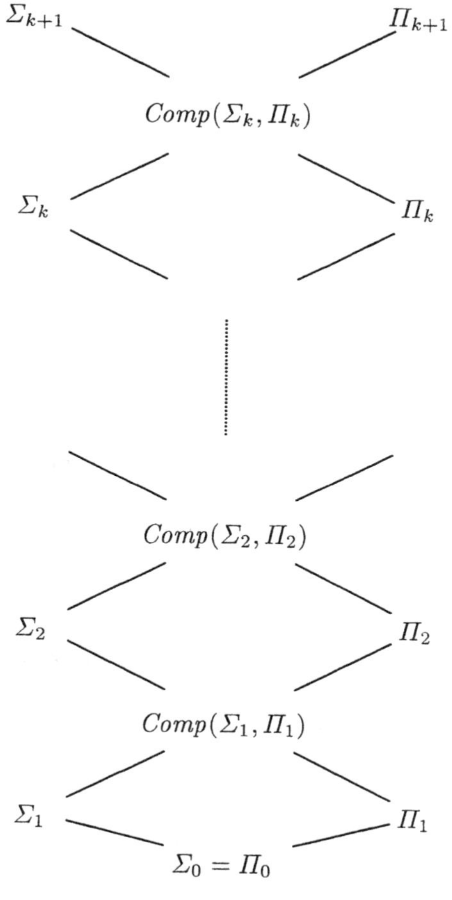

Fig. 2.1. The hierarchy of clones

the terms $\mu z.\nu w.f(x,z,w)$, $\nu y.f(x,y,v)$ and $\nu y.f(x,y,\mu z.\nu w.f(x,z,w))$. On the other hand, our term cannot be obtained by composition of two terms in $\Sigma_2^{EL}(F)$ since the variable x occurs free in $\mu z.\nu w.f(x,z,w)$. This term is actually of alternation depth 3 in the sense of Emerson and Lei [35].

2.7 Vectorial μ-calculi

In Section 1.4, page 24, we have introduced a notation for vectorial fixed-point terms. We now wish to incorporate this issue in the formalism of the present chapter. To this end, it is convenient to introduce the concept of a *vectorial μ-calculus*, i.e., a vectorial extension of the μ-calculus in the sense of Section 2.1 (page 42).

Thus, in particular, vectorial fixed-point terms will be formally presented as vectors of the ordinary (i.e., scalar) fixed-point terms.

By the Bekič principle (Lemma 1.4.2, page 27), vectorial fixed-point terms do not lead to a greater expressive power. However, vectorial notation has several advantages. In particular, it admits a prefix normal form which gives rise to an elegant characterization of the alternation-depth fixed-point hierarchy. Also, vectorial fixed points terms are usually more succinct than the equivalent scalar ones.

2.7.1 Vectorial extension of a μ-calculus

Definition 2.7.1. A *vector $\boldsymbol{v}$ of length n* is any sequence $\langle v_1,\dots,v_n\rangle$ of n elements. For $i\in\{1,\dots,n\}$, $\pi_i(\boldsymbol{v})$ is the ith element of the sequence $\boldsymbol{v}$ (see Section 1.1.7, page 7). Obviously, a vector is defined by its components so that if $\boldsymbol{v}$ and $\boldsymbol{v}'$ are two vectors of length n,

$$\forall i\in\{1,\dots,n\},\pi_i(\boldsymbol{v})=\pi_i(\boldsymbol{v}') \quad \text{implies} \quad \boldsymbol{v}=\boldsymbol{v}'.$$

If $\boldsymbol{v}$ is a vector of length n and $\boldsymbol{v}'$ is a vector of length n' we denote by $\langle\boldsymbol{v},\boldsymbol{v}'\rangle$ the vector of length $n+n'$ defined by

$$\pi_i(\langle\boldsymbol{v},\boldsymbol{v}'\rangle)=\begin{cases}\pi_i(\boldsymbol{v}) & \text{if } 1\le i\le n,\\ \pi_{i-n}(\boldsymbol{v}') & \text{if } n+1\le i\le n+n'.\end{cases}$$

Obviously, this product is associative: $\langle\boldsymbol{v},\langle\boldsymbol{v}',\boldsymbol{v}''\rangle\rangle=\langle\langle\boldsymbol{v},\boldsymbol{v}'\rangle,\boldsymbol{v}''\rangle$ and we may denote this vector by $\langle\boldsymbol{v},\boldsymbol{v}',\boldsymbol{v}''\rangle$.

If $\boldsymbol{v}$ is a vector of length n and if $\boldsymbol{i}=\langle i_1,\dots,i_m\rangle$ is a sequence of elements of $\{1,\dots,n\}$ then $\pi_{\boldsymbol{i}}(\boldsymbol{v})$ is the vector of length m defined by $\pi_j(\pi_{\boldsymbol{i}}(\boldsymbol{v}))=\pi_{i_j}(\boldsymbol{v})$, i.e., $\pi_{\boldsymbol{i}}(\langle v_1,\dots,v_n\rangle)=\langle v_{i_1},\dots,v_{i_m}\rangle$.

Definition 2.7.2. Let $\mathcal{T} = \langle T, id, ar, comp, \mu, \nu \rangle$ be a μ-calculus. The *vectorial extension* of the μ-calculus $\mathcal{T}$ is the structure

$$\mathcal{T}^* = \langle T^*, id^*, ar^*, comp^*, \mu^*, \nu^* \rangle$$

where

- $T^* = \bigcup_{n \geq 1} T^n$,
- $id^* = id$,
- for $\boldsymbol{t} = \langle t_1, \ldots, t_n \rangle \in T^n$, $ar^*(\boldsymbol{t}) = \bigcup_{i=1}^n ar(t_i)$,
- for $\boldsymbol{t} = \langle t_1, \ldots, t_n \rangle \in T^n$, and $\rho : Var \to T$,

$$comp^*(\boldsymbol{t}, \rho) = \langle t_1[\rho], \ldots, t_n[\rho] \rangle \in T^n,$$

- for $\boldsymbol{t} = \langle t_1, \ldots, t_n \rangle \in T^n$, and for $\boldsymbol{x} = \langle x_1, \ldots, x_n \rangle$ a vector of *distinct* variables, $\theta^* \boldsymbol{x}.\boldsymbol{t}$ is the vector $\langle t'_1, \ldots, t'_n \rangle$ of T^n where t'_i is recursively defined as follows.
  - If $n = 1$ then $t'_1 = \theta x_1.t$.
  - If $n > 1$, let $\langle t_1^{(i)}, \ldots, t_{i-1}^{(i)}, t_{i+1}^{(i)}, \ldots, t_n^{(i)} \rangle =$

$$\theta^* \langle x_1, \ldots, x_{i-1}, x_{i+1}, \ldots, x_n \rangle . \langle t_1, \ldots, t_{i-1}, t_{i+1}, \ldots, t_n \rangle.$$

  Then $t'_i = \theta x_i.(t_i[id\{t_1^{(i)}/x_1, \ldots, t_{i-1}^{(i)}/x_{i-1}, t_{i+1}^{(i)}/x_{i+1}, \ldots, t_n^{(i)}/x_n\}])$.

Note that $T = T^1 \subseteq T^*$ and that all the operations on T^* are extensions of the operations on T, therefore in the sequel we will omit the superscript $*$. We will also denote by $\boldsymbol{t}[\rho]$ the object $comp^*(\boldsymbol{t}, \rho)$.

Proposition 2.7.3. *Let T be a μ-calculus. If $\boldsymbol{x} = \langle x_1, \ldots, x_n \rangle$ is a vector of distinct variables, and if $\boldsymbol{t} = \langle t_1, \ldots, t_n \rangle \in T^n$ then $ar(\theta \boldsymbol{x}.\boldsymbol{t}) = ar(\boldsymbol{t}) - \{x_1, \ldots, x_n\}$.*

Proof. The proof is by induction on n. For $n = 1$, this property is just the Axiom 3 of the μ-calculus.

Let $\theta \boldsymbol{x}.\boldsymbol{t} = \langle t'_1, \ldots, t'_n \rangle$. Then $ar(\theta \boldsymbol{x}.\boldsymbol{t}) = \bigcup_{i=1}^n ar(t'_i)$ where

$$t'_i = \theta x_i.t_i[id\{t_1^{(i)}/x_1, \ldots, t_{i-1}^{(i)}/x_{i-1}, t_{i+1}^{(i)}/x_{i+1}, \ldots, t_n^{(i)}/x_n\}]$$

and

$$\begin{aligned} \boldsymbol{t}^{(i)} &= \langle t_1^{(i)}, \ldots, t_{i-1}^{(i)}, t_{i+1}^{(i)}, \ldots, t_n^{(i)} \rangle \\ &= \theta \langle x_1, \ldots, x_{i-1}, x_{i+1}, \ldots, x_n \rangle . \langle t_1, \ldots, t_{i-1}, t_{i+1}, \ldots, t_n \rangle. \end{aligned}$$

Thus,

$$\begin{aligned} ar(t_j^{(i)}) &\subseteq ar(\boldsymbol{t}^{(i)}) \\ &\subseteq ar(\langle t_1, \ldots, t_{i-1}, t_{i+1}, \ldots, t_n \rangle) - \{x_1, \ldots, x_{i-1}, x_{i+1}, \ldots, x_n\} \\ &\subseteq ar(\boldsymbol{t}) - \{x_1, \ldots, x_{i-1}, x_{i+1}, \ldots, x_n\}, \end{aligned}$$

and $ar(t'_i) \subseteq ar(t_i[id\{t_1^{(i)}/x_1, \ldots, t_{i-1}^{(i)}/x_{i-1}, t_{i+1}^{(i)}/x_{i+1}, \ldots, t_n^{(i)}/x_n\}]) - \{x_i\}$. But $ar(t_i[id\{t_1^{(i)}/x_1, \ldots, t_{i-1}^{(i)}/x_{i-1}, t_{i+1}^{(i)}/x_{i+1}, \ldots, t_n^{(i)}/x_n\}])$ $\subseteq (ar(t_i) - \{x_1, \ldots, x_{i-1}, x_{i+1}, \ldots, x_n\}) \cup \bigcup_{j \neq i} ar(t_j^{(i)})$ $\subseteq ar(\boldsymbol{t}) - \{x_1, \ldots, x_{i-1}, x_{i+1}, \ldots, x_n\}$, and the result follows. □

Let $\mathcal{T}_1$ and $\mathcal{T}_2$ be two μ-calculi. A homomorphism h from $\mathcal{T}_1$ to $\mathcal{T}_2$ can be extended into a mapping h^* from the vectorial μ-calculus over $\mathcal{T}_1$ to the vectorial μ-calculus over $\mathcal{T}_2$ by $h^*(\langle t_1, \ldots, t_n \rangle) = \langle h(t_1), \ldots, h(t_n) \rangle$. The following result is a consequence of the definition of a vectorial μ-calculus.

Proposition 2.7.4. $h^*(\theta \boldsymbol{x}.\boldsymbol{t}) = \theta \boldsymbol{x}.h^*(\boldsymbol{t})$.

Proof. The proof is similar to the previous one, by induction on the length of $\boldsymbol{t}$. □

2.7.2 Interpretations of vectorial fixed-point terms

In case $\mathcal{T}$ is the μ-calculus of fixed-point terms (see Section 2.3.2, page 47), the elements of its vectorial extension are called *vectorial fixed-point terms.* Note that an expression like $\theta_k \boldsymbol{x}^{(k)}. \cdots .\theta_1 \boldsymbol{x}^{(1)}.\boldsymbol{t}$, is not, properly speaking, a vectorial fixed-point term, but rather a meta-expression denoting such a term, which, by definition, is a vector of ordinary (i.e., scalar) fixed-point terms. However, we will call such a meta-expression a vectorial fixed-point term, keeping in mind its real meaning.

Since the mapping that sends a term t to its interpretation $[t]_{\mathcal{I}}$ is a homomorphism (see Proposition 2.3.10, page 50) we get, as a consequence of the previous proposition, the following property, which is basic for all applications of vectorial fixed-point terms.

Proposition 2.7.5. *Let* $\tau = \theta_k \boldsymbol{x}^{(k)}. \cdots .\theta_1 \boldsymbol{x}^{(1)}.\boldsymbol{t}$ *be a vectorial fixed-point term. Let* $\mathcal{I}$ *be an interpretation. Then* $[\tau]_{\mathcal{I}}$ *and* $[\boldsymbol{t}]_{\mathcal{I}}$ *are in the vectorial extensions of the functional* μ*-calculus* $\mathcal{F}(D_{\mathcal{I}})$ *and* $[\tau]_{\mathcal{I}} = \theta_k \boldsymbol{x}_k. \cdots .\theta_1 \boldsymbol{x}_1.[\boldsymbol{t}]_{\mathcal{I}}$.

The following proposition shows that $[\theta \boldsymbol{x}.\boldsymbol{t}]_{\mathcal{I}} = \theta \boldsymbol{x}.[\boldsymbol{t}]_{\mathcal{I}}$ is indeed a fixed point and justifies our definition of $\theta \boldsymbol{x}.\boldsymbol{t}$ in a vectorial μ-calculus. It can also be seen as a generalization of the Bekič principle.

Proposition 2.7.6. *Let* $\mathcal{F}(D)$ *be the functional* μ*-calculus over* D*, and let* $v : Var \to D$ *be a valuation. For any vector* $\boldsymbol{x} = \langle x_1, \ldots, x_n \rangle$ *of distinct* *variables and any vector* $\boldsymbol{f} = \langle f_1, \dots, f_n \rangle$ *of elements of* $\mathcal{F}(D)$*, the vector* $(\theta \boldsymbol{x}.\boldsymbol{f})[v]$*, which belongs to the vectorial extension of* $\mathcal{F}(D)$*, is equal to the extremal fixed point of the mapping* $\boldsymbol{g} : D^n \to D^n$ *defined by* $\boldsymbol{g}(d_1, \dots, d_n) = \boldsymbol{f}[v\{d_1/x_1, \dots, d_n/x_n\}]$*, and is also equal to*

$$\theta\langle y_1, \dots, y_n\rangle.\boldsymbol{f}[v\{\hat{y_1}/x_1, \dots, \hat{y_n}/x_n\}].$$

Proof. The proof is by induction on n.

For $n = 1$, the result is a consequence of the definition of $\theta x.f$ in $\mathcal{F}(D)$.

Now, let us assume that $\theta = \mu$ (the proof is similar for $\theta = \nu$), and let $\boldsymbol{d} = \langle d_1, \dots, d_n \rangle$ be the least fixed point of $\boldsymbol{g}$.

Let $\mu \boldsymbol{x}.\boldsymbol{f} = \langle f'_1, \dots, f'_n \rangle$ so that $(\mu \boldsymbol{x}.\boldsymbol{f})[v] = \langle f'_1[v], \dots, f'_n[v] \rangle$. By definition of $\mu \boldsymbol{x}.\boldsymbol{f}$,

$$f'_i[v] = (\mu x_i.f_i[id\{f_1^{(i)}/x_1, \dots, f_{i-1}^{(i)}/x_{i-1}, f_{i+1}^{(i)}/x_{i+1}, \dots, f_n^{(i)}/x_n\}])[v].$$

By Proposition 2.3.10, $f'_i[v] =$

$$\begin{aligned} & \mu y_i.(f_i[id\{f_1^{(i)}/x_1, \dots, f_{i-1}^{(i)}/x_{i-1}, f_{i+1}^{(i)}/x_{i+1}, \dots, f_n^{(i)}/x_n\}][v\{\hat{y_i}/x_i\}]) \\ = \ & f_i[id\{f_1^{(i)}/x_1, \dots, f_{i-1}^{(i)}/x_{i-1}, f_{i+1}^{(i)}/x_{i+1}, \dots, f_n^{(i)}/x_n\}][v\{f'_i[v]/x_i\}]. \end{aligned}$$

Since $id\{f_1^{(i)}/x_1, \dots, f_{i-1}^{(i)}/x_{i-1}, f_{i+1}^{(i)}/x_{i+1}, \dots, f_n^{(i)}/x_n\} \star v\{f'_i[v]/x_i\}$ is equal to

$$v\{h_1^{(i)}/x_1, \dots, h_{i-1}^{(i)}/x_{i-1}, f'_i[v]/x_i, h_{i+1}^{(i)}/x_{i+1}, \dots, h_n^{(i)}/x_n\},$$

where $h_j^{(i)} = f_j^{(i)}[v\{f'_i[v]/x_i\}]$, we get

$$f'_i[v] = f_i[v\{h_1^{(i)}/x_1, \dots, h_{i-1}^{(i)}/x_{i-1}, f'_i[v]/x_i, h_{i+1}^{(i)}/x_{i+1}, \dots, h_n^{(i)}/x_n\}].$$

Applying the induction hypothesis to

$$\begin{aligned} & \langle h_1^{(i)}, \dots, h_{i-1}^{(i)}, h_{i+1}^{(i)}, \dots, h_n^{(i)} \rangle = \langle f_1^{(i)}, \dots, f_{i-1}^{(i)}, f_{i+1}^{(i)}, \dots, f_n^{(i)} \rangle[v\{f'_i[v]/x_i\}] \\ & = (\mu\langle x_1, \dots, x_{i-1}, x_{i+1}, \dots, x_n\rangle.\langle f_1, \dots, f_{i-1}, f_{i+1}, \dots, f_n\rangle)[v\{f'_i[v]/x_i\}] \end{aligned}$$

we get, for $j \neq i$,

$$h_j^{(i)} = f_j[v\{h_1^{(i)}/x_1, \dots, h_{i-1}^{(i)}/x_{i-1}, f'_i[v]/x_i, h_{i+1}^{(i)}/x_{i+1}, \dots, h_n^{(i)}/x_n\}].$$

Hence, $\langle h_1^{(i)}, \dots, h_{i-1}^{(i)}, f'_i[v], h_{i+1}^{(i)}, \dots, h_n^{(i)} \rangle =$

$$\boldsymbol{f}[v\{h_1^{(i)}/x_1, \dots, h_{i-1}^{(i)}/x_{i-1}, f'_i[v]/x_i, h_{i+1}^{(i)}/x_{i+1}, \dots, h_n^{(i)}/x_n\}],$$

by which it follows that $\boldsymbol{d} \leq \langle h_1^{(i)}, \dots, h_{i-1}^{(i)}, f'_i[v], h_{i+1}^{(i)}, \dots, h_n^{(i)} \rangle$. In particular, $d_i \leq f'_i[v]$, thus, $\boldsymbol{d} \leq (\mu \boldsymbol{x}.\boldsymbol{f})[v]$.

Conversely, by definition of $\boldsymbol{d}$, $d_i = f_i[v\{d_1/x_1, \ldots, d_n/x_n\}]$, thus $\langle d_1, \ldots, d_{i-1}, d_{i+1}, \ldots, d_n\rangle =$

$$\langle f_1, \ldots, f_{i-1}, f_{i+1}, \ldots, f_n\rangle[v\{d_1/x_1, \ldots, d_n/x_n\}].$$

By applying the induction hypothesis, we get

$$\begin{array}{ll} & \langle f_1^{(i)}, \ldots, f_{i-1}^{(i)}, f_{i+1}^{(i)}, \ldots, f_n^{(i)}\rangle[v\{d_i/x_i\}] \\ = & (\mu\langle x_1, \ldots, x_{i-1}, x_{i+1}, \ldots, x_n\rangle.\langle f_1, \ldots, f_{i-1}, f_{i+1}, \ldots, f_n\rangle)[v\{d_i/x_i\}] \\ \leq & \langle d_1, \ldots, d_{i-1}, d_{i+1}, \ldots, d_n\rangle \end{array}$$

It follows that

$$\begin{array}{rcl} d_i & = & f_i[v\{d_1/x_1, \ldots, d_n/x_n\}] \\ & \geq & f_i[v\{d_1'/x_1, \ldots, d_{i-1}'/x_{i-1}, d_i/x_i, d_{i+1}'/x_{i+1}, \ldots, d_n'/x_n\}] \end{array}$$

where $d_j' = f_j^{(i)}[v\{d_i/x_i\}]$.

But $v\{d_1'/x_1, \ldots, d_{i-1}'/x_{i-1}, d_i/x_i, d_{i+1}'/x_{i+1}, \ldots, d_n'/x_n\}$ is equal to $id\{f_1^{(i)}/x_1, \ldots, f_{i-1}^{(i)}/x_{i-1}, f_{i+1}^{(i)}/x_{i+1}, \ldots, f_n^{(i)}/x_n\} \star v\{d_i/x_i\}$.
Thus, $d_i \geq f_i[id\{f_1^{(i)}/x_1, \ldots, f_{i-1}^{(i)}/x_{i-1}, f_{i+1}^{(i)}/x_{i+1}, \ldots, f_n^{(i)}/x_n\}][v\{d_i/x_i\}]$, by which it follows that
$d_i \geq (\mu x_i.f_i[id\{f_1^{(i)}/x_1, \ldots, f_{i-1}^{(i)}/x_{i-1}, f_{i+1}^{(i)}/x_{i+1}, \ldots, f_n^{(i)}/x_n\}])[v] = f_i'[v]$.
□

2.7.3 The vectorial hierarchy

Definition 2.7.7. Let T be a μ-calculus and let T^* be its vectorial extension. A subset $C \subseteq T^*$ is called a *vectorial clone* if

- it contains *Var*,
- if $\langle t_1, \ldots, t_n\rangle \in C$ then $t_i \in C$ for any $i \in \{1, \ldots, n\}$,
- if $\boldsymbol{t}$ and $\boldsymbol{t}'$ are in C, so is $\langle \boldsymbol{t}, \boldsymbol{t}'\rangle$,
- it is closed under composition in the following sense: If $\boldsymbol{t} \in C$ and if ρ is a substitution such that $\rho(y) \in C \cap T$ for any $y \in ar(\boldsymbol{t})$ then $\boldsymbol{t}[\rho] \in C$.

A vectorial clone C is a *μ-vectorial clone* if it is additionally closed under the μ operator in the following sense: if $\boldsymbol{t} = \langle t_1, \ldots, t_n\rangle$ is in C, and if $\boldsymbol{x}$ is a vector of n distinct variables, then $\mu\boldsymbol{x}.\boldsymbol{t}$ is in C. Similarly, C is a *ν-vectorial clone* if it is closed under the ν operator and it is a *fixed-point vectorial clone* if it closed under both μ and ν, that is, if it is both a μ-vectorial clone and a ν-vectorial clone.

It is easy to see that the intersection of a nonempty family of vectorial clones of any type is again a vectorial clone of this type. Therefore, for any set $T' \subseteq T^*$, there exists a least vectorial clone containing T'; we shall denote it by $Comp^{\rightarrow}(T')$.

Similarly, there exist a least μ-vectorial clone, a least ν-vectorial clone, and a least fixed-point vectorial clone containing T', which we shall denote respectively by $\mu^{\rightarrow}(T')$, $\nu^{\rightarrow}(T')$, and $fix^{\rightarrow}(T')$.

For $T' \subseteq T^*$, let $pr(T') = \{t \in T \mid \exists \langle t_1, \dots, t_n \rangle \in T', \exists i : t = t_i\}$ be the set of components of vectors of T. It is easy to see that if C is a vectorial clone, then $T' \subseteq C \Leftrightarrow pr(T') \subseteq C$. Therefore, any kind of the above vectorial clones is indeed generated by a subset of T.

Given a subset T' of T we define a hierarchy of elements of T, relative to T', by

$$\Sigma_0^{\rightarrow}(T') = \Pi_0(T') = Comp^{\rightarrow}(T'),$$

and, for $k < \omega$,

$$\begin{aligned} \Sigma_{k+1}^{\rightarrow}(T') &= \mu^{\rightarrow}(\Pi_k^{\rightarrow}(T')) \\ \Pi_{k+1}^{\rightarrow}(T') &= \nu^{\rightarrow}(\Sigma_k^{\rightarrow}(T')) \end{aligned}$$

The following proposition is an immediate consequence of the previous definitions.

Proposition 2.7.8. *For any $k \geq 0$, $pr(\Sigma_k^{\rightarrow}(T')) \subseteq \Sigma_k(T') \subseteq \Sigma_k^{\rightarrow}(T')$ and $pr(\Pi_k^{\rightarrow}(T')) \subseteq \Pi_k(T') \subseteq \Pi_k^{\rightarrow}(T')$.*

2.7.4 Vectorial fixed-point terms in normal form

Let us consider the μ-calculus of fixed point terms defined in Section 2.3.2 and its vectorial extension, as defined in Section 2.7.1. For each integer $n \geq 1$ we define the following classification on the n-tuples of fixed point terms in $fix\mathcal{T}(F)$.

- $\mathcal{S}_0^n(F) = \mathcal{P}_0^n(F)$ is the set of n-tuples of base terms (see Definition 2.3.1, page 47).
- If $\boldsymbol{t} \in \mathcal{S}_0^n(F) = \mathcal{P}_0^n(F)$ then $\mu\boldsymbol{x}.\boldsymbol{t} \in \mathcal{S}_1^n(F)$ and $\nu\boldsymbol{x}.\boldsymbol{t} \in \mathcal{P}_1^n(F)$.
- For $n > 0$,
  - if $\boldsymbol{t} \in \mathcal{S}_k^n(F)$ then $\mu\boldsymbol{x}.\boldsymbol{t} \in \mathcal{S}_k^n(F)$ and $\nu\boldsymbol{x}.\boldsymbol{t} \in \mathcal{P}_{k+1}^n(F)$,
  - if $\boldsymbol{t} \in \mathcal{P}_k^n(F)$ then $\mu\boldsymbol{x}.\boldsymbol{t} \in \mathcal{S}_{k+1}^n(F)$ and $\nu\boldsymbol{x}.\boldsymbol{t} \in \mathcal{P}_k^n(F)$.

Definition 2.7.9. A vectorial fixed-point term of length n is in *normal form* if it belongs to $\mathcal{S}_k^n(F)$ or to $\mathcal{P}_k^n(F)$ for some $k \geq 0$.

Note that this classification does not cover all the n-tuples. Moreover it is purely syntactic as exemplified in the next proposition, and it is not a hierarchy, since, for instance, $\mathcal{S}_k^n(F) \not\subseteq \mathcal{S}_{k+1}^n(F)$.

Proposition 2.7.10. *Let $\boldsymbol{t}$ be a vector of base terms. Then $\theta_1\boldsymbol{x}_1.\cdots.\theta_p.\boldsymbol{x}_p.\boldsymbol{t}$ is in $\mathcal{S}_k^n(F)$ if and only if*

- *$p = k = 0$, or*
- *$p \geq 1$, $k \geq 1$, $\theta_1 = \mu$ and $\theta_2\boldsymbol{x}_2.\cdots.\theta_p.\boldsymbol{x}_p.\boldsymbol{t} \in \mathcal{S}_k^n(F) \cup \mathcal{P}_{k-1}^n(F)$.*

A symmetrical result holds for $\mathcal{P}_k^n(F)$.

By Proposition 2.7.8 (page 63), it is clear that every component of an n-tuple in $\mathcal{S}_k^n(F)$ (resp. $\mathcal{P}_k^n(F)$) is in $\Sigma_k(F)$) (resp. $\Pi_k(F)$)).

In this section we prove a kind of converse, transforming the previous classification into a hierarchy (see Proposition 2.7.14) by using the notion of a variant. This gives rise to a normal form for describing fixed-point terms.

First we introduce the following notation: If $\boldsymbol{x} = \langle x_1, \ldots, x_n\rangle$ then $\{\boldsymbol{x}\} = \{x_1, \ldots, x_n\}$. If ρ is a substitution and $\boldsymbol{t} = \langle t_1, \ldots, t_n\rangle$ an n-tuple of terms then $\rho\{\boldsymbol{t}/\boldsymbol{x}\}$ denotes $\rho\{t_1/x_1, \ldots, t_n/x_n\}$.

For any vector $\boldsymbol{v} = \langle v_1, \ldots, v_n\rangle$ and any $i : 1 \leq i \leq n$, we denote by $\boldsymbol{v}_{-i}$ the vector $\langle v_1, \ldots, v_{i-1}, v_{i+1}, \ldots, v_n\rangle$.

For instance, we can shorten the definition of $\boldsymbol{t}' = \theta\boldsymbol{x}.\boldsymbol{t}$ given in Definition 2.7.2 (page 59) by writing $t'_i = \theta x_i.t_i[id\{\theta\boldsymbol{x}_{-i}.\boldsymbol{t}_{-i}/\boldsymbol{x}_{-i}\}]$

We say that a n-tuple $\boldsymbol{t} = \langle t_1, \ldots, t_n\rangle$ is a *variant* of $\boldsymbol{t}' = \langle t'_1, \ldots, t'_n\rangle$, denoted by $\boldsymbol{t} \approx \boldsymbol{t}'$ if $\forall i : 1 \leq i \leq n,\ t_i \approx t'_i$.

The following proposition generalizes Axiom 7 of a μ-calculus.

Proposition 2.7.11. *Let ρ be a substitution and let $\boldsymbol{t}' = (\theta\boldsymbol{x}.\boldsymbol{t})[\rho]$. Let $\boldsymbol{y}$ be a vector of variables such that $\{\boldsymbol{y}\} \cap ar(\boldsymbol{t}') = \emptyset$. Then $\boldsymbol{t}' \approx \theta\boldsymbol{y}.(\boldsymbol{t}[\rho\{\boldsymbol{y}/\boldsymbol{x}\}])$.*

Proof. The proof is by induction on the length n of the tuples. For $n = 1$, the result is immediate by definition of $\approx$.

Otherwise $t'_i = (\theta x_i.t_i[id\{\boldsymbol{s}^{(i)}/\boldsymbol{x}_{-i}\}])[\rho] = \theta z.t_i[id\{\boldsymbol{s}^{(i)}/\boldsymbol{x}_{-i}\}][\rho\{z/x_i\}]$, where $\boldsymbol{s}^{(i)} = \langle s_1^{(i)}, \ldots, s_{i-1}^{(i)}, s_{i+1}^{(i)}, \ldots, s_n^{(i)}\rangle = \theta\boldsymbol{x}_{-i}.\boldsymbol{t}_{-i}$ Thus,

$$t'_i \approx \theta y_i.t_i[id\{\boldsymbol{s}^{(i)}/\boldsymbol{x}_{-i}\}][\rho\{z/x_i\}][id\{y_i/z\}].$$

Taking into account the choice of z and y_i, we can check that, on $ar(\boldsymbol{t})$,

$$id\{\boldsymbol{s}^{(i)}/\boldsymbol{x}_{-i}\} \star \rho\{z/x_i\} \star id\{y_i/z\} = \rho\{\boldsymbol{s}^{(i)}[\rho\{y_i/x_i\}]/\boldsymbol{x}_{-i}, y_i/x_i\}$$

By the induction hypothesis,

$$s^{(i)}[\rho\{y_i/x_i\}] = (\theta \boldsymbol{x}_{-i}.\boldsymbol{t}_{-i})[\rho\{y_i/x_i\}] \approx \theta \boldsymbol{y}_{-i}.(\boldsymbol{t}_{-i}[\rho\{\boldsymbol{y}/\boldsymbol{x}\}]),$$

thus $t'_i \approx \theta y_i.t_i[\rho\{\theta \boldsymbol{y}_{-i}.(\boldsymbol{t}_{-i}[\rho\{\boldsymbol{y}/\boldsymbol{x}\}])/\boldsymbol{x}_{-i}, y_i/x_i\}]$.

On the other hand, the i-th component of $\theta \boldsymbol{y}.(\boldsymbol{t}[\rho\{\boldsymbol{y}/\boldsymbol{x}\}])$ is

$$\theta y_i.(t_i[\rho\{\boldsymbol{y}/\boldsymbol{x}\}][id\{\theta \boldsymbol{y}_{-i}.(\boldsymbol{t}_{-i}[\rho\{\boldsymbol{y}/\boldsymbol{x}\}])/\boldsymbol{y}_{-i}\}])$$

which is equal to t'_i since, on $ar(t_i)$,

$$\rho\{\boldsymbol{y}/\boldsymbol{x}\} \star id\{\theta \boldsymbol{y}_{-i}.(\boldsymbol{t}_{-i}[\rho\{\boldsymbol{y}/\boldsymbol{x}\}])/\boldsymbol{y}_{-i}\} = \rho\{\theta \boldsymbol{y}_{-i}.(\boldsymbol{t}_{-i}[\rho\{\boldsymbol{y}/\boldsymbol{x}\}])/\boldsymbol{x}_{-i}, y_i/x_i\}.$$

□

Applying the previous proposition with $\rho = id$ we get the vectorial generalization of the property (v3) in the definition of a variant (Section 2.4.2, page 51).

Corollary 2.7.12. *If $\{\boldsymbol{y}\} \cap (ar(\boldsymbol{t}) - \{\boldsymbol{x}\}) = \emptyset$ then $\theta \boldsymbol{x}.\boldsymbol{t} \approx \theta \boldsymbol{y}.(\boldsymbol{t}[id\{\boldsymbol{y}/\boldsymbol{x}\}])$.*

We also have a generalization of (v4).

Proposition 2.7.13. *If $\{\boldsymbol{x}\} \cap ar(\boldsymbol{t}) = \emptyset$ then $\theta \boldsymbol{x}.\boldsymbol{t} \approx \boldsymbol{t}$.*

Proof. Let $\boldsymbol{t} = \langle t_1, \ldots, t_n\rangle$ and $\theta \boldsymbol{x}.\boldsymbol{t} = \langle t'_1, \ldots, t'_n\rangle$. By definition we have $t'_i = \theta x_i.t_i[id\{\theta \boldsymbol{x}_{-i}.\boldsymbol{t}_{-i}/\boldsymbol{x}_{-i}\}]$. Since $\{\boldsymbol{x}_{-i}\} \cap ar(t_i) = \emptyset$, $t_i[id\{\theta \boldsymbol{x}_{-i}.\boldsymbol{t}_{-i}/\boldsymbol{x}_{-i}\}] = t_i[id] \approx t_i$, thus $t'_i \approx \theta x_i.t_i$. Since $x_i \notin ar(t_i)$, $\theta x_i.t_i \approx t_i$. □

As a consequence of this proposition, we get the following

Proposition 2.7.14. *For any $n \geq 0$ and any $n > 0$, we have, up to $\approx$,*

$$\mathcal{S}^n_n(F) \cup \mathcal{P}^n_n(F) \subseteq \mathcal{S}^n_{n+1}(F) \cap \mathcal{P}^n_{n+1}(F).$$

Proposition 2.7.15. *Let $\boldsymbol{x}$, $\boldsymbol{y}$, and $\boldsymbol{t}$ of length n, $\boldsymbol{x}'$, $\boldsymbol{y}'$, and $\boldsymbol{t}'$ of length n'. If the sets $\{\boldsymbol{y}\}$, $\{\boldsymbol{y}'\}$, and $ar(\theta \boldsymbol{x}.\boldsymbol{t}) \cup ar(\theta \boldsymbol{x}'.\boldsymbol{t}')$ are pairwise disjoint, then $\langle \theta \boldsymbol{x}.\boldsymbol{t}, \theta \boldsymbol{x}'.\boldsymbol{t}'\rangle \approx \theta\langle \boldsymbol{y}, \boldsymbol{y}'\rangle.\langle \boldsymbol{t}[id\{\boldsymbol{y}/\boldsymbol{x}\}], \boldsymbol{t}'[id\{\boldsymbol{y}'/\boldsymbol{x}'\}]\rangle$.*

Proof. By Corollary 2.7.12, we have

$$\theta \boldsymbol{x}.\boldsymbol{t} \approx \theta \boldsymbol{y}.\boldsymbol{t}[id\{\boldsymbol{y}/\boldsymbol{x}\}] \quad \text{and} \quad \theta \boldsymbol{x}'.\boldsymbol{t}' \approx \theta \boldsymbol{y}'.\boldsymbol{t}'[id\{\boldsymbol{y}'/\boldsymbol{x}'\}].$$

Thus we have only to show that $\langle \theta \boldsymbol{x}.\boldsymbol{t}, \theta \boldsymbol{x}'.\boldsymbol{t}'\rangle \approx \theta\langle \boldsymbol{x}, \boldsymbol{x}'\rangle.\langle \boldsymbol{t}, \boldsymbol{t}'\rangle$ provided $\boldsymbol{x} \cap \boldsymbol{x}' = \boldsymbol{x} \cap ar(\boldsymbol{t}') = \boldsymbol{x}' \cap ar(\boldsymbol{t}) = \emptyset$.

The proof is by induction on $n + n'$. For $n = n' = 1$, we have $\theta\langle x, x'\rangle.\langle t, t'\rangle = \langle \theta x.(t[id\{\theta x'.t'/x'\}]), \theta x'.(t'[id\{\theta x.t/x\}])\rangle$. Since $x' \notin ar(t)$ (resp. $x \notin ar(t)$), $t[id\{\theta x'.t'/x'\}] = t[id] \approx t$ (resp. $t'[id\{\theta x.t/x\}] = t'[id] \approx t'$). It follows that $\theta\langle x, x'\rangle.\langle t, t'\rangle \approx \langle \theta x.t, \theta x'.t'\rangle$.

Let $\langle \boldsymbol{s}, \boldsymbol{s}'\rangle = \theta\langle \boldsymbol{x}, \boldsymbol{x}'\rangle.\langle \boldsymbol{t}, \boldsymbol{t}'\rangle$. By definition

$$s_i = \theta x_i.(t_i[id\{\theta\langle \boldsymbol{x}_{-i}, \boldsymbol{x}'\rangle.\langle \boldsymbol{t}_{-i}, \boldsymbol{t}'\rangle / \langle \boldsymbol{x}_{-i}, \boldsymbol{x}'\rangle\}]).$$

By the induction hypothesis, $\theta\langle \boldsymbol{x}_{-i}, \boldsymbol{x}'\rangle.\langle \boldsymbol{t}_{-i}, \boldsymbol{t}'\rangle \approx \langle \theta \boldsymbol{x}_{-i}.\boldsymbol{t}_{-i}, \theta \boldsymbol{x}'.\boldsymbol{t}'\rangle$. Thus,

$$s_i \approx \theta x_i.(t_i[id\{\theta \boldsymbol{x}_{-i}.\boldsymbol{t}_{-i}/\boldsymbol{x}_{-i}, \theta \boldsymbol{x}'.\boldsymbol{t}'/\boldsymbol{x}'\}]).$$

Since $ar(\boldsymbol{t}) \cap \{\boldsymbol{x}'\} = \emptyset$ we get

$$t_i[id\{\theta \boldsymbol{x}_{-i}.\boldsymbol{t}_{-i}/\boldsymbol{x}_{-i}, \theta \boldsymbol{x}'.\boldsymbol{t}'/\boldsymbol{x}'\}] = t_i[id\{\theta \boldsymbol{x}_{-i}.\boldsymbol{t}_{-i}/\boldsymbol{x}_{-i}\}].$$

It follows that s_i is a variant of the i-th component of $\theta \boldsymbol{x}.\boldsymbol{t}$, thus, $\boldsymbol{s} \approx \theta \boldsymbol{x}.\boldsymbol{t}$. For similar reasons $\boldsymbol{s}' \approx \theta \boldsymbol{x}'.\boldsymbol{t}'$. □

The next proposition shows how in a vectorial extension we can perform substitution with only fixed-point operators. Before stating this result, let us give a simple example of its intuitive meaning.

Example 2.7.16. Let D be a complete lattice and $f : D^2 \to D$ and $g : D \to D$ be monotonic. Let $\langle f(y,z), g(z)\rangle \in D^{D^{\{y,z\}}} \times D^{D^{\{z\}}}$ and let $\langle a(z), b(z)\rangle = \theta\langle x, y\rangle.\langle f(y,z), g(z)\rangle \in D^{D^{\{z\}}} \times D^{D^{\{z\}}}$. We have

$$\langle a(z), b(z)\rangle = \langle f(y,z), g(z)\rangle[id\{a(z)/x, b(z)/y\}] = \langle f(b(z), z), g(z)\rangle.$$

It follows that $b(z) = g(z)$ and $a(z) = f(g(z), z) = f(y,z)[id\{g(z)/y\}]$. Thus, the functional term $f(g(z), z)$ is the component of a vectorial term built only with base terms. □

Proposition 2.7.17. *Let $\boldsymbol{x} = \langle x_1, \ldots, x_n\rangle$, and $\boldsymbol{t} = \langle t_1, \ldots, t_n\rangle$ be vectors of length n, $\boldsymbol{x}' = \langle x'_1, \ldots, x'_{n'}\rangle$ and $\boldsymbol{t}' = \langle t'_1, \ldots, t'_{n'}\rangle$ be vectors of length n' such that*

$$\{\boldsymbol{x}\} \cap \{\boldsymbol{x}'\} = \{\boldsymbol{x}\} \cap ar(\langle \boldsymbol{t}, \boldsymbol{t}'\rangle) = \emptyset,$$
$$\{\boldsymbol{x}'\} \cap ar(\boldsymbol{t}) = \emptyset.$$

Then $\theta\langle \boldsymbol{x}, \boldsymbol{x}'\rangle.\langle \boldsymbol{t}, \boldsymbol{t}'\rangle \approx \langle \boldsymbol{t}[id\{\boldsymbol{t}'/\boldsymbol{x}'\}], \boldsymbol{t}'\rangle$.

Proof. Let $\langle \boldsymbol{s}, \boldsymbol{s}'\rangle = \theta\langle \boldsymbol{x}, \boldsymbol{x}'\rangle.\langle \boldsymbol{t}, \boldsymbol{t}'\rangle$. First we show that $\boldsymbol{s}' \approx \boldsymbol{t}'$. By definition, the ith component of $\boldsymbol{s}'$ is equal to $\theta x'_i(t'_i[id\{\theta\langle \boldsymbol{x}, \boldsymbol{x}'_{-i}\rangle.\langle \boldsymbol{t}, \boldsymbol{t}'_{-i}\rangle/\langle \boldsymbol{x}, \boldsymbol{x}'_{-i}\rangle\}])$. Since $ar(\boldsymbol{t}') \cap (\boldsymbol{x} \cup \boldsymbol{x}') = \emptyset$, we have $t'_i[id\{\theta\langle \boldsymbol{x}, \boldsymbol{x}'_{-i}\rangle.\langle \boldsymbol{t}, \boldsymbol{t}'_{-i}\rangle/\langle \boldsymbol{x}, \boldsymbol{x}'_{-i}\rangle\}] = t'_i[id] \approx t'_i$ and, by Proposition 2.7.13, $\theta x'_i.t'_i \approx t'_i$.

Now let us compute s_i, the ith component of $\boldsymbol{s}$, which is equal to

$$\theta x_i(t_i[id\{\theta\langle \boldsymbol{x}_{-i}, \boldsymbol{x}'\rangle.\langle \boldsymbol{t}_{-i}, \boldsymbol{t}'\rangle/\langle \boldsymbol{x}_{-i}, \boldsymbol{x}'\rangle\}]).$$

We have just shown that $\theta\langle \boldsymbol{x}_{-i}, \boldsymbol{x}'\rangle.\langle \boldsymbol{t}_{-i}, \boldsymbol{t}'\rangle$ is equivalent to a vector $\langle \boldsymbol{r}, \boldsymbol{t}'\rangle$, thus $s_i \approx \theta x_i(t_i[id\{\langle \boldsymbol{r}, \boldsymbol{t}'\rangle/\langle \boldsymbol{x}_{-i}, \boldsymbol{x}'\rangle\}])$. Since $ar(\boldsymbol{t}) \cap \{\boldsymbol{x}\} = \emptyset$, we have $t_i[id\{\langle \boldsymbol{r}, \boldsymbol{t}'\rangle/\langle \boldsymbol{x}_{-i}, \boldsymbol{x}'\rangle\}] = t_i[id\{\boldsymbol{t}'/\boldsymbol{x}'\}]$.
Finally, since $ar(t_i[id\{\boldsymbol{t}'/\boldsymbol{x}'\}]) \subseteq ar(\boldsymbol{t}) \cup ar(\boldsymbol{t}')$ which does not contain x_i, by Proposition 2.7.13, $s_i \approx t_i[id\{\boldsymbol{t}'/\boldsymbol{x}'\}]$. □

Proposition 2.7.18. *Let t' be a vector of length n' in $\mathcal{S}_k^{n'}(F)$ (resp. $\mathcal{P}_k^{n'}(F)$) and t'' be a vector of length n'' in $\mathcal{S}_k^{n''}(F)$ (resp. $\mathcal{P}_k^{n''}(F)$). Then there exists a vector t of length $n = n'+n''$ in $\mathcal{S}_k^{n}(F)$ (resp. $\mathcal{P}_k^{n}(F)$) such that $\langle t', t''\rangle \approx t$.*

Proof. Let $t' = \theta'_1 x'_1.\cdots.\theta'_{p'} x'_{p'}.s'$ and $t'' = \theta''_1 x''_1.\cdots.\theta''_{p''} x''_{p''}.s''$.

In the first step, we show by induction on $p' + p''$ that there exists $m \geq \max(p, p')$, $\theta_1, \ldots, \theta_m$, $z'_1, \ldots, z'_m, z''_1, \ldots, z''_m$ such that

$$\begin{aligned} t' &\approx \theta_1 z'_1.\cdots.\theta_m z'_m.s' \in \mathcal{S}_k^{n'}(F), \quad (\text{resp. } \mathcal{P}_k^{n'}(F)) \\ t'' &\approx \theta_1 z''_1.\cdots.\theta_m z''_m.s'' \in \mathcal{S}_k^{n''}(F), \quad (\text{resp. } \mathcal{P}_k^{n''}(F)). \end{aligned}$$

If $p' + p'' = 0$, we also have $k = 0$ and we take $m = 0$. If $p' + p'' > 0$, then $p' > 0, p'' > 0, k > 0$. In this case we have $\theta'_1 = \theta''_1 = \theta$ and $t' = \theta x'_1.r'$, $t'' = \theta x''_1.r''$ where, assuming that $\theta = \mu$, $t' \in \mathcal{S}_k^{n'}(F)$, $t'' \in \mathcal{S}_k^{n''}(F)$,

$$\begin{aligned} r' &= \theta'_2 x'_2.\cdots.\theta'_{p'} x'_{p'}.s' \in \mathcal{S}_k^{n'}(F) \cup \mathcal{P}_{k-1}^{n'}(F), \\ r'' &= \theta''_2 x''_2.\cdots.\theta''_{p''} x''_{p''}.s'' \in \mathcal{S}_k^{n''}(F) \cup \mathcal{P}_{k-1}^{n''}(F). \end{aligned}$$

If r' and r'' are both in $\mathcal{S}_k(F)$ or both in $\mathcal{P}_{k-1}(F)$, we can apply the induction hypothesis. Otherwise, assume that $r' \in \mathcal{S}_k^{n'}(F)$ and $r'' \in \mathcal{P}_{k-1}^{n''}(F)$. Then, by Proposition 2.7.13, $t'' \approx \mu x''_1.t''$ and we can apply the induction hypothesis to r' and t''.

The second step is to show that if

$$\begin{aligned} t' &= \theta_1 z'_1.\cdots.\theta_m z'_m.s' \\ t'' &= \theta_1 z''_1.\cdots.\theta_m z''_m.s'', \end{aligned}$$

where s' and s'' are base terms then

$$\langle t', t''\rangle \approx \theta_1\langle y'_1, y''_1\rangle.\cdots.\theta_m\langle y'_m, y''_m\rangle.\langle s'_1, s''_1\rangle,$$

where s'_1 and s''_1 are base terms. Here the proof is by induction on m. If $m = 0$ there is nothing to do. If $m \geq 1$, let $r' = \theta_2 z'_2.\cdots.\theta_m z'_m.s'$ and $r'' = \theta_2 z''_2.\cdots.\theta_m z''_m.s''$ so that $t' = \theta_1 z'_1.r'$ and $t'' = \theta_1 z''_1.r''$. By Proposition 2.7.15, we get

$$\langle t', t''\rangle = \theta_1\langle y', y''\rangle.\langle r'[id\{y'/z'_1\}], r''[id\{y''/z''_1\}]\rangle.$$

Now, by applying Proposition 2.7.11 (page 64) we get that $r'[id\{y'/z'_1\}]$ is equivalent to some vector in the form $\theta_2 v_2.\cdots.\theta_m v_m.s_1$ where s_1 is a vector of base terms, and similarly for $r''[id\{y''/z''_1\}]$, so that we can apply the induction hypothesis to $\langle r'[id\{y'/z'_1\}], r''[id\{y''/z''_1\}]\rangle$. □

Now, we are ready to prove the main theorem of this section.

Theorem 2.7.19. *Let t be a term in $\Sigma_k(F)$ (resp. $\Pi_k(F)$). If $k > 0$, there exists n and a vector $\boldsymbol{t}$ in $\mathcal{S}_k^n(F)$ (resp. $\mathcal{P}_k^n(F)$) such that t is a variant of the first component of $\boldsymbol{t}$. If $k = 0$ then there exist n, a vector $\boldsymbol{t}_1 \in \mathcal{S}_1^n(F)$ and a vector $\boldsymbol{t}_2 \in \mathcal{P}_1^n(F)$ such that t is a variant of the first component of both $\boldsymbol{t}_1$ and $\boldsymbol{t}_2$.*

Moreover, $ar(\boldsymbol{t}) = ar(t)$ in the former case and $ar(\boldsymbol{t}_1) = ar(\boldsymbol{t}_2) = ar(t)$ in the latter.

Proof. The proof is by induction on k.

For $k = 0$ the proof is by induction on t.

If t is a base term, we have $t \approx \theta x.t$ which is in $\mathcal{S}_1^n(F)$ or in $\mathcal{P}_1^n(F)$, according to the value of θ.

If $t = f(t_1, \ldots, t_m)$ then each t_i is in $\Sigma_0(F) = \Pi_0(F)$, and, by the induction hypothesis, t_i is a variant of the first component of a vector $\boldsymbol{t}_i$, which can be chosen in either in $\mathcal{S}_1^{n_i}(F)$ or in $\mathcal{P}_1^{n_i}(F)$. Let us choose $\mathcal{S}_1^{n_i}(F)$ (the proof is similar for $\mathcal{P}_1^{n_i}(F)$). By Proposition 2.7.18, $\boldsymbol{t} = \langle \boldsymbol{t}_1, \ldots, \boldsymbol{t}_m \rangle$ is in $\mathcal{S}_1^n(F)$. Now, $\langle t_1, \ldots, t_m \rangle$ consists of the components of indices $i_1, \ldots, i_m$ of the vector $\boldsymbol{t}$. Let $\boldsymbol{y}$ be a vector of length n such that $\{\boldsymbol{y}\} \cap ar(\boldsymbol{t}) = \emptyset$ and y be a variable neither in $\{\boldsymbol{y}\}$ nor in $ar(\boldsymbol{t})$. Let $\boldsymbol{s} = \mu\langle y, \boldsymbol{y}\rangle.\langle \mu y.f(y_{i_1}, \ldots, y_{i_m}), \boldsymbol{t}\rangle$ which is a variant of a vector in $\mathcal{S}_1^{n+1}(F)$, since $\langle \mu y.f(y_{i_1}, \ldots, y_{i_m}), \boldsymbol{t}\rangle$ is a variant of a vector in $\mathcal{S}_1^{n+1}(F)$ by Proposition 2.7.18. The hypotheses of Proposition 2.7.17 are satisfied, thus the first component of $\langle \mu y.f(y_{i_1}, \ldots, y_{i_m}), \boldsymbol{t}\rangle$ is a variant of $(\mu y.f(y_{i_1}, \ldots, y_{i_m}))[id\{\boldsymbol{t}/\boldsymbol{y}\}]$, and since $\mu y.f(y_{i_1}, \ldots, y_{i_m}) \approx f(y_{i_1}, \ldots, y_{i_m})$, it is also a variant of $f(y_{i_1}, \ldots, y_{i_m})[id\{\boldsymbol{t}/\boldsymbol{y}\}] = f(t_1, \ldots, t_m)$.

For $k > 0$, the proof is by induction on the definition of $\Sigma_k(F)$ (the proof is similar for $\Pi_k(F)$).

Let $t = \mu x.t'$ with t' in $\Sigma_k(F)$ or in $\Pi_{k-1}(F)$. By the induction hypothesis t' is a variant of the first component t'_1 of a vector $\boldsymbol{t}'$ of length n which is in

$\mathcal{S}_k^n(F)$ if t' is in $\Sigma_n(F)$,
$\mathcal{P}_{k-1}^n(F)$ if t' is in $\Pi_{k-1}(F)$ and $k > 1$,
$\mathcal{S}_k^n(F)$ if t' is in $\Pi_{k-1}(F)$ and $k = 1$.

Let $\boldsymbol{y}$ be a vector of length $n-1$ such that $ar(\boldsymbol{t}') \cap \{\boldsymbol{y}\} = \emptyset$. Then $\mu\langle x, \boldsymbol{y}\rangle.\boldsymbol{t}'$ is always a variant of a vector in $\mathcal{S}_k^n(F)$ and its first component s_1 is equal to $\mu x.(t'_1[id\{\mu \boldsymbol{y}.\boldsymbol{t}'_{-1}/\boldsymbol{y}\}])$. By the choice of $\boldsymbol{y}$, $t'_1[id\{\mu \boldsymbol{y}.\boldsymbol{t}'_{-1}/\boldsymbol{y}\}] = t'_1[id] \approx t'_1 \approx t'$, thus $s_1 \approx \mu x.t'$.

Let $t = t'[\rho]$ where $t' \in \Sigma_k(F)$ and $\rho(x) \in \Sigma_k(F)$ for any $x \in ar(t')$. Let $\boldsymbol{x} = \langle x_1, \ldots, x_m\rangle$ be such that $\{\boldsymbol{x}\} = ar(t')$, and let $\boldsymbol{s} = \langle \rho(x_1), \ldots, \rho(x_m)\rangle$, so that $t'[\rho] = t'[id\{\boldsymbol{s}/\boldsymbol{x}\}]$. By the induction hypothesis, t' is a variant of the first component of a vector $\boldsymbol{t}' \in \mathcal{S}_k^{n'}(F)$ and $\rho(x_i)$ is a variant of the first component of a vector $\boldsymbol{t}_i \in \mathcal{S}_k^{n_i}(F)$. Let $\boldsymbol{t} = \langle \boldsymbol{t}_1, \ldots, \boldsymbol{t}_m\rangle$, let i_j be the index of $\rho(x_j)$ in $\boldsymbol{t}$, let $\boldsymbol{y}$ be a vector of length n such that $ar(\boldsymbol{t}) \cap \{\boldsymbol{y}\} = \emptyset = \{\boldsymbol{x}\} \cap \{\boldsymbol{y}\}$

and let $z = \langle y_{i_1}, \ldots, y_{i_m} \rangle$. By Proposition 2.7.11 (page 64), $t'' = t'[id\{z/x\}]$ is a variant of a vector in $\mathcal{S}_k^{n'}(F)$, thus $\langle t'', t \rangle$ is a variant of a vector in $\mathcal{S}_k^{n'+n}(F)$, and so is $\mu\langle x', y \rangle.\langle t'', t \rangle$ where $\{x'\} \cap \{y\} = \{x'\} \cap ar(t) = \{x'\} \cap ar(t'') = \emptyset$. We can apply Proposition 2.7.17 to get that the first component of $\mu\langle x', y\rangle.\langle t'', t\rangle$ is a variant of $t'[id\{z/x\}][id\{t/y\}]$. It is easy to check that this term is equal to $t'[id\{z/x\}][id\{t/y\}] = t'[\rho]$. □

From Proposition 2.7.8 (page 63) and the previous theorem we get:

Corollary 2.7.20. *For a term t the following properties are equivalent.*

- $t \in \Sigma_k(F)$,
- $t \in \Sigma_k^{\rightarrow}(F)$,
- *t is a component of some vectorial term in* $\mathcal{S}_k^n(F)$ *for some* $n \geq 1$

and similarly for Π *and* $\mathcal{P}$.

2.8 Bibliographic notes and sources

We have already mentioned in the bibliographic notes after chapter 1, the prototypes of μ-calculi considered by Park [80] and Emerson and Clarke [36]. The μ-calculus as a general purpose logical system, not confined to a particular application, originated with the work of Kozen [55] who proposed the modal μ-calculus as an extension of the propositional modal logic by a least fixed point operator (by duality, the logic comprises the greatest fixed points as well). Similar ideas appeared slightly earlier in the work of Pratt [82], who based his calculus on the concept of a least root rather than least fixed point. Nearly at the same time, Immerman [42] and Vardi [98] independently gave a model-theoretic characterization of polynomial time complexity, using an extension of first-order logic by fixed-point operators (the subject not covered in this book). The modal μ-calculus of Kozen has subsequently received much study motivated both by the mathematical appeal of the logic, and by its potential usefulness for program verification. These studies revealed that the μ-calculus subsumes most of previously defined logics of programs (see [41] and references therein), is decidable in exponential time (Emerson and Jutla [32]), and admits a natural complete proof system (Walukiewicz [103]).

A slightly different approach to the μ-calculus was taken by the authors of this book [74, 75, 77, 10, 78], who considered an algebraic calculus of fixed-point terms interpretable over arbitrary complete lattices or, more restrictively, over powerset algebras. We will reconcile the two views in Chapter 6, by presenting the modal μ-calculus in our framework of powerset algebras.

The concept of an abstract μ-calculus as presented in Section 2.1 of this chapter has not been presented before; it can be viewed as a remote analogue of the concept of a combinatorial algebra in the λ-calculus (see Barendregt [11]).

3. The Boolean μ-calculus

The simplest nontrivial complete lattice consists only of $\perp$ and $\top$ with $\perp \neq \top$; this is the celebrated Boole algebra. We denote it $\mathbb{B}$ and use the classical notations, 0 for $\perp$, 1 for $\top$, + for $\vee$, and the operator $\wedge$ is denoted by a dot or omitted. We call the functional μ-calculus over $\mathbb{B}$ (see Section 2.2, page 44), $\mathcal{F}(\mathbb{B})$, the *Boolean* μ-calculus.

The importance of the Boolean μ-calculus stems from the role of powerset interpretations (see Section 2.5, page 53) via the obvious isomorphism of $\mathbb{B}^E$ and $\mathcal{P}(E)$. Note that if E is finite, $\mathcal{F}(\mathbb{B}^E)$ is captured by the vectorial extension of $\mathcal{F}(\mathbb{B})$ (see Section 2.7, page 58). As E can be infinite, we will be led also to consideration of infinite vectors of Boolean fixed-point terms (see Sections 3.3.2 and 3.3.3, pages 78 and sq.).

The Boolean μ-calculus is the core of this book. As we will see later in Chapters 10 and 11, most of the algorithmic problems of the μ-calculus amounts to evaluating Boolean vectorial fixed-point terms. Also, the selection property shown in Section 3.3 (page 75) turns out to be crucial for determinacy of the infinite games of Chapter 4. These games in turn are essential for the equivalence between the μ-calculus and automata (Chapter 7), as well as for the hierarchy problem (Chapter 8).

3.1 Monotone Boolean functions

Let $h \in \mathbb{B}^X \to \mathbb{B}$. If $x \in X$ we write $h = h(x)$ to make it clear that h depends on x and we write $h(t)$ for $h[id\{t/x\}]$, if x is known from the context. Then the classical Shannon lemma is written $h(x) = \overline{x}h(0) + xh(1)$.

Indeed, if h is monotonic with respect to x, Shannon's lemma becomes

Lemma 3.1.1 (Shannon's lemma). $h(x) = h(0) + xh(1)$.

Proof. Let $g(x) = h(0) + xh(1)$. We have $g(0) = h(0)$ and $g(1) = h(0) + h(1)$ which is equal to $h(1)$ since $h(0) \leq h(1)$. □

Here are some consequences of this lemma.

Proposition 3.1.2. $\mu x.h(x) = h(0)$; $\nu x.h(x) = h(1)$.

Proof. Since $h(x) = h(0) + xh(1)$ with $h(0) \leq h(1)$, we get $h(h(0)) = h(0) + h(0)h(1) = h(0)$ and $h(1) = h(0) + h(1)h(1) = h(1)$. Therefore $h(0)$ and $h(1)$ are fixed points of $h(x)$. But we know that $h(0) \leq \mu x.h(x) \leq \nu x.h(x) \leq h(1)$, hence the result. □

The following proposition shows that any monotonic mapping is homomorphic with respect to Boolean sum and product.

Proposition 3.1.3. *If $h(x)$ is monotonic with respect to x then $h(x + y) = h(x) + h(y)$ and $h(xy) = h(x)h(y)$.*

Proof. We have to prove that for any $b, b' \in \mathbb{B}$, $h(b + b') = h(b) + h(b')$. Without loss of generality, we may assume that $b \leq b'$ hence $h(b) \leq h(b')$. Then $h(b + b') = h(b') = h(b) + h(b')$.

The proof is similar for the product.

Moreover, it is easy to see that $ar(h(x + y)) = ar(h(xy)) = ar(h(x) + h(y)) = ar(h(x)h(y))$ which is equal to $ar(h(x)) \cup ar(h(y))$. □

Another consequence of the previous lemma is the following corollary.

Corollary 3.1.4. *For any $x \in Var$, if h is monotonic with respect to x, then $h[\rho] = h[\rho\{0/x\}] + \rho(x)h[\rho\{1/x\}]$.*

Proof. If is easy to see that

$$\begin{aligned}(h[id\{0/x\}] + xh[id\{1/x\}])[\rho] &= h[id\{0/x\}][\rho] + \rho(x)(h[id\{1/x\}][\rho]) \\ &= h[\rho\{0/x\}] + \rho(x)h[\rho\{1/x\}].\end{aligned}$$

□

3.2 Powerset interpretations and the Boolean μ-calculus

Due to the obvious isomorphism of the powerset lattice $\langle \mathcal{P}(E), \subseteq \rangle$ and the product lattice $\mathbb{B}^E$, we can identify any powerset interpretation $\mathcal{I}$ of the universe $\mathcal{P}(E)$ (see Section 2.5, page 53) with a μ-interpretation of the universe $\mathbb{B}^E$, i.e., such that the interpretation of any fixed-point term is an object of $\mathcal{F}(\mathbb{B}^E)$. Now, any function in $\mathcal{F}(\mathbb{B}^E)$ can readily be represented by a possibly infinite vector of (possibly infinite) monotone Boolean terms. In particular, with any fixed-point term t we can associate its characteristic Boolean function that is the representation of $[t]_{\mathcal{I}}$. In this section, we fix a notation for these Boolean characteristics, and justify it by some basic results. The purpose of this section is technical[1], as we will later wish to exploit the aforementioned connection between powerset and Boolean interpretations.

[1] This section is not related to the subsequent sections of this chapter.

Let $\mathcal{I}$ be a powerset interpretation for F with domain $\mathcal{P}(E)$. Let h be the canonical bijection from $\mathbb{B}^E$ to $\mathcal{P}(E)$, i.e., if g is a mapping from E to $\mathbb{B}$, $h(g) = \{e \in E \mid g(e) = 1\}$. Note that this bijection is monotonic. This bijection is extended into a bijection from $(\mathbb{B}^E)^D = \mathbb{B}^{E\times D}$ to $\mathcal{P}(E)^D$, still denoted by h. Thus, if g is a mapping from $E \times D$ to $\mathbb{B}$, $h(g)$ is the mapping from D to $\mathcal{P}(E)$ defined by $h(g)(d) = \{e \in E \mid g(e,d) = 1\}$.

Let Var' be a new set of variables, large enough for there to exist an injection $u : E \times Var \to Var'$. For $e \in E$ and $x \in Var$, we denote by $u_{e,x}$ the variable $u(e,x)$ of Var'. If $\boldsymbol{x}$ is a vector of distinct variables of Var, indexed by an arbitrary set J, we denote by $u_{\boldsymbol{x}}$ the vector of variables of Var', indexed by $E \times J$, such that if the component of index j of $\boldsymbol{x}$ is y, the component of index (e,j) of $u_{\boldsymbol{x}}$ is $u_{e,y}$, which we also denote by $\langle u_{e,x_j}\rangle_{e\in E, j\in J}$. In particular, if x is a variable of Var, $u_{\langle x\rangle}$ is the vector indexed by E whose component of index e is $u_{e,x}$. Likewise, if X is a subset of Var, u_X is the subset $\{u_{e,x} \mid e \in E, x \in X\}$ of Var'.

For any valuation $w : Var' \to \mathbb{B}$, $h^*(w)$ is the mapping from Var to $\mathcal{P}(E)$ defined by $h^*(w)(x) = \{e \in E \mid w(u_{e,x}) = 1\}$. If we see $w(u_{\langle x\rangle})$ as an element of $\mathbb{B}^E$, we can also write $h^*(w)(x) = h(w(u_{\langle x\rangle}))$.

From these definitions it follows immediately.

Lemma 3.2.1. *Let $\boldsymbol{x}$ be a vector of length n of distinct variables in Var. Let $w : Var' \to \mathbb{B}$ and $\boldsymbol{a}$ be a vector in $(\mathbb{B}^E)^n$, that can also be viewed as a vector of Boolean values indexed by $E \times \{1,\dots,n\}$. Then $h^*(w\{\boldsymbol{a}/u_{\boldsymbol{x}}\}) = h^*(w)\{h(\boldsymbol{a})/\boldsymbol{x}\}$.*

Proof. By the remark above $h^*(w\{\boldsymbol{a}/u_{\boldsymbol{x}}\})(y) = h(w\{\boldsymbol{a}/u_{\boldsymbol{x}}\}(u_{\langle y\rangle}))$. Let $\boldsymbol{x}$ be $\langle x_1,\dots,x_n\rangle$ and $\boldsymbol{a}$ be $\langle a_1,\dots,a_n\rangle$. Then $h(w\{\boldsymbol{a}/u_{\boldsymbol{x}}\}(u_{\langle x_i\rangle})) = h(a_i) = h^*(w)\{h(\boldsymbol{a})/\boldsymbol{x}\}(x_i)$, and for any y which is not a component of $\boldsymbol{x}$,

$$h(w\{\boldsymbol{a}/u_{\boldsymbol{x}}\}(u_{\langle y\rangle})) = h(w(u_{\langle y\rangle})) = h^*(w)(y) = h^*(w)\{h(\boldsymbol{a})/\boldsymbol{x}\}(y).$$

□

Definition 3.2.2. For any valuation $v : Var \to \mathcal{P}(E)$ we define $char(v) : \mathbb{B}^{Var'} \to \mathbb{B}$, the *characteristic of* v, by

$$char(v) = \prod_{x\in Var}\prod_{e\in v(x)} u_{e,x}.$$

so that for any $w : Var' \to \mathbb{B}$,

$$char(v)(w) = \prod_{x\in Var}\prod_{e\in v(x)} w(u_{e,x}).$$

Lemma 3.2.3. *Let $v : Var \to \mathcal{P}(E)$ and $w : Var' \to \mathbb{B}$ be two valuations. Then $char(v)(w) = 1$ if and only if $\forall x \in Var, v(x) \subseteq h^*(w)(x)$.*

Proof. We have $char(v)(w) = 1$ if and only if $\forall x \in Var, \forall e \in v(x)$, $w(u_{e,x}) = 1$. But $w(u_{e,x}) = 1$ if and only if $e \in h^*(w)(x)$. □

Definition 3.2.4. Let t be any fixed-point term over F and let $\mathcal{I}$ be a powerset interpretation over $\mathcal{P}(E)$. With t and $\mathcal{I}$ we associate the characteristic function $\chi_{\mathcal{I}}(t) : \mathbb{B}^{Var'} \to \mathbb{B}^E$ such that the component of index e of $\chi_{\mathcal{I}}(t)$ is equal to $\sum_{v:e\in[\![t]\!]_{\mathcal{I}}[v]} char(v)$.

Proposition 3.2.5. *Let t be a term and let $\mathcal{I}$ be a powerset interpretation.*
For any $w : Var' \to \mathbb{B}$, $h(\chi_{\mathcal{I}}(t)(w)) = [\![t]\!]_{\mathcal{I}}[h^*(w)]$.
In particular, if t is closed, $h(\chi_{\mathcal{I}}(t)) = [\![t]\!]_{\mathcal{I}}$.

Proof. By definition of h and $\chi_{\mathcal{I}}$, $e \in h(\chi_{\mathcal{I}}(t)(w))$ if and only if the component of index e of $\chi_{\mathcal{I}}(t)(w)$ is equal to 1 if and only if there exists $v : Var \to \mathcal{P}(E)$ such that $e \in [\![t]\!]_{\mathcal{I}}[v]$ and $char(v)(w) = 1$. But, by the previous lemma, $char(v)(w) = 1$ if and only if $\forall x \in Var, v(x) \subseteq h^*(w)(x)$.

Therefore, we have $e \in h(\chi_{\mathcal{I}}(t)(w))$ if and only if there exists v such that $e \in [\![t]\!]_{\mathcal{I}}[v]$ and $\forall x \in Var, v(x) \subseteq h^*(w)(x)$.

If such a v exists, by monotonicity of $[\![t]\!]_{\mathcal{I}}$, $e \in [\![t]\!]_{\mathcal{I}}[h^*(w)]$. Conversely, if $e \in [\![t]\!]_{\mathcal{I}}[h^*(w)]$, we can take $v = h^*(w)$ and thus $e \in h(\chi_{\mathcal{I}}(t)(w))$. □

As we have seen before, $h : \mathbb{B}^E \to \mathcal{P}(E)$ can be extended componentwise into a mapping from $(\mathbb{B}^E)^n$ to $\mathcal{P}(E)^n$, and the mapping $\chi_{\mathcal{I}}$ is extended to vectorial terms by $\chi_{\mathcal{I}}(\langle t_1, \ldots, t_n\rangle) = \langle \chi_{\mathcal{I}}(t_1), \ldots, \chi_{\mathcal{I}}(t_n)\rangle$. It is easy to see that the previous proposition is still valid for vectorial terms.

Proposition 3.2.6. *Let $\boldsymbol{t}$ be a vectorial term and let $\mathcal{I}$ be a powerset interpretation.*
For any $w : Var' \to \mathbb{B}$, $h(\chi_{\mathcal{I}}(\boldsymbol{t})(w)) = [\![\boldsymbol{t}]\!]_{\mathcal{I}}[h^*(w)]$.

As a consequence, $\chi_{\mathcal{I}}$ preserves fixed-point operators in the following sense.

Proposition 3.2.7. *Let $\boldsymbol{t}$ be a vectorial term of length n, and let $\boldsymbol{x}$ be a vector of variables of Var of length n. Then $\chi_{\mathcal{I}}(\theta\boldsymbol{x}.\boldsymbol{t}) = \theta u_{\boldsymbol{x}}.\chi_{\mathcal{I}}(\boldsymbol{t})$.*

Proof. By Proposition 2.2.2 (page 45), it is sufficient to prove that for any $w : Var' \to \mathbb{B}$, $\chi_{\mathcal{I}}(\theta\boldsymbol{x}.\boldsymbol{t})(w) = (\theta u_{\boldsymbol{x}}.\chi_{\mathcal{I}}(\boldsymbol{t}))(w)$.

As usual, we prove the result for $\theta = \mu$, the case $\theta = \nu$ is symmetric.

Let $\boldsymbol{a} = \chi_{\mathcal{I}}(\mu\boldsymbol{x}.\boldsymbol{t})(w)$ and $\boldsymbol{b} = (\mu u_{\boldsymbol{x}}.\chi_{\mathcal{I}}(\boldsymbol{t}))(w)$.

By the previous proposition $h(\boldsymbol{a}) = [\![\mu\boldsymbol{x}.\boldsymbol{t}]\!]_{\mathcal{I}}[h^*(w)] = (\mu\boldsymbol{x}.[\![\boldsymbol{t}]\!]_{\mathcal{I}})[h^*(w)]$ which is equal, by Proposition 2.2.4 (page 46), to $\mu\boldsymbol{x}.([\![\boldsymbol{t}]\!]_{\mathcal{I}}[h^*(w)\{\boldsymbol{x}/\boldsymbol{x}\}])$. It follows that $h(\boldsymbol{a}) = [\![\boldsymbol{t}]\!]_{\mathcal{I}}[h^*(w)\{h(\boldsymbol{a})/\boldsymbol{x}\}]$ which is also equal, by Lemma 3.2.1, to $[\![\boldsymbol{t}]\!]_{\mathcal{I}}[h^*(w\{\boldsymbol{a}/u_{\boldsymbol{x}}\}]$. Again by the previous proposition,

$$[\![t]\!]_{\mathcal{I}}[h^*(w\{\boldsymbol{a}/u_{\boldsymbol{x}}\})] = h(\chi_{\mathcal{I}}(\boldsymbol{t})(w\{\boldsymbol{a}/u_{\boldsymbol{x}}\}))$$

and since h is a bijection, this implies $\boldsymbol{a} = \chi_{\mathcal{I}}(\boldsymbol{t})(w\{\boldsymbol{a}/u_{\boldsymbol{x}}\})$. It follows that $\boldsymbol{a} \geq \mu u_{\boldsymbol{x}}.(\chi_{\mathcal{I}}(\boldsymbol{t})(w\{u_{\boldsymbol{x}}/u_{\boldsymbol{x}}\})) = (\mu u_{\boldsymbol{x}}.\chi_{\mathcal{I}}(\boldsymbol{t}))(w) = \boldsymbol{b}$.

On the other hand, $\boldsymbol{b} = (\mu u_{\boldsymbol{x}}.\chi_{\mathcal{I}}(\boldsymbol{t}))(w) = \mu u_{\boldsymbol{x}}.(\chi_{\mathcal{I}}(\boldsymbol{t})(w\{u_{\boldsymbol{x}}/u_{\boldsymbol{x}}\})) = \chi_{\mathcal{I}}(\boldsymbol{t})(w\{\boldsymbol{b}/u_{\boldsymbol{x}}\})$. It follows that

$$h(\boldsymbol{b}) = h(\chi_{\mathcal{I}}(\boldsymbol{t})(w\{\boldsymbol{b}/u_{\boldsymbol{x}}\})) = [\![t]\!]_{\mathcal{I}}[h^*(w\{\boldsymbol{b}/u_{\boldsymbol{x}}\})] = [\![t]\!]_{\mathcal{I}}[h^*(w)\{h(\boldsymbol{b})/\boldsymbol{x}\})].$$

Hence, $h(\boldsymbol{b}) \geq \mu\boldsymbol{x}.([\![t]\!]_{\mathcal{I}}[h^*(w)\{\boldsymbol{x}/\boldsymbol{x}\})]) = (\mu\boldsymbol{x}.[\![t]\!]_{\mathcal{I}})[h^*(w)] = [\![\mu\boldsymbol{x}.\boldsymbol{t}]\!]_{\mathcal{I}}[h^*(w)]$. By the previous proposition, $[\![\mu\boldsymbol{x}.\boldsymbol{t}]\!]_{\mathcal{I}}[h^*(w)]$ is equal to $h(\chi_{\mathcal{I}}(\mu\boldsymbol{x}.\boldsymbol{t})(w)) = h(\boldsymbol{a})$ and since h is a monotonic bijection, $h(\boldsymbol{b}) \geq h(\boldsymbol{a})$ implies $\boldsymbol{b} \geq \boldsymbol{a}$. □

By iterated applications of the previous proposition, we get

Proposition 3.2.8. *If τ is the vectorial term $\theta_k\boldsymbol{x}_k.\cdots.\theta_1\boldsymbol{x}_1.\boldsymbol{t}$. then $\chi_{\mathcal{I}}(\tau) = \theta_k u_{\boldsymbol{x}_k}.\cdots.\theta_1 u_{\boldsymbol{x}_1}.\chi_{\mathcal{I}}(\boldsymbol{t})$.*

3.3 The selection property

It is obvious that for any closed Boolean terms b_1, b_2, the disjunction $b_1 + b_2$ is equivalent to one of the disjuncts, b_i. Although this property is not true for non-closed terms, it generalizes, somewhat surprisingly, to closed terms with arbitrary fixed-point prefix. That means, for any closed fixed-point term $\theta_1x_1.\theta_2x_2.\dots.\theta_kx_k.(b_1 + b_2)$ (where b_1, b_2 need not, of course, to be closed) we can select $i \in \{1, 2\}$ such that the original term is equivalent to $\theta_1x_1.\theta_2x_2.\dots.\theta_kx_k.b_i$. This property further generalizes to vectorial fixed-point terms, and also to infinite vectors of infinite disjunctions. We call it *selection property* since it allows the replacement of the sum by an adequately selected summand. The selection property underlies several important results in the μ-calculus. In particular, it is at the basis of the determinacy result for parity games proved in the next chapter (Theorem 4.3.8, page 92).

We give a simple proof of this result for vectorial terms of finite length. Then we generalize it for vectorial terms indexed by sets of arbitrary cardinality like those used in the previous section.

3.3.1 Finite vectorial fixed-point terms

First we need some lemmas.

As a consequence of the Gauss elimination principle (Proposition 1.4.7, page 30), we get

Lemma 3.3.1. *Let $\boldsymbol{x} = \langle x_1, \dots, x_n \rangle$ and let x be a variable not in $\boldsymbol{x}$. Let $f = f(x, x_1, \dots, x_n)$ be a monotonic function over a complete lattice and let $\boldsymbol{f} = \langle f_1, \dots, f_n \rangle$ be a vector of such functions. Let us denote by $\langle x, \boldsymbol{x} \rangle$ the sequence $\langle x, x_1, \dots, x_n \rangle$ and by $\langle f, \boldsymbol{f} \rangle$ the sequence $\langle f, f_1, \dots, f_n \rangle$. Let $\boldsymbol{g} = \theta \boldsymbol{x}.\boldsymbol{f}$ and $g = \theta x.f[id\{\boldsymbol{g}/\boldsymbol{x}\}]$. Then $\theta\langle x, \boldsymbol{x}\rangle.\langle f, \boldsymbol{f}\rangle = \langle g, \boldsymbol{g}[id\{g/x\}]\rangle$.*

Remark We will use the obvious notation $f(\boldsymbol{g})$ for $f[id\{\boldsymbol{g}/\boldsymbol{x}\}]$ and $\boldsymbol{g}(g)$ for $\boldsymbol{g}[id\{g/x\}]$.

The next lemma is a particular case of Proposition 1.4.5, page 29.

Lemma 3.3.2. *Let $f(x, \boldsymbol{x})$ be a function and $\boldsymbol{f}(x, \boldsymbol{x})$ be a vector of functions. Then*

$$\theta\langle x, \boldsymbol{x}\rangle.\langle f(x, \boldsymbol{x}), \boldsymbol{f}(f(x, \boldsymbol{x}), \boldsymbol{x})\rangle = \theta\langle x, \boldsymbol{x}\rangle.\langle f(x, \boldsymbol{x}), \boldsymbol{f}(x, \boldsymbol{x})\rangle.$$

Lemma 3.3.3. *Let for $i = 0, 1, 2$, $f_i(x, \boldsymbol{x})$ be monotonic functions such that $f_0 = f_1 + f_2$, and let $\boldsymbol{f}$ be a vector of functions of the same length as $\boldsymbol{x}$. Then there exists $g_0, g_1, g_2, \boldsymbol{g}$ such that $\theta\langle x, \boldsymbol{x}\rangle.\langle f_i, \boldsymbol{f}\rangle = \langle g_i, \boldsymbol{g}(g_i)\rangle$ and $g_0 = g_1 + g_2$.*

Proof. Let $\boldsymbol{g} = \theta \boldsymbol{x}.\boldsymbol{f}$ and $g_i = \theta x.f_i(\boldsymbol{g})$. By Lemma 3.3.1

$$\theta\langle x, \boldsymbol{x}\rangle.\langle f_i, \boldsymbol{f}\rangle = \langle g_i, \boldsymbol{g}(g_i)\rangle.$$

Let Ω be 0 or 1 according to whether θ is μ or ν. Then, by Proposition 3.1.2 (page 71), $g_i = f_i(\Omega, \boldsymbol{g}(\Omega))$. Since $f_0 = f_1 + f_2$, we get $g_0 = g_1 + g_2$ by Proposition 3.1.3 (page 72). □

Proposition 3.3.4. *For $i = 0, 1, 2$ and $k > 0$, let*

$$\boldsymbol{k}_i = \theta_1\langle x_1, \boldsymbol{x}_1\rangle.\theta_2\langle x_2, \boldsymbol{x}_2\rangle.\cdots.\theta_k\langle x_k, \boldsymbol{x}_k\rangle.\langle f_i, \boldsymbol{f}\rangle$$

with $f_0 = f_1 + f_2$.

Then there exist $g_0, g_1, g_2, \boldsymbol{g}$ such that $\boldsymbol{k}_i = \langle g_i, \boldsymbol{g}(g_i)\rangle$ and $g_0 = g_1 + g_2$.

Proof. The proof is by induction on k. For $k = 1$, the result is proved in Lemma 3.3.3.

By the induction hypothesis, $\boldsymbol{k}_i = \theta_1\langle x_1, \boldsymbol{x}_1\rangle.\langle g_i, \boldsymbol{g}(g_i)\rangle$ with $g_0 = g_1 + g_2$. By Lemma 3.3.2, $\boldsymbol{k}_i = \theta_1\langle x_1, \boldsymbol{x}_1\rangle.\langle g_i, \boldsymbol{g}(x_1)\rangle$, and, again by Lemma 3.3.3, $\boldsymbol{k}_i = \langle h_i, \boldsymbol{h}(h_i)\rangle$ with $h_0 = h_1 + h_2$. □

Theorem 3.3.5. *Let $\boldsymbol{f} : \mathbb{B}^k \times \mathbb{B}^{kn} \to \mathbb{B}^n$ (with $k > 0$), and, for $i = 0, 1, 2$ let $f_i : \mathbb{B}^k \times \mathbb{B}^{kn} \to \mathbb{B}$, and*

$$\boldsymbol{k}_i = \theta_1\langle x_1, \boldsymbol{x}_1\rangle.\theta_2\langle x_2, \boldsymbol{x}_2\rangle.\cdots.\theta_k\langle x_k, \boldsymbol{x}_k\rangle.\langle f_i, \boldsymbol{f}\rangle \in \mathbb{B}^{n+1}.$$

If $f_0 = f_1 + f_2$ then $\boldsymbol{k}_0 = \boldsymbol{k}_1 + \boldsymbol{k}_2$, and, moreover, there exists $i \in \{1, 2\}$ such that $\boldsymbol{k}_0 = \boldsymbol{k}_i$.

Proof. By the above proposition, $\boldsymbol{k}_i = \langle g_i, \boldsymbol{g}(g_i)\rangle$ with $g_0 = g_1 + g_2 \in \mathbb{B}$. By Proposition 3.1.3 (page 72), $\boldsymbol{g}(g_0) = \boldsymbol{g}(g_1) + \boldsymbol{g}(g_2)$, hence $\boldsymbol{k}_0 = \boldsymbol{k}_1 + \boldsymbol{k}_2$.

Since $g_0 = g_1 + g_2 \in \mathbb{B}$, there exists i such that $g_0 = g_i$, hence the result. □

The following "dual" theorem is proved in exactly the same way.

Theorem 3.3.6. *Let* $f : \mathbb{B}^k \times \mathbb{B}^{kn} \to \mathbb{B}^n$ *(with* $k > 0$*), and, for* $i = 0, 1, 2$ *let* $f_i : \mathbb{B}^k \times \mathbb{B}^{kn} \to \mathbb{B}$, *and*

$$\boldsymbol{k}_i = \theta_1\langle x_1, \boldsymbol{x}_1\rangle.\theta_2\langle x_2, \boldsymbol{x}_2\rangle.\cdots.\theta_n\langle x_k, \boldsymbol{x}_k\rangle.\langle f_i, \boldsymbol{f}\rangle \in \mathbb{B}^{n+1}.$$

If $f_0 = f_1.f_2$ *then* $k_0 = k_1.k_2$.

As a consequence of Theorem 3.3.5, we get the following result.

Theorem 3.3.7. *Let, for* $i = 1, \ldots, n$, $f_i = f_i^{(1)} + f_i^{(2)}$ *where* $f_i^{(j)} : \mathbb{B}^{kn} \to \mathbb{B}$. *Then, for each* $i = 1, \ldots, n$, *there exists* $s(i) \in \{1, 2\}$ *such that*
$\theta_1\boldsymbol{x}_1.\cdots.\theta_k\boldsymbol{x}_k.\langle f_1, \ldots, f_i, \ldots, f_n\rangle =$

$$\theta_1\boldsymbol{x}_1.\cdots.\theta_k\boldsymbol{x}_k.\langle f_1^{(s(1))}, \ldots, f_i^{(s(i))}, \ldots, f_n^{(s(n))}\rangle.$$

Proof. Apply n times Theorem 3.3.5. □

We can express this theorem another way. Add $2n$ Boolean variables $z_1, \ldots, z_n$ and $z'_1, \ldots, z'_n$ and replace $f_i = f_i^{(0)} + f_i^{(1)}$ by $f'_i = z_i.f_i^{(0)} + z'_i.f_i^{(1)}$.

Theorem 3.3.8. *Let* $\boldsymbol{g}(z_1, \ldots, z_i, \ldots, z_n, z'_1, \ldots, z'_i, \ldots, z'_n) =$

$$\theta_1\boldsymbol{x}_1.\theta_2\boldsymbol{x}_2.\cdots.\theta_k\boldsymbol{x}_k.\langle f'_1, \ldots, f'_i, \ldots, f'_k\rangle.$$

Then there exist $b_1, \ldots, b_n \in \mathbb{B}$ *such that*

$$\boldsymbol{g}(1, \ldots, 1, \ldots, 1, \ldots, 1, \ldots, 1) = \boldsymbol{g}(b_1, \ldots, b_i, \ldots, b_n, \overline{b_1}, \ldots, \overline{b_i}, \ldots, \overline{b_n}).$$

Example 3.3.9. Let

$$\boldsymbol{f}(z_1, z_2, z'_1, z'_2) = \nu\langle x_1, x_2, x_3\rangle.\mu\langle y_1, y_2, y_3\rangle.\langle z_1.y_2 + z'_1.x_3, z_2.y_1 + z'_2.x_3, x_3\rangle.$$

We have $\mu\langle y_1, y_2, y_3\rangle.\langle z_1.y_2 + z'_1.x_3, z_2.y_1 + z'_2.x_3, x_3\rangle =$

$$\langle (z_1.z'_2 + z'_1)x_3, (z_2.z'_1 + z'_2)x_3, x_3\rangle$$

and

$$\boldsymbol{f}(z_1, z_2, z'_1, z'_2) = \langle z_1.z'_2 + z'_1, z_2.z'_1 + z'_2, 1\rangle.$$

Hence,

$$
\begin{aligned}
\boldsymbol{f}(1,1,1,1) &= \langle 1,1,1\rangle,\\
\boldsymbol{f}(1,1,0,0) &= \langle 0,0,1\rangle,\\
\boldsymbol{f}(0,0,1,1) &= \langle 1,1,1\rangle,\\
\boldsymbol{f}(1,0,0,1) &= \langle 1,1,1\rangle,\\
\boldsymbol{f}(0,1,1,0) &= \langle 1,1,1\rangle.
\end{aligned}
$$

It follows that $\nu\langle x_1,x_2,x_3\rangle.\mu\langle y_1,y_2,y_3\rangle.\langle y_2+x_3,y_1+x_3,x_3\rangle$ is also equal to $\nu\langle x_1,x_2,x_3\rangle.\mu\langle y_1,y_2,y_3\rangle.\langle y_2,x_3,x_3\rangle$, to $\nu\langle x_1,x_2,x_3\rangle.\mu\langle y_1,y_2,y_3\rangle.\langle x_3,y_1,x_3\rangle$, and to $\nu\langle x_1,x_2,x_3\rangle.\mu\langle y_1,y_2,y_3\rangle.\langle x_3,x_3,x_3\rangle$, but not to $\nu\langle x_1,x_2,x_3\rangle.\mu\langle y_1,y_2,y_3\rangle.\langle y_2,y_1,x_3\rangle$. □

3.3.2 Infinite vectors of fixed-point terms

Theorem 3.3.8 continues to hold for infinite systems of equations, i.e., when we consider that the sequences $\boldsymbol{x}_j$ and $\boldsymbol{f}$ are infinite instead of being of finite length n.

Let, for $j=1,\ldots,k$, $\boldsymbol{x}_j$ be a vector of variables, indexed by some set I of arbitrary cardinality, and let $\boldsymbol{f}$ be a vector of monotonic Boolean functions over the variables in the vectors $\boldsymbol{x}_j$, indexed by the same set I. Indeed, $\boldsymbol{f}$ can be viewed as a monotonic mapping from $(\mathbb{B}^I)^k$ to $\mathbb{B}^I$. Then, since $\mathbb{B}^I$ is a complete lattice, $\theta_1\boldsymbol{x}_1.\theta_2\boldsymbol{x}_2.\cdots.\theta_k\boldsymbol{x}_k.\boldsymbol{f}(\boldsymbol{x}_1,\boldsymbol{x}_2\ldots,\boldsymbol{x}_k)$ exists.

Let us consider two vectors of Boolean variables $\boldsymbol{z}$ and $\boldsymbol{z}'$, indexed by I, and let us assume that $\boldsymbol{f}=\langle f_i\rangle_{i\in I}$ with $f_i = z_i.f_i^{(0)} + z_i' f_i^{(1)}$.

For any vector $\boldsymbol{u}\in\mathbb{B}^I$ let us denote by $\overline{\boldsymbol{u}}$ the vector $\langle\overline{u_i}\rangle_{i\in I}$ Then the following result is a generalization of Theorem 3.3.8, where $\mathbf{1}$ denotes the infinite vector of 1's of appropriate length.

Theorem 3.3.10. *Let $g(\boldsymbol{z},\boldsymbol{z}')=\theta_1\boldsymbol{x}_1.\theta_2\boldsymbol{x}_2.\cdots.\theta_k\boldsymbol{x}_k.\boldsymbol{f}$. Then there exists $\boldsymbol{u}\in\mathbb{B}^\omega$ such that $g(\mathbf{1},\mathbf{1})=g(\boldsymbol{u},\overline{\boldsymbol{u}})$.*

To prove this theorem we need a definition.

Definition 3.3.11. We say that a monotonic mapping

$$\boldsymbol{f}(\boldsymbol{z},\boldsymbol{z}',\boldsymbol{x}) : \mathbb{B}^I\times\mathbb{B}^I\times(\mathbb{B}^I)^m\to\mathbb{B}^I$$

has property S (S for *Selection*) if
$\forall\boldsymbol{u},\boldsymbol{u}',\boldsymbol{v},\boldsymbol{v}'\in\mathbb{B}^I$ such that $\boldsymbol{u}\le\boldsymbol{v}$ and $\boldsymbol{u}'\le\boldsymbol{v}'$,
$\forall\boldsymbol{e}_1,\boldsymbol{e}_2\in(\mathbb{B}^I)^m$ such that $\boldsymbol{e}_1\le\boldsymbol{e}_2$,
if $\boldsymbol{u}+\boldsymbol{u}'\le\boldsymbol{f}(\boldsymbol{u},\boldsymbol{u}',\boldsymbol{e}_1)$ then there exist $\boldsymbol{w},\boldsymbol{w}'\in\mathbb{B}^I$ such that

- $\boldsymbol{u}\le\boldsymbol{w}\le\boldsymbol{v}$ and $\boldsymbol{u}'\le\boldsymbol{w}'\le\boldsymbol{v}'$,
- $\boldsymbol{u}\cdot\boldsymbol{u}'=\boldsymbol{w}\cdot\boldsymbol{w}'$,

– $\boldsymbol{w} + \boldsymbol{w}' = \boldsymbol{f}(\boldsymbol{w}, \boldsymbol{w}', \boldsymbol{e}_2) = \boldsymbol{f}(\boldsymbol{v}, \boldsymbol{v}', \boldsymbol{e}_2)$,

where $\boldsymbol{u} + \boldsymbol{u}'$ and $\boldsymbol{u} \cdot \boldsymbol{u}'$ are the pointwise extensions to $\mathbb{B}^I$ of the sum and product of $\mathbb{B}$.

Lemma 3.3.12. *If $\boldsymbol{f}(\boldsymbol{z}, \boldsymbol{z}', \boldsymbol{x}) = \langle f_i \rangle_{i \in I}$ is such that for all i in I, $f_i = z_i.f_i^{(0)} + z_i' f_i^{(1)}$ for some $f_i^{(0)}$ and $f_i^{(1)}$, then it has property S.*

Proof. It is sufficient to show that $f(z, z', \boldsymbol{x}) = z.g_1(\boldsymbol{x}) + z'.g_2(\boldsymbol{x})$ has property S in the following restricted sense: $\forall u, u', v, v' \in \mathbb{B}$ such that $u \leq v$ and $u' \leq v'$, $\forall \boldsymbol{e}_1, \boldsymbol{e}_2 \in (\mathbb{B}^I)^m$ such that $\boldsymbol{e}_1 \leq \boldsymbol{e}_2$, if $u + u' \leq f(u, u', \boldsymbol{e}_1)$ then there exist $w, w' \in \mathbb{B}^I$ such that

– $u \leq w \leq v$ and $u' \leq w' \leq v'$,
– $uu' = ww'$,
– $w + w' = f(w, w', \boldsymbol{e}_2) = f(v, v', \boldsymbol{e}_2)$.

Let us remark that $u + u' \leq f(u, u', \boldsymbol{e}_1) \leq f(u, u', \boldsymbol{e}_2) \leq f(v, v', \boldsymbol{e}_2)$.

If $u + u' = 1$, we have $1 = f(u, u', \boldsymbol{e}_2) = f(v, v', \boldsymbol{e}_2) = u + u'$ and we take $w = u, w' = u'$. If $f(v, v', \boldsymbol{e}_2) = 0$ then we have $0 = u + u' = f(u, u', \boldsymbol{e}_2) = f(v, v', \boldsymbol{e}_2) = u + u'$ and, again, we take $w = u, w' = u'$.

It remains to consider the case $u + u' = 0$ (hence, $u = u' = uu' = 0$) and $f(v, v', \boldsymbol{e}_2) = 1$. We cannot have $v + v' = 0$, for $f(0, 0, \boldsymbol{e}) = 0$. If $v.v' = 0$ we can take $w = v, w' = v'$ since $f(v, v', \boldsymbol{e}_2) = v + v' = 1$. If $v.v' = 1$, (i.e., $v = v' = 1$), then $f(v, v', \boldsymbol{e}_2) = f(1, 1, \boldsymbol{e}_2) = g_1(\boldsymbol{e}_2) + g_2(\boldsymbol{e}_2) = 1$. Thus $g_i(\boldsymbol{e}_2) = 1$ for some $i \in \{1, 2\}$ and we take $w = 1, w' = 0$ or $w = 0, w' = 1$ according to the value of i. □

Lemma 3.3.13. *Let us assume that $\boldsymbol{f}(\boldsymbol{z}, \boldsymbol{z}', \boldsymbol{x}, \boldsymbol{y}) : \mathbb{B}^I \times \mathbb{B}^I \times (\mathbb{B}^I)^{m+1} \to \mathbb{B}^I$ has property S. Then $\theta \boldsymbol{x}.\boldsymbol{f}(\boldsymbol{z}, \boldsymbol{z}', \boldsymbol{x}, \boldsymbol{y}) : \mathbb{B}^I \times \mathbb{B}^I \times (\mathbb{B}^I)^m \to \mathbb{B}^I$ has property S.*

Proof. Let $\boldsymbol{g}(\boldsymbol{z}, \boldsymbol{z}', \boldsymbol{y}) = \theta \boldsymbol{x}.\boldsymbol{f}(\boldsymbol{z}, \boldsymbol{z}', \boldsymbol{x}, \boldsymbol{y})$. Let $\boldsymbol{u}, \boldsymbol{u}', \boldsymbol{v}, \boldsymbol{v}' \in \mathbb{B}^I$ such that $\boldsymbol{u} \leq \boldsymbol{v}$ and $\boldsymbol{u}' \leq \boldsymbol{v}'$, let $\boldsymbol{e}_1, \boldsymbol{e}_2 \in (\mathbb{B}^I)^m$ such that $\boldsymbol{e}_1 \leq \boldsymbol{e}_2$, and let us assume that $\boldsymbol{u} + \boldsymbol{u}' \leq \boldsymbol{g}(\boldsymbol{u}, \boldsymbol{u}', \boldsymbol{e}_1)$. We have to show that there exist $\boldsymbol{w}, \boldsymbol{w}' \in \mathbb{B}^I$ such that

– $\boldsymbol{u} \leq \boldsymbol{w} \leq \boldsymbol{v}$ and $\boldsymbol{u}' \leq \boldsymbol{w}' \leq \boldsymbol{v}'$,
– $\boldsymbol{u}.\boldsymbol{u}' = \boldsymbol{w}.\boldsymbol{w}'$,
– $\boldsymbol{w} + \boldsymbol{w}' = \boldsymbol{g}(\boldsymbol{w}, \boldsymbol{w}', \boldsymbol{e}_2) = \boldsymbol{g}(\boldsymbol{v}, \boldsymbol{v}', \boldsymbol{e}_2)$.

Let $\boldsymbol{X} = \boldsymbol{g}(\boldsymbol{u}, \boldsymbol{u}', \boldsymbol{e}_1)$ and $\boldsymbol{Y} = \boldsymbol{g}(\boldsymbol{v}, \boldsymbol{v}', \boldsymbol{e}_2)$.
Obviously, $\boldsymbol{X} = \boldsymbol{f}(\boldsymbol{u}, \boldsymbol{u}', \boldsymbol{X}, \boldsymbol{e}_1) \leq \boldsymbol{Y} = \boldsymbol{f}(\boldsymbol{v}, \boldsymbol{v}', \boldsymbol{Y}, \boldsymbol{e}_2)$.

We have two different proofs according to whether $\theta = \mu$ or $\theta = \nu$.

Case $\theta = \mu$ Let us consider the vector $\boldsymbol{Y}_\beta$ of elements of $\mathbb{B}^I$, indexed by ordinal numbers, and defined by

$\boldsymbol{Y}_0 = \boldsymbol{X}$,
$\boldsymbol{Y}_{\alpha+1} = \boldsymbol{f}(\boldsymbol{v}, \boldsymbol{v}', \boldsymbol{Y}_\alpha, \boldsymbol{e_2})$,
$\boldsymbol{Y}_\beta = \sum_{\alpha<\beta} \boldsymbol{Y}_\alpha$.

Since $\boldsymbol{f}$ is monotonic and

$$\boldsymbol{Y}_0 = \boldsymbol{X} = \boldsymbol{f}(\boldsymbol{u}, \boldsymbol{u}', \boldsymbol{X}, \boldsymbol{e}_1) \leq \boldsymbol{f}(\boldsymbol{v}, \boldsymbol{v}', \boldsymbol{Y}_0, \boldsymbol{e}_2) = \boldsymbol{Y}_1,$$

by Proposition 1.2.13 (page 12), this sequence is nondecreasing and $\boldsymbol{Y} = \boldsymbol{Y}_\gamma = \boldsymbol{Y}_{\gamma+1}$ for some ordinal γ.

Now, we construct, inductively, two nondecreasing sequences $\boldsymbol{w}_\alpha$ and $\boldsymbol{w}'_\alpha$, for $0 \leq \alpha \leq \gamma + 1$, that satisfy

- $\forall \alpha \leq \gamma + 1, \boldsymbol{u} \leq \boldsymbol{w}_\alpha \leq \boldsymbol{v}, \boldsymbol{u}' \leq \boldsymbol{w}'_\alpha \leq \boldsymbol{v}'$,
- $\forall \alpha \leq \gamma + 1, \boldsymbol{u}.\boldsymbol{u}' = \boldsymbol{w}_\alpha.\boldsymbol{w}'_\alpha$,
- $\forall \alpha : 1 \leq \alpha \leq \gamma + 1, \boldsymbol{w}_\alpha + \boldsymbol{w}'_\alpha = \boldsymbol{Y}_\alpha \leq \boldsymbol{f}(\boldsymbol{w}_\alpha, \boldsymbol{w}'_\alpha, \boldsymbol{Y}_\alpha, \boldsymbol{e}_2)$,
- $\forall \alpha \leq \gamma, \boldsymbol{Y}_{\alpha+1} = \boldsymbol{f}(\boldsymbol{w}_{\alpha+1}, \boldsymbol{w}'_{\alpha+1}, \boldsymbol{Y}_\alpha, \boldsymbol{e}_2)$.

The definition is as follows:
$\boldsymbol{w}_0 = \boldsymbol{u}$, $\boldsymbol{w}'_0 = \boldsymbol{u}'$.
Since $\boldsymbol{w}_0 + \boldsymbol{w}'_0 = \boldsymbol{u} + \boldsymbol{u}' \leq \boldsymbol{X} \leq \boldsymbol{Y}_1 = \boldsymbol{f}(\boldsymbol{v}, \boldsymbol{v}', \boldsymbol{Y}_0, \boldsymbol{e}_2)$, and since $\boldsymbol{f}$ has property S, there exists $\boldsymbol{w}_1$ and $\boldsymbol{w}'_1$ such that

$\boldsymbol{w}_1.\boldsymbol{w}'_1 = \boldsymbol{w}_0.\boldsymbol{w}'_0$,
$\boldsymbol{w}_1 + \boldsymbol{w}'_1 = \boldsymbol{Y}_1 = \boldsymbol{f}(\boldsymbol{w}_1, \boldsymbol{w}'_1, \boldsymbol{Y}_0, \boldsymbol{e}_2) \leq \boldsymbol{f}(\boldsymbol{w}_1, \boldsymbol{w}'_1, \boldsymbol{Y}_1, \boldsymbol{e}_2)$.

Similarly, if $\boldsymbol{w}_\alpha + \boldsymbol{w}'_\alpha = \boldsymbol{Y}_\alpha \leq \boldsymbol{f}(\boldsymbol{w}_\alpha, \boldsymbol{w}'_\alpha, \boldsymbol{Y}_\alpha, \boldsymbol{e}_2)$, we can find $\boldsymbol{w}_{\alpha+1}$ and $\boldsymbol{w}'_{\alpha+1}$ such that

$$\boldsymbol{w}_{\alpha+1}.\boldsymbol{w}'_{\alpha+1} = \boldsymbol{w}_\alpha.\boldsymbol{w}'_\alpha = \boldsymbol{u}.\boldsymbol{u}',$$

$$\begin{aligned} \boldsymbol{Y}_{\alpha+1} = \boldsymbol{w}_{\alpha+1} + \boldsymbol{w}'_{\alpha+1} &= \boldsymbol{f}(\boldsymbol{w}_{\alpha+1}, \boldsymbol{w}'_{\alpha+1}, \boldsymbol{Y}_\alpha, \boldsymbol{e}_2) \\ &\leq \boldsymbol{f}(\boldsymbol{w}_{\alpha+1}, \boldsymbol{w}'_{\alpha+1}, \boldsymbol{Y}_{\alpha+1}, \boldsymbol{e}_2). \end{aligned}$$

For limit ordinals, we set $\boldsymbol{w}_\beta = \sum_{\alpha<\beta} \boldsymbol{w}_\alpha$ and $\boldsymbol{w}'_\beta = \sum_{\alpha<\beta} \boldsymbol{w}'_\alpha$. Since $\boldsymbol{w}_\beta.\boldsymbol{w}'_\beta = \sum_{\alpha<\beta} \boldsymbol{w}_\alpha.\boldsymbol{w}'_\alpha$, we get $\boldsymbol{w}_\beta.\boldsymbol{w}'_\beta = \boldsymbol{u}.\boldsymbol{u}'$. Now, we have $\boldsymbol{Y} = \boldsymbol{Y}_{\gamma+1} = \boldsymbol{f}(\boldsymbol{w}_{\gamma+1}, \boldsymbol{w}'_{\gamma+1}, \boldsymbol{Y}_\gamma, \boldsymbol{e}_2) = \boldsymbol{f}(\boldsymbol{w}_{\gamma+1}, \boldsymbol{w}'_{\gamma+1}, \boldsymbol{Y}, \boldsymbol{e}_2)$. We take $\boldsymbol{w} = \boldsymbol{w}_{\gamma+1}$, $\boldsymbol{w}' = \boldsymbol{w}'_{\gamma+1}$. Since $\boldsymbol{Y} = \boldsymbol{f}(\boldsymbol{w}, \boldsymbol{w}', \boldsymbol{Y}, \boldsymbol{e}_2)$, we have

$$\boldsymbol{Y} \geq \mu \boldsymbol{x}.\boldsymbol{f}(\boldsymbol{w}, \boldsymbol{w}', \boldsymbol{x}, \boldsymbol{e}_2) = \boldsymbol{g}(\boldsymbol{w}, \boldsymbol{w}', \boldsymbol{e}_2).$$

It remains to prove the reverse inequality. Let $\boldsymbol{Z}$ be any element of $\mathbb{B}^I$ such that $\boldsymbol{Z} = \boldsymbol{f}(\boldsymbol{w}, \boldsymbol{w}', \boldsymbol{Z}, \boldsymbol{e}_2)$. We prove by induction that $\boldsymbol{Y}_\alpha \leq \boldsymbol{Z}$. Firstly, $\boldsymbol{Y}_0 = \boldsymbol{X} = \mu \boldsymbol{x}.\boldsymbol{f}(\boldsymbol{u}, \boldsymbol{u}', \boldsymbol{x}, \boldsymbol{e}_1) \leq \mu \boldsymbol{x}.\boldsymbol{f}(\boldsymbol{w}, \boldsymbol{w}', \boldsymbol{x}, \boldsymbol{e}_2) \leq \boldsymbol{Z}$. If $\boldsymbol{Y}_\alpha \leq \boldsymbol{Z}$ then $\boldsymbol{Y}_{\alpha+1} = \boldsymbol{f}(\boldsymbol{w}_{\alpha+1}, \boldsymbol{w}'_{\alpha+1}, \boldsymbol{Y}_\alpha, \boldsymbol{e}_2) \leq \boldsymbol{f}(\boldsymbol{w}, \boldsymbol{w}', \boldsymbol{Z}, \boldsymbol{e}_2) = \boldsymbol{Z}$ and $\boldsymbol{Y}_\beta = \sum_{\alpha<\beta} \boldsymbol{Y}_\alpha \leq \boldsymbol{Z}$.

Case $\theta = \nu$ Since $\boldsymbol{f}$ has property S, there exist $\boldsymbol{w}_0$ and $\boldsymbol{w}'_0$ such that

$$\begin{aligned} &\boldsymbol{u} \leq \boldsymbol{w}_0 \leq \boldsymbol{v} \text{ and } \boldsymbol{u}' \leq \boldsymbol{w}'_0 \leq \boldsymbol{v}' \\ &\boldsymbol{w}_0.\boldsymbol{w}'_0 = \boldsymbol{u}.\boldsymbol{u}', \\ &\boldsymbol{w}_0 + \boldsymbol{w}'_0 = \boldsymbol{Y} = \boldsymbol{f}(\boldsymbol{w}_0, \boldsymbol{w}'_0, \boldsymbol{Y}, \boldsymbol{e}_2). \end{aligned}$$

Now, consider the nonincreasing sequence $\boldsymbol{Y}_\alpha$, indexed by ordinal numbers, defined by

$$\begin{aligned} &\boldsymbol{Y}_0 = \mathbf{1} \\ &\boldsymbol{Y}_{\alpha+1} = \boldsymbol{f}(\boldsymbol{v}, \boldsymbol{v}', \boldsymbol{Y}_\alpha, \boldsymbol{e_2}), \\ &\boldsymbol{Y}_\beta = \prod_{\alpha<\beta} \boldsymbol{Y}_\alpha. \end{aligned}$$

Then $\boldsymbol{Y} = \boldsymbol{Y}_\gamma = \boldsymbol{Y}_{\gamma+1}$ for some ordinal γ.

Since $\boldsymbol{f}$ has property S, for each successor ordinal $\alpha + 1 \leq \gamma + 1$ there exist $\boldsymbol{w}_{\alpha+1}$ and $\boldsymbol{w}'_{\alpha+1}$ such that

$$\begin{aligned} &\boldsymbol{w}_0 \leq \boldsymbol{w}_{\alpha+1} \leq \boldsymbol{v} \text{ and } \boldsymbol{w}'_0 \leq \boldsymbol{w}'_{\alpha+1} \leq \boldsymbol{v}' \\ &\boldsymbol{w}_{\alpha+1}.\boldsymbol{w}'_{\alpha+1} = \boldsymbol{w}_0.\boldsymbol{w}'_0 = \boldsymbol{u}.\boldsymbol{u}', \\ &\boldsymbol{w}_{\alpha+1} + \boldsymbol{w}'_{\alpha+1} = \boldsymbol{Y}_{\alpha+1} = \boldsymbol{f}(\boldsymbol{w}_{\alpha+1}, \boldsymbol{w}'_{\alpha+1}, \boldsymbol{Y}_\alpha, \boldsymbol{e}_2). \end{aligned}$$

Let $\boldsymbol{w} = \prod_{\alpha+1\leq\gamma+1} \boldsymbol{w}_{\alpha+1}$ and $\boldsymbol{w}' = \prod_{\alpha+1\leq\gamma+1} \boldsymbol{w}'_{\alpha+1}$. We claim that

1. $\boldsymbol{u} \leq \boldsymbol{w} \leq \boldsymbol{v}$ and $\boldsymbol{u}' \leq \boldsymbol{w}' \leq \boldsymbol{v}'$,
2. $\boldsymbol{w}.\boldsymbol{w}' = \boldsymbol{w}_0.\boldsymbol{w}'_0 = \boldsymbol{u}.\boldsymbol{u}'$,
3. $\boldsymbol{w} + \boldsymbol{w}' = \boldsymbol{Y}$,
4. $\boldsymbol{Y} = \nu\boldsymbol{x}.\boldsymbol{f}(\boldsymbol{w}, \boldsymbol{w}', \boldsymbol{x}, \boldsymbol{e}_2)$.

Since $\boldsymbol{u} \leq \boldsymbol{w}_0 \leq \boldsymbol{w}_{\alpha+1} \leq \boldsymbol{v}$ and $\boldsymbol{u}' \leq \boldsymbol{w}'_0 \leq \boldsymbol{w}'_{\alpha+1} \leq \boldsymbol{v}'$, point 1 above is satisfied.

Since $\boldsymbol{w}_0 \leq \boldsymbol{w} \leq \boldsymbol{w}_{\alpha+1}$ and $\boldsymbol{w}'_0 \leq \boldsymbol{w}' \leq \boldsymbol{w}'_{\alpha+1}$, we have $\boldsymbol{u}.\boldsymbol{u}' = \boldsymbol{w}_0.\boldsymbol{w}'_0 \leq \boldsymbol{w}.\boldsymbol{w}' \leq \boldsymbol{w}_{\alpha+1}.\boldsymbol{w}'_{\alpha+1} = \boldsymbol{u}.\boldsymbol{u}'$ and point 2 above is satisfied.

For point 3, we have $\boldsymbol{Y} = \boldsymbol{w}_0 + \boldsymbol{w}'_0 \leq \boldsymbol{w} + \boldsymbol{w}' \leq \boldsymbol{w}_{\gamma+1} + \boldsymbol{w}'_{\gamma+1} = \boldsymbol{Y}$.

For point 4, we have

$$\boldsymbol{Y} = \boldsymbol{f}(\boldsymbol{w}_0, \boldsymbol{w}'_0, \boldsymbol{Y}, \boldsymbol{e}_2) \leq \boldsymbol{f}(\boldsymbol{w}, \boldsymbol{w}', \boldsymbol{Y}, \boldsymbol{e}_2) \leq \boldsymbol{f}(\boldsymbol{w}_{\gamma+1}, \boldsymbol{w}'_{\gamma+1}, \boldsymbol{Y}, \boldsymbol{e}_2) = \boldsymbol{Y}$$

hence $\boldsymbol{Y} \leq \nu\boldsymbol{x}.\boldsymbol{f}(\boldsymbol{w}, \boldsymbol{w}', \boldsymbol{x}, \boldsymbol{e}_2)$. Let $\boldsymbol{Z} = \nu\boldsymbol{x}.\boldsymbol{f}(\boldsymbol{w}, \boldsymbol{w}', \boldsymbol{x}, \boldsymbol{e}_2) = \boldsymbol{f}(\boldsymbol{w}, \boldsymbol{w}', \boldsymbol{Z}, \boldsymbol{e}_2)$ and let us show, inductively, that $\boldsymbol{Z} \leq \boldsymbol{Y}_\alpha$ for any $\alpha \leq \gamma+1$. Obviously, $\boldsymbol{Z} \leq \boldsymbol{Y}_0 = \mathbf{1}$. If $\boldsymbol{Z} \leq \boldsymbol{Y}_\alpha$ then $\boldsymbol{Z} = \boldsymbol{f}(\boldsymbol{w}, \boldsymbol{w}', \boldsymbol{Z}, \boldsymbol{e}_2) \leq \boldsymbol{f}(\boldsymbol{w}_{\alpha+1}, \boldsymbol{w}'_{\alpha+1}, \boldsymbol{Y}_\alpha, \boldsymbol{e}_2) = \boldsymbol{Y}_{\alpha+1}$, and $\boldsymbol{Z} \leq \boldsymbol{Y}_\alpha$, for all $\alpha < \beta$, implies $\boldsymbol{Z} \leq \prod_{\alpha<\beta} \boldsymbol{Y}_\alpha = \boldsymbol{Y}_\beta$. □

Proof of Theorem 3.3.10 (page 78). It is a direct consequence of the two previous lemmas. Let $g(\boldsymbol{z}, \boldsymbol{z}') = \theta_1\boldsymbol{x}_1.\theta_2\boldsymbol{x}_2.\cdots.\theta_k\boldsymbol{x}_k.\boldsymbol{f}$ where $\boldsymbol{f}$ has the form explained at the beginning of this section (page 78). Then $\boldsymbol{g}$ has property S. Thus, since $\mathbf{0} = \mathbf{0} + \mathbf{0} \leq \boldsymbol{g}(\mathbf{0}, \mathbf{0}) \leq \boldsymbol{g}(\mathbf{1}, \mathbf{1})$, there exist $\boldsymbol{w}$ and $\boldsymbol{w}'$ such that $\mathbf{0} = \boldsymbol{w}.\boldsymbol{w}'$ and $\boldsymbol{w} + \boldsymbol{w}' = \boldsymbol{g}(\boldsymbol{w}, \boldsymbol{w}') = \boldsymbol{g}(\mathbf{1}, \mathbf{1})$. But $\mathbf{0} = \boldsymbol{w}.\boldsymbol{w}'$ implies $\boldsymbol{w}' \leq \overline{\boldsymbol{w}}$, hence $\boldsymbol{g}(\mathbf{1}, \mathbf{1}) = \boldsymbol{g}(\boldsymbol{w}, \boldsymbol{w}') \leq \boldsymbol{g}(\boldsymbol{w}, \overline{\boldsymbol{w}}) \leq \boldsymbol{g}(\mathbf{1}, \mathbf{1})$.

3.3.3 Infinite vectors of infinite fixed-point terms

We can further generalize the selection property shown in the previous section (Theorem 3.3.10) by considering $\boldsymbol{f}$ whose components are arbitrary sums of monotonic function from $(\mathbb{B}^I)^k$ to $\mathbb{B}^I$, rather than sums of only two of them.

Thus we assume that the i-th component of $\boldsymbol{f}$ is $f_i = \sum_{j\in J_i} f_i^{(j)}$ where J_i is any set of indices. Without loss of generality, we may assume that the sets J_i are disjoint and we set $J = \bigcup_{i\in I} J_i$.

Let us consider fresh Boolean variables z_j for $j \in J$ and write f_i in the form $\sum_{j\in J_i} z_j.f_i^{(j)}$. Now $\boldsymbol{f}$ is a mapping from $\mathbb{B}^J \times (\mathbb{B}^I)^k$ to $\mathbb{B}^I$ and

$$\boldsymbol{g}(\boldsymbol{z}) = \theta_1\boldsymbol{x}_1.\theta_2\boldsymbol{x}_2.\cdots.\theta_k\boldsymbol{x}_k.\boldsymbol{f}(\boldsymbol{z}, \boldsymbol{x}_1, \boldsymbol{x}_2 \ldots, \boldsymbol{x}_k)$$

is a mapping from $\mathbb{B}^J$ to $\mathbb{B}^I$.

Then the generalization of Theorem 3.3.10 (page 78) becomes the following.

Theorem 3.3.14. *Let* $\boldsymbol{g}(\boldsymbol{z}) = \theta_1\boldsymbol{x}_1.\theta_2\boldsymbol{x}_2.\cdots.\theta_k\boldsymbol{x}_k.\boldsymbol{f}$. *Then there exists* $\boldsymbol{u} = \langle u_j\rangle_{j\in J} \in \mathbb{B}^J$ *such that*

$\boldsymbol{g}(\boldsymbol{1}) = \boldsymbol{g}(\boldsymbol{u})$ *and*
for any $i \in I$ *there is at most one* j *in* J_i *such that* $u_j = 1$.

Proof. The proof is quite similar to the proof of Theorem 3.3.10. As above we define the *property* S for $\boldsymbol{f}(\boldsymbol{z}, \boldsymbol{x}) : \mathbb{B}^J \times (\mathbb{B}^I)^m \to \mathbb{B}^I$. The two functions $\boldsymbol{u} + \boldsymbol{u}'$ and $\boldsymbol{u}.\boldsymbol{u}'$ which appear in this definition have to be replaced by the two monotonic functions S and P from $\mathbb{B}^J$ to $\mathbb{B}^I$ defined as follows. For $\boldsymbol{u} = (u_j)_{j\in J} \in \mathbb{B}^J$

- the i-th component of $S(\boldsymbol{u})$ is equal to $\sum_{j\in J_i} u_j$,
- the i-th component of $P(\boldsymbol{u})$ is equal to 0 if and only if there is at most one j in in J_i such that $u_j = 1$.

Now, $\boldsymbol{f}(\boldsymbol{z}, \boldsymbol{x})$ has property S if
$\forall \boldsymbol{u}, \boldsymbol{v} \in \mathbb{B}^J$ such that $\boldsymbol{u} \le \boldsymbol{v}$, $\forall \boldsymbol{e}_1, \boldsymbol{e}_2 \in (\mathbb{B}^I)^m$ such that $\boldsymbol{e}_1 \le \boldsymbol{e}_2$, if $S(\boldsymbol{u}) \le \boldsymbol{f}(\boldsymbol{u}, \boldsymbol{e}_1)$ then there exist $\boldsymbol{w} \in \mathbb{B}^J$ such that

- $\boldsymbol{u} \le \boldsymbol{w} \le \boldsymbol{v}$,
- $P(\boldsymbol{u}) = P(\boldsymbol{w})$,
- $S(\boldsymbol{w}) = \boldsymbol{f}(\boldsymbol{w}, \boldsymbol{e}_2) = \boldsymbol{f}(\boldsymbol{v}, \boldsymbol{e}_2)$,

To prove the basis of the induction, it is enough to prove that

$$f(\boldsymbol{z}, \boldsymbol{x}) = \sum_{j\in J} z_j.f_j(\boldsymbol{x})$$

has property S, with $S(\boldsymbol{u}) = \sum_{j \in J} u_j$ and $P(\boldsymbol{u}) = 0$ if and only if there is at most one j such that $u_j = 1$. Let $\boldsymbol{u} \leq \boldsymbol{v}$, $\boldsymbol{e}_1 \leq \boldsymbol{e}_2$ and $S(\boldsymbol{u}) \leq f(\boldsymbol{u}, \boldsymbol{e}_1)$. Then $S(\boldsymbol{u}) \leq f(\boldsymbol{u}, \boldsymbol{e}_1) \leq f(\boldsymbol{v}, \boldsymbol{e}_2) \leq S(\boldsymbol{v})$. If $S(\boldsymbol{u}) = 1$ or $f(\boldsymbol{v}, \boldsymbol{e}_2) = 0$, we can take $\boldsymbol{w} = \boldsymbol{u}$. If $S(\boldsymbol{u}) = 0$ and $f(\boldsymbol{v}, \boldsymbol{e}_2) = 1$, then $\forall j \in J, u_j = 0$ and, since $f(\mathbf{v}, \boldsymbol{e}_2) = \sum_{j \in J} v_j . f_j(\boldsymbol{e}_2) = 1$, there exists j_0 such that $v_{j_0} = 1$ and $f_{j_0}(\boldsymbol{e}_2) = 1$. Then, we take $\boldsymbol{w}$ defined by $w_j = 1$ if and only if $j = j_0$.

To prove the induction step, we proceed exactly as in Lemma 3.3.13. The only difference is that since sum and product are replaced by S and P, we need, for the case $\theta = \mu$, the following property: if $(\boldsymbol{w}_\alpha)_{\alpha<\beta}$ is an increasing sequence of elements of $\mathbb{B}^J$ such that $\boldsymbol{u} \leq \boldsymbol{w}_\alpha$ and $P(\boldsymbol{u}) = P(\boldsymbol{w}_\alpha)$ then $P(\boldsymbol{u}) = P(\sum_{\alpha<\beta} \boldsymbol{w}_\alpha)$ and $S(\sum_{\alpha<\beta} \boldsymbol{w}_\alpha) = \sum_{\alpha<\beta} S(\boldsymbol{w}_\alpha)$. This property is proved exactly like the similar one with sum and product. □

As a corollary of the previous theorem we get the following:

Corollary 3.3.15. *Let $\tau = \theta_1 \boldsymbol{x}_1 . \theta_2 \boldsymbol{x}_2 . \cdots . \theta_k \boldsymbol{x}_k . \boldsymbol{t}$ be a closed fixed-point term of the Boolean μ-calculus such that for any $i \in I$, the component of t of index i is either 0, 1, or $\sum_{j \in J_i} t_i^j$ (with $J_i \neq \emptyset$). Then there exists $\tau' = \theta_1 \boldsymbol{x}_1 . \theta_2 \boldsymbol{x}_2 . \cdots . \theta_k \boldsymbol{x}_k . \boldsymbol{t}'$ such that $[\tau] = [\tau']$ where the component of t' of index i is*

0 if the component of t of index i is 0,
1 if the component of t of index i is 1,
t_i^j, with $j \in J_i$, if the component of t of index i is $\sum_{j \in J_i} t_i^j$.

By duality we also have

Corollary 3.3.16. *Let $\tau = \theta_1 \boldsymbol{x}_1 . \theta_2 \boldsymbol{x}_2 . \cdots . \theta_n \boldsymbol{x}_k . \boldsymbol{t}$ be a closed fixed-point term of the Boolean μ-calculus such that for any $i \in I$, the component of t of index i is either 0, 1, or $\prod_{j \in J_i} t_i^j$ (with $J_i \neq \emptyset$). Then there exists $\tau' = \theta_1 \boldsymbol{x}_1 . \theta_2 \boldsymbol{x}_2 . \cdots . \theta_k \boldsymbol{x}_k . \boldsymbol{t}'$ such that $[\tau] = [\tau']$ where the component of t' of index i is*

0 if the component of t of index i is 0,
1 if the component of t of index i is 1,
t_i^j, with $j \in J_i$, if the component of t of index i is $\prod_{j \in J_i} t_i^j$.

3.4 Bibliographic notes and sources

The selection property of the Boolean μ-calculus was discovered by revealing the connection between the μ-calculus and infinite games (see the next chapter). It is implicit in the works by Emerson and Jutla [33] and Walukiewicz [102]. The purely μ-calculus proof given here was originally presented in [4]. The same result was also proved, independently, by A. Mader [59].

4. Parity games

Suppose two persons, let us call them Eva and Adam, argue about the value of a closed Boolean term, for example $((0 \vee 1) \wedge ((0 \wedge (1 \vee 0)) \vee (1 \vee (0 \wedge 0))))$. Suppose Eva affirms the value is 1 while Adam claims the opposite. One way to resolve the controversy is to play a game in which the players successively select the subterms, and the choice belongs to Eva or Adam, depending on whether the leading connective is $\vee$ or $\wedge$, respectively. Finally, Eva wins if the last term is 1, otherwise Adam is the winner. A possible play can proceed as follows: Adam: $((0 \wedge (1 \vee 0)) \vee (1 \vee (0 \wedge 0)))$, Eva: $(1 \vee (0 \wedge 0))$, Eva: 1. It is easy to see that, in general, Eva has a winning strategy in this game if and only if the value of the term is indeed 1.

Now if instead of an ordinary term, Eva and Adam wish to consider a closed Boolean fixed-point term, they can still embark on playing the above game, assuming that whenever the play reaches the subterm x of $\theta x.t$, it resumes from t. But now the resulting play can be infinite. (The rules of the game generalize easily to the case of a vectorial fixed-point term.) It turns out to be possible to define an infinite winning condition in such a way that, again, Eva wins if and only if the value of the term (or, respectively, a component of the vectorial term) is 1. It will be the parity condition announced in the title of this chapter. By the correspondence between the Boolean μ-calculus and powerset interpretations (see Section 3.2 of the previous chapter), the games will provide a new, "dynamic" semantics of the fixed-point terms.

Infinite games is a well–established subject of descriptive set theory; their relevance for automata theory was first recognized by Büchi, and a close connection to the μ-calculus was discovered by Emerson and Jutla [33], who also introduced the parity winning condition that we adopt here (see also Bibliographical notes at the end of this chapter). One major advantage of the games is that an alternative semantics they provide is often more intuitive than the one based on the Knaster–Tarski theorem. It is particularly helpful in case of complicated terms, with nested alternations. Indeed, games are behind the properties which witness the infinity of the alternation-depth hierarchy (see Chapter 8). Games are also useful as a mathematical tool, not least because they allow the replacement of a universal quantifier by an existential one. (By determinacy, the non-existence of a strategy of one player implies the existence of the strategy of the other.) In particular, in Chapter 7, we use parity games to define the semantics of some general–purpose alternating automata, and to establish the equivalence between those automata and the μ-calculus.

4.1 Games and strategies

A *parity game* is a game of possibly infinite duration played by two players, whom we call *Eva* and *Adam*, in connection to existential and universal quantifiers, respectively. The game will be specified by a tuple $G = (Pos_a, Pos_e, Mov, rank)$ satisfying the following conditions.

- Pos_a and Pos_e are disjoint sets whose elements we call *positions* of Adam and Eva, respectively. We also write $Pos = Pos_a \cup Pos_e$. We assume $Pos \neq \emptyset$, although Pos_a, or Pos_e may be empty.
- Mov is a subset of $Pos \times Pos$ whose elements are called *moves*. Thus, the pair $(Pos\ ,\ Mov)$ forms a directed graph; we will sometimes refer to this graph as to the arena of the game.
- $rank : Pos \to \mathbb{N}$ is a mapping such that the set of values $rank(Pos)$ is *finite*.

A position v is a *deadlock* if there is no v' such that $(v, v') \in Mov$.

A *play* starting in a position v is a sequence, finite or infinite, $v_0, v_1, \ldots,$ such that $v_0 = v$, and for all i, whenever v_i is defined and there exists w, such that $(v_i, w) \in Mov$, then v_{i+1} is defined, and $(v_i, v_{i+1}) \in Mov$. In other words, a play is a *maximal* path in the arena: either it is infinite or it is finite and its last position is a deadlock. We think of a play as of a sequence of positions subsequently selected by the players consistently with the Mov relation. That is, e.g., if v_i is a position of Eva then Eva selects a position v_{i+1} which can be a position of Adam, who will move in his turn, but can also be again a position of Eva. It is often the case that a graph is bipartite (with the partition, actually being Pos_e and Pos_a), and then the players play in alternation. But in view of further applications, we prefer not to have such a restriction. Note that, in our game, it is even possible that a single player plays all the time which, of course, is not a guarantee to win.

A *finite* play $v_0, v_1, \ldots, v_m$ is *won* by Eva, or by Adam, if v_m is a position of the opposite player, that is, of Adam, or of Eva, respectively. That is, Eva wins the play if Adam cannot make any move (and *vice–versa*). An *infinite* play $v_0, v_1, \ldots$ is won by Eva if $\limsup_{i\to\infty} rank(v_i)$ is *even*, otherwise, Adam is the winner. Note that the above limit is always defined because the set of values of the function *rank* is finite, and it is equal to the greatest rank occurring infinitely often in the play.

A *strategy* for Eva is a partial mapping $s : (Pos)^* Pos_e \to Pos$ such that $s(v_0 v_1 \dots v_i)$ is defined whenever v_i is not a deadlock, and in this case $(v_i, s(v_0 v_1 \dots v_i)) \in Mov$. Intuitively, a strategy tells Eva the next move according to the current configuration of the play, if there is any possible move. A play (finite or infinite) $v_0 v_1 \dots$ is *consistent* with a strategy s if, whenever an initial segment $v_0 v_1 \dots v_i$ is in the domain of s then $v_{i+1} = s(v_0 v_1 \dots v_i)$. Given a position $v \in Pos$ (of Eva or of Adam), we say that a strategy s is *winning* for Eva *at the position* v if any play starting from v and consistent with s is won by Eva. We say that v is a *winning position* of Eva if there is a winning strategy for Eva at this position. We emphasize that v itself need not be a position of Eva.

We say that a strategy is *globally winning* for Eva if it is winning at any winning position of Eva. That is, whenever Eva can win, she can use this strategy (independently of starting position).

The analogous concepts for Adam are defined similarly. Obviously, a position cannot be winning both for Adam and Eva. But we shall prove below (Theorem 4.3.10, page 92), that a position is always winning for one of the two players.

4.2 Positional strategies

We say that a strategy $s : (Pos)^* Pos_e \to Pos$ for Eva is *positional* if for any $uv, u'v' \in (Pos)^* Pos_e$, $v = v' \Rightarrow s(uv) = s(u'v')$. In other words the next move of Eva depends only on her current position, and the positional strategy can be seen as a mapping $s : Pos_e \to Pos$ such that $(v, s(v)) \in Mov$. Such a strategy is called positional, as in [30]; in the literature it is sometimes called *memoryless* because Eva does not have to remember the sequence of moves played so far to decide on the next move.

Let s be any positional strategy for Eva. If Eva plays according to this strategy, she always plays the same move in any position of Pos_e. If Adam knows the Eva's strategy, he can play all the Eva's moves as well and the game becomes a one–player game. This is explained by the following lemma.

Lemma 4.2.1. *Let s be a positional strategy for Eva for the game*

$$G = (Pos_a, Pos_e, Mov, rank).$$

Let $G(s)$ be the game $(Pos, \emptyset, Mov', rank)$ where

$$Mov' = (Mov \cap (Pos_a \times Pos)) \cup \{(v, s(v)) \mid v \in Pos_e\}.$$

The strategy s is winning in v_0 if and only if for every maximal path $v_0 v_1 \cdots$ in $G(s)$, either this path is infinite and $\limsup_{n \to \infty} rank(v_n)$ *is even, or it is finite and its last position is in Pos_a.*

Proof. The set of plays that Adam can play on $G(s)$ is exactly the set of plays consistent with s that Eva and Adam can play on G. □

In case Eva's strategy s is not positional , we can still get a similar result. We consider the game $G' = (Pos', Mov', rank')$ defined by

- $Pos'_e = (Pos)^* Pos_e$, $Pos'_a = (Pos)^* Pos_a$,
- $Mov' = \{(uv, uvv') \mid u \in (Pos)^*, (v, v') \in Mov\}$,
- $rank'(uv) = rank(v)$.

It is easy to see that with any strategy s for Eva on G, one can associate a positional strategy s' on G', and vice-versa, such that s is winning at v on G if and only if s' is winning at v on G'.

Corollary 4.2.2. *Let s be any strategy for Eva. Let $G'(s) = (Pos', \emptyset, Mov'', rank')$ where*

$$\begin{aligned} Mov'' = \quad & \{(uv, uvv') \mid u \in (Pos)^*, (v, v') \in Mov \cap (Pos_a \times Pos)\} \\ \cup \quad & \{(uv, uvs(uv)) \mid u \in (Pos)^*, v \in Pos_e\}. \end{aligned}$$

Then s is winning at v_0 on G if and only if for every maximal path $v'_0 v'_1 \cdots$ in $G'(s)$, either this path is infinite and $\limsup_{n \to \infty} r'(v'_n)$ is even, or it is finite and its last position is in Pos_a.

4.3 The μ-calculus of games

4.3.1 Boolean terms for games

Let $G = (Pos_a, Pos_e, Mov, rank)$ be a parity game, and let k be such that $rank(Pos) \subseteq \{1, \ldots, k\}$.(By a translation of 2, we may always assume that $rank(v)$ is never 0.)

Let $\boldsymbol{w} \in \mathbb{B}^{Pos}$ be defined by $\boldsymbol{w}_v = 1$ if and only if v is a winning position for Eva. It is clear that the components of $\boldsymbol{w}$ satisfy the equalities

$$\boldsymbol{w}_v = \begin{cases} \sum_{(v,v') \in Mov} \boldsymbol{w}_{v'} & \text{if } v \in Pos_e, \\ \prod_{(v,v') \in Mov} \boldsymbol{w}_{v'} & \text{if } v \in Pos_a. \end{cases}$$

Indeed, we prove in this section, that $\boldsymbol{w}$ is the value of a vectorial Boolean fixed-point term built up from the right-hand sides of these equalities.

Definition 4.3.1. For $i = 1, \ldots, k$ let $\boldsymbol{x}^{(i)}$ be vectors of variables indexed by Pos such that the sets $\{\boldsymbol{x}^{(i)}\}$ $(i = 1, \ldots, k)$ are pairwise disjoint. Let $\boldsymbol{f} : (\mathbb{B}^{Pos})^k \to \mathbb{B}^{Pos}$ be the monotonic mapping whose component $\boldsymbol{f}_v$ of index v is defined by

$$f_v(\boldsymbol{x}^{(1)},\dots,\boldsymbol{x}^{(k)}) = \begin{cases} \sum_{(v,v')\in Mov} \boldsymbol{x}_{v'}^{(rank(v'))} & \text{if } v \in Pos_e, \\ \prod_{(v,v')\in Mov} \boldsymbol{x}_{v'}^{(rank(v'))} & \text{if } v \in Pos_a. \end{cases}$$

Note that if there is a deadlock $v \in Pos_e$ (resp. Pos_a) then $f_v(\boldsymbol{x}^{(1)},\dots,\boldsymbol{x}^{(k)}) = 0$ (resp. 1), because $\sum \emptyset = 0$ and $\prod \emptyset = 1$.

Finally, let $\boldsymbol{b} = \theta_k \boldsymbol{x}^{(k)}.\cdots.\mu\boldsymbol{x}^{(1)}.\boldsymbol{f}(\boldsymbol{x}^{(1)},\dots,\boldsymbol{x}^{(k)})$ where θ_i is equal to μ if i is odd, and to ν if i is even.

In view of Lemma 4.2.1 we need some results on one-player games

Let $G = (V, E)$ be a directed graph and let $r : V \to \{1,\dots,k\}$. Moreover let β be a mapping that assigns 0 or 1 to each deadlock in G.

A maximal path $v_0v_1\dots$ in G is said to be an *even path* (resp. *odd path*) if it is infinite and $\limsup_{n\to\infty} r(v_n)$ is even (resp. odd), or it is finite and its last vertices is assigned the value 1 (resp. 0) by β. Let $P(v)$ be the set of all maximal paths in G starting in v.

Definition 4.3.2. Let $\boldsymbol{x}^{(1)},\dots,\boldsymbol{x}^{(k)}$ be disjoint vectors of variables, indexed by V. Let $\boldsymbol{g} : (\mathbb{B}^V)^k \to \mathbb{B}^V$ where, for each $v \in V$, the v-th component $\boldsymbol{g}_v$ of $\boldsymbol{g}$ is defined by $\boldsymbol{g}_v(\boldsymbol{x}^{(1)},\dots,\boldsymbol{x}^{(k)}) = \prod_{v'\in V_v} \boldsymbol{x}_{v'}^{(r(v'))}$ if $V_v = \{v' \mid (v,v') \in E\}$ is a nonempty subset of V, otherwise, it is the Boolean value assigned to v by β.

The following property is a direct consequence of the definition of $\boldsymbol{g}$.

Lemma 4.3.3. *Let $(v,v') \in E$ with and $r(z') = m'$. Let $\boldsymbol{b}^{(m)} \in \mathbb{B}^V$ for $m = 1,\dots,k$. Then $\boldsymbol{g}_v(\boldsymbol{b}^{(1)},\dots,\boldsymbol{b}^{(k)}) \le \boldsymbol{b}_{v'}^{(m')}$.*

Proposition 4.3.4. *Let $\boldsymbol{c} = \theta_k \boldsymbol{x}^{(k)}.\cdots.\mu\boldsymbol{x}^{(1)}.\boldsymbol{g}(\boldsymbol{x}^{(1)},\dots,\boldsymbol{x}^{(k)})$ where θ_i is equal to μ if i is odd, and to ν if i is even.*

$\boldsymbol{c}_v = 0$ if and only $P(v)$ contains an odd path.

Proof. First, let us remark that without loss of generality, we may assume that for all v, V_v is not empty, so that all maximal paths are infinite. Indeed if V_v is empty, we add to E the edge (v,v) and we modify $r(v)$ so that $r(v)$ is even if v is assigned the value 1, and $r(v)$ is odd otherwise. If a maximal path ends in v it can now be extended into an infinite path $v, v, v\dots$, and this path is even if and only if $r(v)$ is even if and only if v was assigned the value 1.

If: Let us assume that there is an odd path $v_0, v_1, \dots$ in $P(v_0)$. By definition of $\boldsymbol{c}$ we have $\boldsymbol{c} = \boldsymbol{g}(\boldsymbol{c},\dots,\boldsymbol{c})$, and, by the previous lemma, $\boldsymbol{c}_{v_n} \le \boldsymbol{c}_{v_{n+1}}$.

Let $m = \limsup_{n\to\infty} r(v_n)$, which is odd, so that $\theta_m = \mu$. There exists an increasing sequence $j_0,\dots,j_p,\dots$ of natural numbers such that

- $\forall p \ge 0$, $r(v_{j_p}) = m$,
- $\forall p \ge 0$, $\forall j \ge 0 : j_p < j < j_{p+1} \Rightarrow r(v_j) < m$.

Let $J_0, J_1, \ldots$ be the sequence of subsets of $\mathbb{N}$ defined by

$$J_0 = \{j_p \mid p \geq 1\}, \quad J_{\ell+1} = \{j \geq j_0 \mid j+1 \in J_\ell\}.$$

Obviously $\bigcup_{\ell \geq 1} J_\ell = \{j \in \mathbb{N} \mid j \geq j_0\}$.

Finally, let $V'_\ell = \{v_j \mid j \in J_\ell\}$, $V' = \bigcup_{\ell \geq 1} V'_\ell$, and $\boldsymbol{d} \in \mathbb{B}^V$ be defined by $\boldsymbol{d}_v = 0 \Leftrightarrow v \in V'$. We claim that $\boldsymbol{c} \leq \boldsymbol{d}$.

Let

$$\boldsymbol{h}(\boldsymbol{x}^{(m)}) = \theta_{m-1}\boldsymbol{x}^{(m-1)}.\cdots.\mu\boldsymbol{x}^{(1)}.\boldsymbol{g}(\boldsymbol{x}^{(1)}, \ldots, \boldsymbol{x}^{(m-1)}, \boldsymbol{x}^{(m)}, \boldsymbol{c}, \ldots, \boldsymbol{c}).$$

We have $\boldsymbol{c} = \mu\boldsymbol{x}^{(m)}.\boldsymbol{h}(\boldsymbol{x}^{(m)})$, so that to prove $\boldsymbol{c} \leq \boldsymbol{d}$, it is sufficient to show that $\boldsymbol{h}(\boldsymbol{d}) \leq \boldsymbol{d}$. Indeed we show by induction on $\ell \geq 1$ that $v \in V'_\ell \Rightarrow \boldsymbol{h}_v(\boldsymbol{d}) = 0$.

By definition of $\boldsymbol{h}$,

$$\boldsymbol{h}_v(\boldsymbol{d}) = \boldsymbol{g}_v(\boldsymbol{h}(\boldsymbol{d}), \ldots, \boldsymbol{h}(\boldsymbol{d}), \boldsymbol{d}, \boldsymbol{c}, \ldots, \boldsymbol{c}).$$

Now, assume that $v \in V'_1$. Then there exists $p \geq 1$ such that $v = v_{j_p-1}$. Let $v' = v_{j_p}$. Since $r(v') = m$, we get, by Lemma 4.3.3, $\boldsymbol{h}_v(\boldsymbol{d}) \leq \boldsymbol{d}_{v'}$. Since $v' \in V'$, $\boldsymbol{d}_{v'} = 0$. Assume that $v \in V'_{\ell+1} - V'_1$. Then there exists $j \in J_{\ell+1}$ such that $v = v_j$ with $j+1 \in J_\ell$ and $r(v_{j+1}) < m$. It follows that $\boldsymbol{h}_v(\boldsymbol{d}) \leq \boldsymbol{h}_{v'}(\boldsymbol{d})$ where $v' = v_{j+1} \in V'_\ell$. By the induction hypothesis $\boldsymbol{h}_{v'}(\boldsymbol{d}) = 0$.

Since $\boldsymbol{c} \leq \boldsymbol{d}$ and since j_0 is in some J_ℓ, $\boldsymbol{c}_{v_{j_0}} = 0$. Using Lemma 4.3.3 again, we get $\boldsymbol{c}_{v_0} = 0$.

Only if: We show that if $\boldsymbol{c}_v = 0$ then $P(v)$ contains an odd path.

By Corollary 3.3.16 (page 83), there exists $\boldsymbol{g}'$ such that

$$\begin{aligned} \boldsymbol{c} &= \theta_k\boldsymbol{x}^{(k)}.\theta_{k-1}\boldsymbol{x}^{(k-1)}.\cdots.\mu\boldsymbol{x}^{(1)}.\boldsymbol{g}(\boldsymbol{x}^{(1)}, \ldots, \boldsymbol{x}^{(k)}) \\ &= \theta_k\boldsymbol{x}^{(k)}.\theta_{k-1}\boldsymbol{x}^{(k-1)}.\cdots.\mu\boldsymbol{x}^{(1)}.\boldsymbol{g}'(\boldsymbol{x}^{(1)}, \ldots, \boldsymbol{x}^{(k)}) \end{aligned}$$

where $\boldsymbol{g}'_v(\boldsymbol{x}^{(1)}, \ldots, \boldsymbol{x}^{(k)}) = \boldsymbol{x}_{v'}^{(r(v'))}$ with $v' \in V_v$.

Let us consider the graph $G' = (V, E')$ associated with $\boldsymbol{g}'$: the unique successor of v is $v' \in V_v$ chosen in the definition of $\boldsymbol{g}'$. For each $v \in V$ there is one and only one infinite path $p(v)$ in G', which is also a path in G. Therefore it is enough to show that if $\boldsymbol{c}_{v_0} = 0$ the unique path $v_0, v_1, \ldots$ in G' is an odd path.

Due to the form of $\boldsymbol{g}'$, Lemma 4.3.3 can be strengthened.

Lemma 4.3.5. *Let $(v, v') \in E'$ with $r(v') = m'$. Let $\boldsymbol{b}^{(m)} \in \mathbb{B}^V$ for $m = 1, \ldots, k$. Then $\boldsymbol{g}'_v(\boldsymbol{b}^{(1)}, \ldots, \boldsymbol{b}^{(k)}) = \boldsymbol{b}_{v'}^{(m')}$.*

It follows that for any $j \in \mathbb{N}$, $\boldsymbol{c}_{v_j} = 0$. We show that the path $p(v_0)$ cannot be even.

Assume it is even, and let $m = \limsup_{n\to\infty} r(v_n)$ which is even, thus $\theta_m = \nu$. There exists an increasing sequence $j_0, \ldots, j_k, \ldots$ of natural numbers such that

– $\forall p \geq 0$, $r(v_{j_p}) = m$,
– $\forall p \geq 0, \forall j \geq 0, j_k p < j < j_{p+1} \Rightarrow r(v_j) < m$.

Let $\boldsymbol{d'} \in \mathbb{B}^V$ be defined by $\boldsymbol{d'}_v = 1$ if and only if there is $j \geq j_0$ such that $v = v_j$ and let us show that $\boldsymbol{d'} \leq \boldsymbol{c}$ that is a contradiction.
Let

$$\boldsymbol{h'}(\boldsymbol{x}^{(m)}) = \theta_{m-1}\boldsymbol{x}^{(m-1)}.\cdots.\mu\boldsymbol{x}^{(1)}.\boldsymbol{g'}(\boldsymbol{x}^{(1)},\ldots,\boldsymbol{x}^{(m-1)},\boldsymbol{x}^{(m)},\boldsymbol{c},\ldots,\boldsymbol{c}).$$

Since $\boldsymbol{c} = \nu\boldsymbol{x}^{(m)}.\boldsymbol{h'}(\boldsymbol{x}^{(m)})$, $\boldsymbol{d'} \leq \boldsymbol{c}$ whenever $\boldsymbol{d'} \leq \boldsymbol{h'}(\boldsymbol{d'})$, i.e, if $j \geq j_0$, $\boldsymbol{h'}_{v_j}(\boldsymbol{d'}) = 1$. The proof is similar to the previous one, using Lemma 4.3.5 □

Proposition 4.3.6. *Let $\boldsymbol{b}$ be the Boolean vector defined in Definition 4.3.1 (page 89). If Eva has a positional strategy winning on v, then $\boldsymbol{b}_v = 1$.*

Proof. By Lemma 4.2.1 (page 87), if Eva has a positional strategy winning in v then there is no odd path from v in $G(s)$. By Proposition 4.3.4 (page 89), this implies $\boldsymbol{c}_v = 1$ where $\boldsymbol{c}$ is defined in Definition 4.3.2 (page 89) with $G = G(s)$. But the mapping $\boldsymbol{g}$, of which $\boldsymbol{c}$ is a fixed point, is, by construction, less than the mapping $\boldsymbol{f}$, of which $\boldsymbol{b}$ is a fixed point. Hence, $\boldsymbol{c} \leq \boldsymbol{b}$ and $\boldsymbol{b}_v = 1$. □

Proposition 4.3.7. *If Eva has a strategy winning at v, then $\boldsymbol{b}_v = 1$.*

Proof. By the previous proposition, and by Corollary 4.2.2 (page 88), we have $\boldsymbol{b}'_v = 1$ with $\boldsymbol{b'} = \theta_k\boldsymbol{y}^{(k)}.\cdots.\mu\boldsymbol{y}^{(1)}.\boldsymbol{f'}(\boldsymbol{y}^{(1)},\ldots,\boldsymbol{y}^{(k)})$, where

$$\boldsymbol{f'}_{uv'}(\boldsymbol{y}^{(1)},\ldots,\boldsymbol{y}^{(k)}) = \begin{cases} \sum_{(v',v'')\in Mov} \boldsymbol{y}^{(r(v''))}_{uv'v''} & \text{if } v' \in Pos_e, \\ \prod_{(v',v'')\in Mov} \boldsymbol{y}^{(r(v''))}_{uv'v''} & \text{if } v' \in Pos_a. \end{cases}$$

Let us define $h : \mathbb{B}^{Pos} \to \mathbb{B}^{(Pos)^+}$ by $h(\boldsymbol{x}) = \boldsymbol{y}$ if and only if $\boldsymbol{y}_{uv'} = \boldsymbol{x}_{v'}$. If $v' \in Pos_e$,

$$\begin{aligned} \boldsymbol{f}'_{uv'}(h(\boldsymbol{x}^{(1)}),\ldots,h(\boldsymbol{x}^{(k)})) &= \sum_{(v',v'')\in Mov} \boldsymbol{x}^{(r(v''))}_{v''} \\ &= \boldsymbol{f}_{v'}(\boldsymbol{x}^{(1)},\ldots,\boldsymbol{x}^{(k)}) \end{aligned}$$

For similar reasons, if $v' \in Pos_a$, then

$$\boldsymbol{f}'_{uv'}(h(\boldsymbol{x}^{(1)}),\ldots,h(\boldsymbol{x}^{(k)})) = \boldsymbol{f}_{v'}(\boldsymbol{x}^{(1)},\ldots,\boldsymbol{x}^{(k)}).$$

On the other hand, the component of index uv' of $h(\boldsymbol{f}(\boldsymbol{x}^{(1)},\ldots,\boldsymbol{x}^{(k)}))$ is equal to $\boldsymbol{f}_{v'}(\boldsymbol{x}^{(1)},\ldots,\boldsymbol{x}^{(k)})$. It follows that

$$h(\boldsymbol{f}(\boldsymbol{x}^{(1)},\ldots,\boldsymbol{x}^{(k)})) = \boldsymbol{f'}(h(\boldsymbol{x}^{(1)}),\ldots,h(\boldsymbol{x}^{(k)})).$$

Obviously, h has the continuity properties that allow us to apply Lemma 1.2.15 (page 13), and we get $h(\boldsymbol{b}) = \boldsymbol{b}'$. Thus, $\boldsymbol{b}'_v = \boldsymbol{b}_v = 1$. □

The selection theorem (Theorem 3.3.14, page 82), applied to the Boolean fixed-point term given in Definition 4.3.1 (page 88), amounts to stating the existence of a positional strategy for Eva in all positions v such that $\boldsymbol{b}_v = 1$.

Theorem 4.3.8. *The position v is winning for Eva if and only if $\boldsymbol{b}_v = 1$. Moreover, there exists a positional strategy s for Eva that is winning at all winning positions.*

Proof. By the previous proposition, we know that if v is a winning position then $\boldsymbol{b}_v = 1$. To complete the proof of this theorem it is enough to prove the existence of s. By the selection theorem, for each $v \in Pos_e$, there exists v' with $(v, v') \in Mov$, denoted by $s(v)$, such that

$$\boldsymbol{b} = \theta_k \boldsymbol{x}^{(k)}. \cdots .\mu \boldsymbol{x}^{(1)}.\boldsymbol{f}''(\boldsymbol{x}^{(1)}, \ldots, \boldsymbol{x}^{(k)})$$

where $\boldsymbol{f}''_v(\boldsymbol{x}^{(1)}, \ldots, \boldsymbol{x}^{(k)}) = \begin{cases} \boldsymbol{x}_{s(v)}^{(r(s(v)))} & \text{if } v \in Pos_e, \\ \prod_{(v,v') \in Mov} \boldsymbol{x}_{v'}^{(r(v'))} & \text{if } v \in Pos_a. \end{cases}$

Obviously, $s : Pos_e \to Pos$ is a positional strategy for Eva, and the Boolean fixed-point term associated with the game $G(s)$ is precisely

$$\boldsymbol{b} = \theta_k \boldsymbol{x}^{(k)}. \cdots .\mu \boldsymbol{x}^{(1)}.\boldsymbol{f}''(\boldsymbol{x}^{(1)}, \ldots, \boldsymbol{x}^{(k)}).$$

By Lemma 4.2.1 (page 87) and Proposition 4.3.4 (page 89), $\boldsymbol{b}_v = 1$ implies that the strategy s is winning in v. □

Using the bijection $h : \mathbb{B}^{Pos} \to \mathcal{P}(Pos)$ we can state this result in the following way.

Corollary 4.3.9. *The set of winning positions for Eva is equal to*

$$h(\theta_k \boldsymbol{x}^{(k)}. \cdots .\mu \boldsymbol{x}^{(1)}.\boldsymbol{f}(\boldsymbol{x}^{(1)}, \ldots, \boldsymbol{x}^{(k)})).$$

Theorem 4.3.10. *If v is not a winning position for Eva, then it is a winning position for Adam.*

Proof. Let us consider the game G from the point of view of Adam. All we did for Eva can be done for Adam, with the unique difference that Adam wins a play if $\limsup_{n \to \infty} r(v_n)$ is odd. This difference disappears if we set $r'(v) = r(v) + 1$.

It is easy to see that the Boolean fixed-point term that characterizes the winning positions for Adam is nothing but

$$\overline{\boldsymbol{b}} = \overline{\theta_k} \boldsymbol{x}^{(k)}. \cdots .\nu \boldsymbol{x}^{(1)}.\widetilde{\boldsymbol{f}}(\boldsymbol{x}^{(1)}, \ldots, \boldsymbol{x}^{(k)})$$

with $\overline{\mu} = \nu$ and $\overline{\nu} = \mu$, since

$$\widetilde{\boldsymbol{f}}_v(\boldsymbol{x}^{(1)}, \ldots, \boldsymbol{x}^{(k)}) = \begin{cases} \prod_{(v,v') \in Mov} \boldsymbol{x}_{v'}^{(r(v')+1)} & \text{if } v \in Pos_e, \\ \sum_{(v,v') \in Mov} \boldsymbol{x}_{v'}^{(r(v')+1)} & \text{if } v \in Pos_a. \end{cases}$$ □

In terms of game theory, the above result states *determinacy* of parity games: each position is winning for some player. Note that in general infinite games with perfect information are not always determined [73].

4.3.2 Game terms

In Section 4.3.1, we have shown that each parity game can be characterized by a Boolean vectorial fixed-point term depending on this game. In this section we consider μ-interpretations (in fact, powerset interpretations) naturally induced by parity games. Then, for each k, we construct a term, depending only on k, which defines the set of winning positions of Eva in (the interpretation induced by) any parity game such that the values of *rank* are in $\{1, \ldots, k\}$. An interested reader can easily find the analogous term for Adam by symmetry.

We first fix $k \geq 1$, and then the language, i.e. a set of function symbols, $\{A, E, rank_1, \ldots, rank_k, Pred_A, Pred_E, \cup, \cap\}$. We assume that $ar(A) = ar(E) = ar(rank_1) = ar(rank_k) = 0$, $ar(Pred_A) = ar(Pred_E) = 1$, and $ar(\cup) = ar(\cap) = 2$.

Now let $G = (Pos_a, Pos_e, Mov, rank)$ be a parity game such that the set $rank(Pos)$ is included in $\{1, \ldots, k\}$ (see the remark at the beginning of Section 4.3.1, page 88). We define the powerset interpretation, say $\mathcal{G}$, over the universe $\mathcal{P}(Pos)$, such that $A^{\mathcal{G}} = Pos_a$, $E^{\mathcal{G}} = Pos_e$, $rank_j^{\mathcal{G}} = \{p \in Pos : rank(p) = j\}$, $\cup$ and $\cap$ are interpreted as binary set union and intersection, respectively, and, for $X \subseteq Pos$,

$$\begin{aligned} Pred_E^{\mathcal{G}}(X) &= \{p \in Pos_e \mid \exists p' \in X : Mov(p, p')\} \\ Pred_A^{\mathcal{G}}(X) &= \{p \in Pos_a \mid \forall p', Mov(p, p') \Rightarrow p' \in X\} \end{aligned}$$

Note that $Pred_E^{\mathcal{G}}$ and $Pred_A^{\mathcal{G}}$ are dual since

$$Pred_A^{\mathcal{G}}(X) = Pos - Pred_E^{\mathcal{G}}(Pos - X).$$

Now we let $Play^{(k)}$ be the term

$$\begin{aligned} Play^{(k)}(x_1, \ldots, x_k) &= (E \cap Pred_E((rank_1 \cap x_1) \cup \ldots \cup (rank_k \cap x_k))) \\ &\quad \cup (A \cap Pred_A((rank_1 \cap x_1) \cup \ldots \cup (rank_k \cap x_k))) \end{aligned}$$

If $P_1, \ldots, P_k$ are subsets of Pos, and if p is a position, it follows from the definition of $Play^{(k)}$ that $p \in [Play^{(k)}]_{\mathcal{G}}(P_1, \ldots, P_k)$ if and only if

– if $p \in Pos_e$ then there is $p' \in P_{rank(p')}$ such that $(p, p') \in Mov$,
– if $p \in Pos_a$ then for all p' such that $(p, p') \in Mov$, $p' \in P_{rank(p')}$.

Finally, we define $W^{(k)}$ by

$$W^{(k)} = \theta_k x_k. \cdots .\nu x_2.\mu x_1.Play^{(k)}(x_1, \ldots, x_k)$$

where θ_k is equal to μ if k is odd, and to ν if k is even.

Then we claim the following.

Proposition 4.3.11. *For any parity game G with $rank(Pos) \subseteq \{1, \ldots, k\}$, the interpretation of $W^{(k)}$, $[W^{(k)}]_{\mathcal{G}} \subseteq Pos$, is precisely the set of positions winning for Eva.*

Proof. Let h be the canonical bijection from $\mathbb{B}^{Pos}$ to $\mathcal{P}(Pos)$.
By Corollary 4.3.9 (page 92), we have to prove that

$$[W^{(k)}]_{\mathcal{G}} = h(\theta_k \boldsymbol{x}^{(k)}. \cdots .\mu \boldsymbol{x}^{(1)}.\boldsymbol{f}(\boldsymbol{x}^{(1)}, \ldots, \boldsymbol{x}^{(k)})).$$

By Proposition 3.2.6 (page 74), $[W^{(k)}]_{\mathcal{G}} = h(\chi(W^{(k)}))$, which is also equal, by Proposition 3.2.7 (page 74) to $h(\theta_k u_{\langle x_k \rangle}. \cdots .\nu u_{\langle x_2 \rangle}.\mu u_{\langle x_1 \rangle}.\chi(Play^{(k)}))$.

Since h is a bijection, it is enough to show that

$$\theta_k u_{\langle x_k \rangle}. \cdots .\nu u_{\langle x_2 \rangle}.\mu u_{\langle x_1 \rangle}.\chi(Play^{(k)}) = \theta_k \boldsymbol{x}^{(k)}. \cdots .\mu \boldsymbol{x}^{(1)}.\boldsymbol{f}(\boldsymbol{x}^{(1)}, \ldots, \boldsymbol{x}^{(k)}).$$

Up to a renaming of the variables, we can identify the vector $u_{\langle x_i \rangle}$ with $\boldsymbol{x}^{(i)}$ (they are both indexed by Pos), so that we have to prove

$$\chi(Play^{(k)})(\boldsymbol{x}^{(1)}, \ldots, \boldsymbol{x}^{(k)}) = \boldsymbol{f}(\boldsymbol{x}^{(1)}, \ldots, \boldsymbol{x}^{(k)}).$$

Let $\boldsymbol{b}_1, \ldots, \boldsymbol{b}_k \in \mathbb{B}^{Pos}$ and let us show that

$$\chi(Play^{(k)})(\boldsymbol{b}_1, \ldots, \boldsymbol{b}_k) = \boldsymbol{f}(\boldsymbol{b}_1, \ldots, \boldsymbol{b}_k).$$

Let us compute the value of the component of index $p \in Pos$ of these two Boolean vectors, observing that, again by Proposition 3.2.6, and by Lemma 3.2.1 (page 73), the component of index p of $\chi(Play^{(k)})(\boldsymbol{b}_1, \ldots, \boldsymbol{b}_k)$ is equal to 1 if and only if $p \in [Play^{(k)}]_{\mathcal{G}}(h(\boldsymbol{b}_1), \ldots, h(\boldsymbol{b}_k))$. We consider the two cases where p is in Pos_e and in Pos_a.

Case 1: p is a position of Eva. Then $p \in [Play^{(k)}]_{\mathcal{G}}(h(\boldsymbol{b}_1), \ldots, h(\boldsymbol{b}_k))$ if and only if there exists p' such that $Mov(p, p')$, and $p' \in h(\boldsymbol{b}_j)$, where $j = rank(p')$, i.e., the component of index p' of $\boldsymbol{b}_j$ is equal to 1. This is exactly the case when the component of index p of $\boldsymbol{f}(\boldsymbol{b}_1, \ldots, \boldsymbol{b}_k)$ is equal to 1.

Case 2: p is a position of Adam. Now, $p \in [Play^{(k)}]_{\mathcal{G}}(h(\boldsymbol{b}_1), \ldots, h(\boldsymbol{b}_k))$ if and only if for all p' such that $Mov(p, p')$, $p' \in h(\boldsymbol{b}_j)$, where $j = rank(p')$. Again, this is exactly the case when the component of index p of $\boldsymbol{f}(\boldsymbol{b}_1, \ldots, \boldsymbol{b}_k)$ is equal to 1. □

4.4 Games for the μ-calculus

We have shown in the previous section that the μ-calculus is a suitable language for parity games since the set of winning positions can be described by a fixed-point term. In this section we show that conversely, the parity games provide a kind of "dynamic" semantics for the μ-calculus. We first consider Boolean terms and then show how the results on these can be extended to arbitrary powerset interpretations.

4.4.1 Games for Boolean terms

We have shown in Theorem 4.3.8 (page 92) that with any game one can associate a closed Boolean vectorial term that characterizes the winning position for Eva. We are going to prove that, conversely, the value of a closed Boolean vectorial term can be characterized by the winning positions of a game. Let $\boldsymbol{b} = \theta_k \boldsymbol{x}^{(k)}. \cdots .\theta_1 \boldsymbol{x}^{(1)}.\boldsymbol{f}(\boldsymbol{x}^{(k)}, \ldots, \boldsymbol{x}^{(1)})$ be a vector indexed by a set I. Let $X = \bigcup_{1 \leq i \leq k} \{\boldsymbol{x}^{(i)}\}$. We denote by $\mathcal{Y}_i$ the set of all subsets Y of X such that $\boldsymbol{f}_i$ gets the value 1 if the value 1 is assigned to all the variables of Y and the value 0 to all other variables. Therefore, $\boldsymbol{f}_i(\boldsymbol{x}^{(k)}, \ldots, \boldsymbol{x}^{(1)}) = \sum_{Y \in \mathcal{Y}_i} \prod_{x \in Y} x$.

Moreover, for each $j \in \{1, \ldots, k\}$, let $r(j) = 2j - 1$ if $\theta_j = \mu$, and $2j$ if $\theta_i = \nu$.

Definition 4.4.1. We define the game $G = (Pos_a, Pos_e, Mov, rank)$ by

- $Pos_e = X$, $Pos_a = \mathcal{P}(X)$,
- for any subset X' of X in Pos_a, $rank(X') = 0$,
- for any $\boldsymbol{x}_i^{(j)} \in X = Pos_e$, $rank(\boldsymbol{x}_i^{(j)}) = r(j)$,
- $Mov = \{(X, x) \mid x \in X\} \cup \{(\boldsymbol{x}_i^{(j)}, Y) \mid Y \in \mathcal{Y}_j\}$.

This game is obviously bipartite.

Proposition 4.4.2. *For all $i \in I$, for all $j \in \{1, \ldots, k\}$, $\boldsymbol{b}_i = 1$ if and only if $\boldsymbol{x}_i^{(j)}$ is a winning position for Eva.*

Proof. The Boolean vectorial term associated with G (see Definition 4.3.1, page 89) has the form $\langle \boldsymbol{A}, \boldsymbol{a} \rangle =$

$$\nu\langle \boldsymbol{Y}^{(2k)}, \boldsymbol{y}^{(2k)}\rangle.\mu\langle \boldsymbol{Y}^{(2k-1)}, \boldsymbol{y}^{(2k-1)}\rangle. \cdots .\mu\langle \boldsymbol{Y}^{(1)}, \boldsymbol{y}^{(1)}\rangle.\nu\langle \boldsymbol{Y}^{(0)}, \boldsymbol{y}^{(0)}\rangle.\langle \boldsymbol{G}, \boldsymbol{g}\rangle$$

where the vectors $\boldsymbol{Y}^{(j)}$, $\boldsymbol{G}$ and $\boldsymbol{A}$ are indexed by $Pos_a = \mathcal{P}(X)$, and $\boldsymbol{y}^{(j)}$, $\boldsymbol{a}$ and $\boldsymbol{g}$ by $Pos_e = X$.

By Theorem 4.3.8 (page 92), x is a winning position for Eva if and only if $\boldsymbol{a}_x = 1$.

Now, from the definition of G, the component $\boldsymbol{G}_Y$ of index $Y \in Pos_a$ of $\boldsymbol{G}$ is equal to $\prod_{x \in Y} \boldsymbol{y}_x^{(rank(x))}$. In particular, $\boldsymbol{G}$ does not depend on $\boldsymbol{Y}^{(j)}$ nor on the variables $\boldsymbol{y}^{(j)}$ when $j \notin r(\{1,\ldots,k\})$, and, therefore, we can write $\boldsymbol{G} = \boldsymbol{G}(\boldsymbol{y}^{(r(k))},\ldots,\boldsymbol{y}^{(r(1))})$. For any $x = x_i^{(j)} \in X$, $\boldsymbol{g}_x = \sum_{Y\in\mathcal{Y}_i} \boldsymbol{Y}_Y^0$, and thus, $\boldsymbol{g} = \boldsymbol{g}(\boldsymbol{Y}^{(0)})$.

Since $\boldsymbol{G}$ and $\boldsymbol{g}$ depend only on $\langle \boldsymbol{Y}^{(j)}, \boldsymbol{y}^{(j)}\rangle$ for $j = 0$ or $j \in r(\{1,\ldots,k\})$, and since in this last case, $r(j)$ is even if and only if $\theta_i = \nu$, we get, by renaming $\boldsymbol{Y}^{(r(j))}$ and $\boldsymbol{y}^{(r(j))}$ in $\boldsymbol{Y}^{(j)}$ and $\boldsymbol{y}^{(j)}$, $\langle \boldsymbol{A}, \boldsymbol{a}\rangle =$

$$\theta_k\langle \boldsymbol{Y}^{(k)}, \boldsymbol{y}^{(k)}\rangle.\cdots.\theta_1\langle \boldsymbol{Y}^{(1)}, \boldsymbol{y}^{(1)}\rangle.\nu\langle \boldsymbol{Y}^{(0)}, \boldsymbol{y}^{(0)}\rangle.\langle \boldsymbol{G}(\boldsymbol{y}^{(k)},\ldots,\boldsymbol{y}^{(1)}), \boldsymbol{g}(\boldsymbol{Y}^{(0)})\rangle.$$

By Proposition 1.4.4 (page 28) $\langle \boldsymbol{A}, \boldsymbol{a}\rangle = \langle \boldsymbol{H}(\boldsymbol{h},\ldots,\boldsymbol{h}), \boldsymbol{h}\rangle$ where $\boldsymbol{H}(\boldsymbol{y}^{(k)},\ldots,\boldsymbol{y}^{(1)}) = \nu \boldsymbol{Y}^{(0)}.\boldsymbol{G}(\boldsymbol{y}^{(k)},\ldots,\boldsymbol{y}^{(1)}) = \boldsymbol{G}(\boldsymbol{y}^{(k)},\ldots,\boldsymbol{y}^{(1)})$, and $\boldsymbol{h} = \theta_k \boldsymbol{y}^{(k)}.\cdots.\theta_1 \boldsymbol{y}^{(1)}.\boldsymbol{g}(\boldsymbol{G}(\boldsymbol{y}^{(k)},\ldots,\boldsymbol{y}^{(1)}))$.

But the component of index $\boldsymbol{x}_i^{(j)}$ of $\boldsymbol{g}(\boldsymbol{G}(\boldsymbol{y}^{(k)},\ldots,\boldsymbol{y}^{(1)}))$ is equal to

$$\sum_{Y\in\mathcal{Y}_i} \boldsymbol{G}_Y = \sum_{Y\in\mathcal{Y}_i} \prod_{\boldsymbol{x}_j^{(p)}\in Y} \boldsymbol{y}^{(p)}_{\boldsymbol{x}_j^{(p)}} = \boldsymbol{f}_i[id\{\boldsymbol{y}^{(p)}_{\boldsymbol{x}_j^{(p)}}/\boldsymbol{x}_j^{(p)}\}].$$

If we consider each vector $\boldsymbol{v}$ indexed by X as a sequence of vectors $\boldsymbol{v}(1),\ldots,\boldsymbol{v}(k)$, each one indexed by I, such that the component of index $\boldsymbol{x}_i^{(j)}$ of $\boldsymbol{v}$ is the component of index i of $\boldsymbol{v}(j)$, we get

$$\boldsymbol{g}(\boldsymbol{G}(\boldsymbol{y}^{(k)},\ldots,\boldsymbol{y}^{(1)})) = \langle \boldsymbol{f}(\boldsymbol{y}^{(k)}(k),\ldots,\boldsymbol{y}^{(1)}(1)),\ldots,\boldsymbol{f}(\boldsymbol{y}^{(k)}(k),\ldots,\boldsymbol{y}^{(1)}(1))\rangle$$

which is a mapping from $(\mathbb{B}^I)^k \to (\mathbb{B}^I)^k$. If we define $\delta : \mathbb{B}^I \to (\mathbb{B}^I)^k$ by $\delta(\boldsymbol{c}) = \langle \boldsymbol{c}, \boldsymbol{c}, \ldots, \boldsymbol{c}\rangle$, the previous equality implies that for any $\boldsymbol{c}_1,\ldots,\boldsymbol{c}_k \in \mathbb{B}^I$,

$$\begin{aligned}\boldsymbol{g}(\boldsymbol{G}(\delta(\boldsymbol{c}_k),\ldots,\delta(\boldsymbol{c}_1))) &= \langle \boldsymbol{f}(\boldsymbol{c}_k,\ldots,\boldsymbol{c}_1),\ldots,\boldsymbol{f}(\boldsymbol{c}_k,\ldots,\boldsymbol{c}_1)\rangle \\ &= \delta(\boldsymbol{f}(\boldsymbol{c}_k,\ldots,\boldsymbol{c}_1)).\end{aligned}$$

By repeated use of Lemma 1.2.15 (page 13), we get

$$\begin{aligned}\boldsymbol{a} &= \theta_k \boldsymbol{y}^{(k)}.\cdots.\theta_1 \boldsymbol{y}^{(1)}.\boldsymbol{g}(\boldsymbol{G}(\boldsymbol{y}^{(k)},\ldots,\boldsymbol{y}^{(1)})) \\ &= \delta(\theta_k \boldsymbol{x}^{(k)}.\cdots.\theta_1 \boldsymbol{x}^{(1)}.\boldsymbol{f}(\boldsymbol{x}^{(k)},\ldots,\boldsymbol{x}^{(1)})) \\ &= \delta(\boldsymbol{b}).\end{aligned}$$

Since $\boldsymbol{a} = \delta(\boldsymbol{b})$, for any i and j, the component of index $\boldsymbol{x}_i^{(j)}$ of $\boldsymbol{a}$ is equal to the component of index i of $\boldsymbol{b}$, and we already know, by definition of $\boldsymbol{a}$, that this component is 1 if and only if $\boldsymbol{x}_i^{(j)}$ is a winning position for Eva in G. □

Moreover, we know, by Theorem 3.3.14 (page 82), that $\boldsymbol{b}$ is also equal to $\theta_k \boldsymbol{x}^{(k)}.\cdots.\theta_1 \boldsymbol{x}^{(1)}.\boldsymbol{f}'(\boldsymbol{x}^{(k)},\ldots,\boldsymbol{x}^{(1)})$ where $\boldsymbol{f}'_i(\boldsymbol{x}^{(k)},\ldots,\boldsymbol{x}^{(1)}) = \prod_{x\in Y} x$ for some $Y \in \mathcal{Y}_i$. This allows us to define a positional winning strategy for Eva: in position $\boldsymbol{x}_i^{(j)}$ choose Y. Thus we get

Corollary 4.4.3. *In the game G associated with*

$$\theta_k \boldsymbol{x}^{(k)}.\cdots.\theta_1 \boldsymbol{x}^{(1)}.\boldsymbol{f}(\boldsymbol{x}^{(k)},\ldots,\boldsymbol{x}^{(1)})$$

there is a winning positional strategy s for Eva such that for any i, j, j', $s(\boldsymbol{x}_i^{(j)}) = s(\boldsymbol{x}_i^{(j')})$.

4.4.2 Games for powerset interpretations

We have seen in Proposition 4.4.2 above that the value of any closed Boolean vectorial fixed-point term can be characterized by a parity game associated with the term. Together with the construction of a Boolean characteristic (Definition 3.2.2, page 73), this gives in two steps a game characterization of the semantics of closed fixed-point terms under an arbitrary powerset interpretation. That is, given a closed vectorial term $\boldsymbol{\tau}$ and a powerset interpretation $\mathcal{I}$, we can construct the closed Boolean term characterizing $[\boldsymbol{\tau}]_{\mathcal{I}}$ in the sense of Proposition 3.2.8 (page 75), and then a game characterizing this Boolean term, as in Proposition 4.4.2. Since the semantics of fixed point terms in any interpretation can be transferred to the induced powerset interpretation (see Corollary 2.5.3, page 55) we can give a game characterization of any closed fixed-point term for any interpretation. However, we think it worthwhile to show a direct construction of this game associated with a term and a powerset interpretation. We will see that positions and moves have a natural meaning in terms of elements of E and valuations.

To make it precise, we will need an obvious concept of *isomorphism of games*. Let $G = (Pos_a, Pos_e, Mov, rank)$ and $G' = (Pos'_a, Pos'_e, Mov', rank')$ be two parity games. We call a mapping $h : (Pos_a \cup Pos_e) \to (Pos'_a \cup Pos'_e)$ an isomorphism from G to G' if it maps Pos_a onto Pos'_a and Pos_e onto Pos'_e bijectively, and, for any positions $p, q \in Pos_a \cup Pos_e$, $Mov(p,q)$ if and only if $Mov'(h(p), h(q))$, and $rank(p) = rank'(h(p))$. Clearly, the inverse mapping h^{-1} is an isomorphism from G' to G. As usual, we call two games *isomorphic* if there is an isomorphism between them.

Now, as in Section 3.2, let us consider a powerset interpretation $\mathcal{I}$ with a domain $\mathcal{P}(E)$, and a closed vectorial fixed-point term in normal form, $\boldsymbol{\tau} = \theta_k \boldsymbol{x}_k.\cdots.\theta_1 \boldsymbol{x}_1.\boldsymbol{t}$. As before, let n be the length of $\boldsymbol{\tau}$ and let $X = \{x_i^{(j)} : i = 1,\ldots,n;\ j = 1,\ldots,k\}$.

Also, as in Definition 4.4.1, we let, for each $j \in \{1,\ldots,k\}$, $r(j) = 2j - 1$ if $\theta_j = \mu$, and $2j$ if $\theta_i = \nu$.

We are ready to define a game for $\boldsymbol{\tau}$ and $\mathcal{I}$, analogous to the one associated with a Boolean term by Definition 4.4.1.

Definition 4.4.4. Let $G(\boldsymbol{\tau}, \mathcal{I}) = (Pos_a, Pos_e, Mov, rank)$ be the game specified by the following items.

- $Pos_e = E \times X$,
- $Pos_a = \mathcal{P}(E)^X$,
- the rank of all Adam's positions is 0,
- The rank of an Eva's position $\langle e, x_i^{(j)} \rangle$ is $r(j)$,
- the moves of Eva consist of all pairs $(\langle e, x_i^{(j)} \rangle, \rho)$ such that $e \in [\![t_i]\!]_{\mathcal{I}}[\rho]$
- the moves of Adam are the pairs $(\rho, \langle e, y \rangle)$ such that $e \in \rho(y)$.

Now, let $\boldsymbol{f} = \chi_{\mathcal{I}}(\boldsymbol{t})$ be the Boolean characteristic of $\boldsymbol{t}$ in the sense of Definition 3.2.4 (page 74) Let $\boldsymbol{b}$ be the Boolean vectorial fixed-point term $\theta_k u_{\boldsymbol{x}^{(k)}}. \cdots .\theta_1 u_{\boldsymbol{x}^{(1)}}.\boldsymbol{f}$ which is indexed by $E \times \{1, \ldots, n\}$.

We define the game $G' = (Pos'_a, Pos'_e, Mov', rank')$ associated with $\boldsymbol{b}$ according to Definition 4.4.1:

- $Pos'_e = u_X = \{u_{e,x_i^{(j)}} \mid e \in E, x_i^{(j)} \in X\}$,
- $Pos'_a = \mathcal{P}(Pos'_e)$,
- any position of Pos'_a has rank 0 and the rank of $u_{e,x_i^{(j)}}$ is $r(j)$,
- Mov' is the set $\{(Y, u_{e,x}) \mid u_{e,x} \in Y \subseteq Pos'_e\}$ of Adam's moves, together with the Eva's moves: all pairs $(u_{e,x_i^{(j)}}, Y)$ such that the component of index (e,i) of $\boldsymbol{f}$ gets the value 1 if the value 1 is assigned to the variables in Y and 0 to the other variables.

Proposition 4.4.5. *The games G' and $G = G(\tau, \mathcal{I})$ are isomorphic.*

Proof. Obviously, the mapping $u : E \times X \to Pos'_e$ that associates $u_{e,x}$ with $\langle e, x \rangle$ is a bijection. Moreover, there is a natural bijection between the subsets of u_X and the mappings of $\mathcal{P}(E)^X$ that associates with $Y \subseteq u_X$ the mapping ρ_Y defined by $\rho_Y(x) = \{e \in E \mid u_{e,x} \in Y\}$. More precisely, with any set $Y \subseteq u_X$ we associate $w_Y : Var' \to \mathbb{B}$ defined by $w_Y(u_{e,x}) = 1 \Leftrightarrow u_{e,x} \in Y$, and let $\rho_Y = h^*(w_Y) : Var \to \mathcal{P}(E)$. Conversely, with any $\rho : Var \to \mathcal{P}(E)$ we associate the set $Y_\rho = \{u_{e,x} \mid e \in \rho(x)\}$. It is clear that $\rho = \rho_{Y_\rho}$.

The Adam's positions have rank 0, in G' as well as in G. The rank of the Eva's position $u_{e,x_i^{(j)}}$ is $r(i)$, that is also the rank of the position $\langle e, x_i^{(j)} \rangle$ in G.

The pair $(Y, u_{e,x})$ is an Adam's move of G' if and only if $u_{e,x} \in Y$. The image in G of this pair is $(\rho_Y, \langle e, x \rangle)$ and it is an Adam's move if and only if $e \in \rho_Y(x)$. But, by definition of ρ_Y, $e \in \rho_Y(x)$ if and only if $u_{e,x} \in Y$.

Finally, $(u_{e,x_i^{(j)}}, Y)$ is an Eva's move in G' if and only if the component of index (e,i) of $\boldsymbol{f}(w_Y)$ is equal to 1, if and only if $e \in h(\boldsymbol{f}_i(w_Y))$. By Proposition 3.2.5 (page 74), $h(\boldsymbol{f}_i(w_Y)) = \chi_{\mathcal{I}}(\boldsymbol{t}_i)(w_Y) = [\![\boldsymbol{t}_i]\!]_{\mathcal{I}}[\rho_Y]$. Thus, $e \in h(\boldsymbol{f}_i(w_Y))$ if and only if $e \in [\![\boldsymbol{t}_i]\!]_{\mathcal{I}}[\rho_Y]$, i.e., if and only if the pair $(\langle e, x_i^{(j)} \rangle, \rho_Y)$ is an Eva's move in G. □

From Proposition 3.2.8, Proposition 4.4.2, and Proposition 4.4.5, we immediately get the following.

Theorem 4.4.6. *Let* $\boldsymbol{\tau} = \theta_k \boldsymbol{x}_k. \cdots .\theta_1 \boldsymbol{x}_1.\boldsymbol{t}$ *be a vectorial fixed-point term in normal form, let* $\mathcal{I}$ *be a powerset interpretation with a domain* $\mathcal{P}(E)$*, and* $val : Var \to \mathcal{P}(E)$ *a valuation. Then, for any* $e \in E$*,* $i = 1, \ldots, n$*,* $j = 1, \ldots, k$*,* $e \in [\![\tau_i]\!]_{\mathcal{I}}[val]$ *if, and only if,* $(e, x_i^{(j)})$ *is a winning position for Eva in the game* $G(\boldsymbol{\tau}, \mathcal{I}, val)$*.*

Moreover, Eva has a globally winning positional strategy, whose value on a position $(e, x_i^{(j)})$ *does not depend on* j*.*

From the above theorem and Proposition 4.3.11 (page 94), we can conclude that the set of winning positions (of either player) in a parity game is definable by a fixed-point term, and conversely, the interpretation of any closed fixed-point term in any powerset coincides with a winning set in some parity game. Moreover, these results indicate a close correspondence between the level of a term in the alternation-depth hierarchy and the range of the rank function of the respective parity game. We do not describe this last correspondence in detail here. Indeed, it can be established similarly to the correspondence between the alternating–depth hierarchy and automata of Chapter 7, *via* the concept of the Mostowski index.

4.5 Weak parity games

In this section, we consider parity games with a certain hierarchical structure of the game graph which forces the highest rank encountered in the play to repeat infinitely often, so that it determines the result of the play. We show that these games, without any further restrictions on the function *rank*, provide a characterization of the alternation-free fixed-point terms, analogous to the one given for all terms by Proposition 4.3.11 (page 94) and Theorem 4.4.6 above. Later on in Chapter 8 we will see a connection between these games and the so-called weak alternating automata (see Section 8.3, page 201).

Definition 4.5.1. A parity game $G = (Pos_a, Pos_e, Mov, rank)$ is a *weak parity game* if for each move $(v, v') \in Mov$, $rank(v) \leq rank(v')$. That implies that for each infinite play $v_0, v_1, \ldots, v_m, \ldots$ $n \leq n' \Rightarrow rank(v_n) \leq rank(v_{n'})$ and therefore there exists N such that $\forall n \geq N$, $rank(v_n) = \limsup_{i \to \infty} rank(v_i)$.

When a game is a weak parity game, the vector $\boldsymbol{b}$ defined in Definition 4.3.1 (page 88), that characterizes the winning positions for Eva, has a simpler definition.

Let us assume that $rank(Pos) \subseteq \{1, \ldots, k\}$. For $j \in \{1, \ldots, k\}$, let $Pos^{(j)}$ be the set of positions of rank j, and let $\boldsymbol{b}_j$ be the vector of Boolean values indexed by $Pos^{(j)}$ such that its component of index v is 1 if and only if v is a winning position for Eva.

By taking into account the partition of Pos onto the sets $Pos^{(j)}$, we can rewrite $\boldsymbol{b} = \theta_k \boldsymbol{x}^{(k)}.\cdots.\mu\boldsymbol{x}^{(1)}.\boldsymbol{f}(\boldsymbol{x}^{(1)},\ldots,\boldsymbol{x}^{(k)})$ defined in Definition 4.3.1 (page 88) in the form

$$\begin{bmatrix} \boldsymbol{b}_1 \\ \vdots \\ \boldsymbol{b}_j \\ \vdots \\ \boldsymbol{b}_k \end{bmatrix} = \theta_k \begin{bmatrix} \boldsymbol{x}_1^{(k)} \\ \vdots \\ \boldsymbol{x}_j^{(k)} \\ \vdots \\ \boldsymbol{x}_k^{(k)} \end{bmatrix} \cdots . \mu \begin{bmatrix} \boldsymbol{x}_1^{(1)} \\ \vdots \\ \boldsymbol{x}_j^{(1)} \\ \vdots \\ \boldsymbol{x}_k^{(1)} \end{bmatrix} . \begin{bmatrix} \boldsymbol{f}_1(\boldsymbol{x}_1^{(1)},\ldots,\boldsymbol{x}_j^{(j)},\ldots,\boldsymbol{x}_k^{(k)}) \\ \vdots \\ \boldsymbol{f}_j(\boldsymbol{x}_1^{(1)},\ldots,\boldsymbol{x}_j^{(j)},\ldots,\boldsymbol{x}_k^{(k)}) \\ \vdots \\ \boldsymbol{f}_k(\boldsymbol{x}_1^{(1)},\ldots,\boldsymbol{x}_j^{(j)},\ldots,\boldsymbol{x}_k^{(k)}) \end{bmatrix} .$$

Since G is a weak parity game, $\boldsymbol{f}_j$ depends only on $\boldsymbol{x}_j^{(j)}, \boldsymbol{x}_{j+1}^{(j+1)}, \ldots, \boldsymbol{x}_k^{(k)}$, hence,

$$\begin{bmatrix} \boldsymbol{b}_1 \\ \vdots \\ \boldsymbol{b}_j \\ \vdots \\ \boldsymbol{b}_k \end{bmatrix} = \theta_k \begin{bmatrix} \boldsymbol{x}_1^{(k)} \\ \vdots \\ \boldsymbol{x}_j^{(k)} \\ \vdots \\ \boldsymbol{x}_k^{(k)} \end{bmatrix} \cdots . \mu \begin{bmatrix} \boldsymbol{x}_1^{(1)} \\ \vdots \\ \boldsymbol{x}_j^{(1)} \\ \vdots \\ \boldsymbol{x}_k^{(1)} \end{bmatrix} . \begin{bmatrix} \boldsymbol{f}_1(\boldsymbol{x}_1^{(1)},\ldots,\boldsymbol{x}_j^{(j)},\ldots,\boldsymbol{x}_k^{(k)}) \\ \vdots \\ \boldsymbol{f}_j(\boldsymbol{x}_j^{(j)},\ldots,\boldsymbol{x}_k^{(k)}) \\ \vdots \\ \boldsymbol{f}_k(\boldsymbol{x}_k^{(k)}) \end{bmatrix} .$$

By Corollary 1.4.3 (page 28), we get

$$\begin{aligned} \boldsymbol{b}_k &= \theta_k \boldsymbol{x}_k^{(k)}.\boldsymbol{f}_k(\boldsymbol{x}_k^{(k)}) \\ \ldots \\ \boldsymbol{b}_i &= \theta_j \boldsymbol{x}_j^{(j)}.\boldsymbol{f}_j(\boldsymbol{x}_j^{(j)}, \boldsymbol{b}_{j+1},\ldots,\boldsymbol{b}_k) \\ \ldots \\ \boldsymbol{b}_1 &= \mu\boldsymbol{x}_1^{(1)}.\boldsymbol{f}_1(\boldsymbol{x}_1^{(1)}, \boldsymbol{b}_2,\ldots,\boldsymbol{b}_j,\ldots,\boldsymbol{b}_k). \end{aligned}$$

It follows that $\boldsymbol{b}$ is in $Comp^{\rightarrow}(\Sigma_1^{\rightarrow} \cup \Pi_1^{\rightarrow})$ if we extend the notion of a vectorial hierarchy (see Definition 2.7.7, page 62) to vectors of arbitrary length in an obvious way.

We are going now to show the converse: namely if a closed vectorial fixed-point term $\boldsymbol{\tau}$ is in $Comp^{\rightarrow}(\Sigma_1^{\rightarrow} \cup \Pi_1^{\rightarrow})$ then, for any interpretation $\mathcal{I}$ whose domain is a powerset $\mathcal{P}(E)$, there exists a weak parity game G associated with $[\boldsymbol{\tau}]_{\mathcal{I}}$ such that for each index i and each element $e \in E$, e is in the component of index i of $[\boldsymbol{\tau}]_{\mathcal{I}}$ if and only if there exists a variable x_i such that (e, x_i) is a winning position for Eva in G.

The construction is by induction on the definition of $\boldsymbol{\tau}$. Because of the associativity of composition, we may assume that $\boldsymbol{\tau}$ has the form $(\boldsymbol{\theta x.t})[\sigma]$

where for each $z \in Z = ar(\theta\boldsymbol{x}.\boldsymbol{t})$, $\sigma(z)$ is a closed fixed-point (scalar, not vectorial)) term in $Comp^{\rightarrow}(\Sigma_1^{\rightarrow} \cup \Pi_1^{\rightarrow})$. Note that if $Z = \emptyset$, $\boldsymbol{\tau} \in \Sigma_1^{\rightarrow} \cup \Pi_1^{\rightarrow}$, and that if $\{\boldsymbol{x}\} \cap ar(\boldsymbol{t}) = \emptyset$, we can consider that $\boldsymbol{\tau} \approx \boldsymbol{t}[\sigma]$ (see Section 2.4.2, page 51).

By the induction hypothesis, for each z in Z, there exists a weak parity game G_z and a variable z such that (e, z) is a winning position if and only if $e \in [\sigma(z)]_{\mathcal{I}}$. Moreover, without loss of generality, we may assume that the ranks of the positions of this game are all strictly greater than 2. We may also assume that the sets of positions of all the games G_z are pairwise disjoint.

Now, we define G as follows. We take the disjoint union of the G_z's to which we add the following positions and the following moves.

- the set Pos_e of Eva's positions (e, x) for $e \in E$ and $x \in \{\boldsymbol{x}\}$ and the set Pos_a of Adam's positions $v : Var \to \mathcal{P}(E)$. These positions are of rank 1 if $\theta = \mu$ and of rank 2 otherwise.
- Eva's moves $((e, x), v)$ for v such that $e \in [\boldsymbol{t}_x]_{\mathcal{I}}[v]$,
- Adam's moves $(v, (e, y))$ for $y \in ar(\boldsymbol{t})$ and $e \in v(y)$.

It it easy to see that a position of G that is in G_z is winning in G if and only if it is winning in G_z. In particular, (e, z) is a winning position if and only if $e \in [\sigma(z)]_{\mathcal{I}}$. Therefore, the positions in Pos_e are winning in G if and only if they are winning in the sub-game G' of G obtained by restricting G to the following set of positions

- the set of Eva's positions Pos_e,
- the subset of Pos_a consisting of all v such that $\forall z \in Z, v(z) = [\sigma(z)]_{\mathcal{I}}$.

Then, it is clear that a position (e, x) is winning in G' if and only if e is in the component of index x of $\theta\boldsymbol{x}.([\boldsymbol{t}]_{\mathcal{I}}[id\{[\sigma(z)]_{\mathcal{I}}/z\}_{z \in Z}]) = [(\theta\boldsymbol{x}.\boldsymbol{t})[\sigma]]_{\mathcal{I}}$.

The following theorem summarizes the results of this section.

Theorem 4.5.2. *If a game G is a weak parity game, its set of winning positions is definable by a fixed-point term in $Comp^{\rightarrow}(\Sigma_1^{\rightarrow} \cup \Pi_1^{\rightarrow})$. Conversely, if a term is in $Comp^{\rightarrow}(\Sigma_1^{\rightarrow} \cup \Pi_1^{\rightarrow})$, its interpretation in any powerset is the set of winning positions of a weak parity game.*

4.6 Bibliographic notes and sources

Parity games on graphs presented in this chapter were introduced in the work of Emerson and Jutla [33]. These games form a special case of a more general situation, namely the infinite–duration games with perfect information. We refer the reader to the chapter by J. Mycielski in the Handbook of Game Theory [73] for a detailed presentation of this issue, which is relevant both to theory of games and to descriptive set theory. Recall that, in the general case, the game is played by two players who sequentially select elements from some set A, thus eventually creating a sequence (possibly infinite) in $A^\infty = A^* \cup A^\omega$. The winning condition is given by a partition of A^∞ into two sets (we can think of them as of "plays won by Eva", and "plays won by Adam", respectively). It is easy to present the parity games in this framework. Note that a play never ends in a draw but, in contrast to finite games with perfect information, an infinite game need not be determined, i.e., the absence of a winning strategy for one player does not guarantee the existence of a winning strategy for the opponent. The determinacy of games with a winning condition given by a Borel set is a celebrated result in descriptive set theory, established by M. Davis [26] (see also [73]). This result guarantees, in particular, the determinacy of parity games. Büchi and Landweber were the first to consider the computational aspect of the strategies. In their work [23] they considered infinite games whose winning conditions were specified by finite automata (i.e., by ω-regular sets, see Chapter 5), and showed that in these games the winning strategies can be specified by finite automata as well. The essential point is that such a strategy does not rely on the whole history of the play but only on a current position, and perhaps some finite amount of memory (e.g., a state of an automaton). This idea gave rise to a number of deep results which we will now mention using the graph-theoretical framework, that we have generally adopted in this book.

Readily, any game with perfect information can be presented by a graph of positions (attributed to the players), so that the play consists of moving a token along the edges of the graph. The winning condition is then specified by a set of paths or, more abstractly, infinite strings over some finite alphabet of colours (or ranks) associated with the positions in the graph. For infinite games, this framework was first used by Emerson and Jutla [33] and, with restriction to finite graphs, also by McNaughton [62]. In such presentation, the result of Büchi and Landweber [23] reads as follows: If the graph of positions is finite and the winning condition is ω-regular then one of the players always has a winning strategy using only a finite amount of memory. The Forgetful Determinacy Theorem of Gurevich and Harrington [40] extends this result to infinite graphs, keeping the winning condition ω-regular. (The work of Gurevich and Harrington was motivated by the search for a simplification of the proof of the Rabin Complementation Lemma, which we will discuss in Chapter 9.) Clearly, the parity criterion can be presented by an ω-regular set (over the alphabet of ranks), but it turns out that it makes strategies even simpler. The fact that in parity games the winner has always a strategy which depends only on current positions, and so does not need any memory at all (Theorem 4.3.8, page 92), was discovered independently by Emerson and Jutla [33], A. W. Mostowski [66], and, with the restriction to finite graphs, by McNaughton [62]. (Another, very transparent proof of the memoryless determinacy of parity games was given later by Walukiewicz [102].) We note that [62, 66] used different formulations of the parity criterion. We refer the reader to the study by Zielonka [105] for a unified treatment of the determinacy results in graph-theoretic framework, together with an explanation of the structure of the memory needed. The introduction to that paper gives much insight into the genetic development of the ideas.

The equivalence between the μ-calculus and games was essentially shown by Emerson and Jutla [33], in terms of the modal μ-calculus. The translation from the μ-formulas to games in their approach is not direct, but uses alternating tree automata as an intermediate step.

5. The μ-calculus on words

It is well known that languages of finite words recognized by finite automata can be defined as least fixed points of some monotonic operators on sets of finite words. Indeed the star operation that appears in regular expressions can be seen as a least fixed-point operator, since L^* is the least language X that satisfies the equation $X = \{\varepsilon\} \cup LX$.

Moreover, the recognizable languages of infinite words can be described by using in addition the operator $L \mapsto L^\omega$ where L^ω is the greatest set of infinite words that satisfies $X = LX$ [81].

In this chapter, we present the μ-calculus on sets of words as an alternative presentation of the notion of recognizability and rationality, and we illustrate the relationship between definability in the μ-calculus and recognizability by finite automata. This relationship will be generalized in Chapter 7.

5.1 Rational languages

5.1.1 Preliminary definitions

Let A be a finite alphabet.

Definition 5.1.1. A *word* u *of length* $\ell = |u|$ *over* A, where $\ell \in \mathbb{N}$ or $\ell = \omega$ is a mapping $u : \{n \in \mathbb{N} \mid 0 \leq n < \ell\} \to A$. A finite word u can be therefore written $u(0)u(1) \cdots u(|u|-1)$ while an infinite word u is written $u(0)u(1) \cdots$. Moreover, there exists a unique word of length 0, called the *empty word*, denoted by ε.

The concatenation of two words u and v, denoted by uv, is the word w of length $|u| + |v|$ defined by $w(i) = u(i)$ if $i < |u|$ and $w(i) = v(i - |u|)$ if $i \geq |u|$. This operation is associative. Obviously, $\varepsilon u = u\varepsilon = u$ and if $|u| = \omega$ then $uv = u$.

More generally let u_i be a word for any $i \in \mathbb{N}$. We denote by $u_0 u_1 \cdots$, the word w of length $|w| = \sum_{i \geq 0} |u_i|$ defined as follows. For $j \geq 0$, let $k_j = \sum_{i<j} |u_i|$, so that $k_0 = 0$ and $k_{j+1} = k_j + |u_j|$. Note that if some of the u_i are empty, one can have $k_{j+1} = k_j$. For $0 \leq i < |w|$, $w(i) = u_j(i - k_j)$ where j is the unique natural number such that $k_j \leq i < k_{j+1}$.

Indeed if $u_i = \varepsilon$ for all $i \geq n$ then the word w defined above is the usual finite product $u_0u_1 \cdots u_n$. It is easy to see that if $w = u_0u_1 \cdots$ then $uw = (uu_0)u_1 \cdots$.

We denote by A^* the set of finite words over A, by A^ω the set of infinite words, and by A^∞ the set $A^* \cup A^\omega$. Obviously, the three powersets $\mathcal{P}(A^\infty)$, $\mathcal{P}(A^*)$, and $\mathcal{P}(A^\omega)$, ordered by inclusion, are complete lattices.

Definition 5.1.2. For each letter a of A, we define the unary mapping a on $\mathcal{P}(A^\infty)$ by $a(K) = \{av \in A^\infty \mid v \in K\}$.

More generally, for any subset L of A^∞ we define the mapping $L : \mathcal{P}(A^\infty) \to \mathcal{P}(A^\infty)$ by $L(K) = \{uv \in A^\infty \mid u \in L, v \in K\}$. Note that if u is in A^ω then $uv = u$, so that $L(K) = L \cap A^\omega \cup \{uv \in A^\infty \mid u \in L \cap A^*, v \in K\}$.

When no ambiguity arises, we write aK and LK for $a(K)$ and $L(K)$

Proposition 5.1.3. *The mapping from $\mathcal{P}(A^\infty)^2$ to $\mathcal{P}(A^\infty)$ that associates LK with L and K is additive with respect to both arguments.*

Proof. Let $L = \bigcup_{i \in I} L_i$ and $K = \bigcup_{j \in J} K_j$. We have to prove that $LK = \bigcup_{i \in I, j \in J} L_iK_j$.

Let $w \in L_iK_j$. Then there is $u \in L_i$ and $v \in K_j$ such that $w = uv$. Obviously $u \in L$ and $v \in K$ so that $w \in LK$. Conversely Let $w \in LK$. Then $w = uv$ with $u \in L$ and $v \in K$. Hence, there exist $i \in I$ and $j \in K$ such that $u \in L_i$ and $v \in K_j$. □

Now, we are ready for defining fixed-point terms for the μ-calculus on words.

Let *Var* be a set of variables. Let A be a finite alphabet. For any subset Ω of $\{\vee, \wedge\}$, we define the set $\mathit{fix}\mathcal{T}(A_\Omega)$ of fixed-point terms by

- each variable is a term, ε is a (constant) term,
- if $a \in A$, and if t is a term, so is at,
- if $\lozenge \in \Omega$, and if t, t' are terms, so is $t \lozenge t'$.
- if t is a term so is $\theta x.t$.

Although these terms can receive any interpretation, we shall consider only the *standard interpretation* whose domain is the complete lattice $\mathcal{P}(A^\infty)$ ordered by inclusion, $\vee$ and $\wedge$ are interpreted as union and intersection, and each unary symbol $a \in A$ is interpreted as the mapping $a : \mathcal{P}(A^\infty) \to \mathcal{P}(A^\infty)$ defined above. In particular, if v is a valuation from *Var* to $\mathcal{P}(A^\infty)$ then $[ax][v] = av(x)$.

The interpretation of each functional symbol is monotonic in all its arguments, and thus any term t has a standard interpretation, denoted by $[t]$, that maps $\mathcal{P}(A^\infty)^{ar(t)}$ into $\mathcal{P}(A^\infty)$.

Indeed, out of the four possible sets $\mathit{fix}\mathcal{T}(A_\Omega)$ of fixed-point terms, we will consider only the sets $\mathit{fix}\mathcal{T}(A_{\{\vee\}})$ and $\mathit{fix}\mathcal{T}(A_{\{\vee,\wedge\}})$. We consider the set $\mathit{fix}\mathcal{T}(A_{\{\vee\}})$ separately, although it is a subset of $\mathit{fix}\mathcal{T}(A_{\{\vee,\wedge\}})$, because it can be studied with much simpler methods than the latter one.

From now on, the terms in $fix\mathcal{T}(A_{\{\vee\}})$ will be called *intersection-free terms.*

5.1.2 Rational languages

The definition of the family of rational languages as the closure of finite languages of finite words under union, concatenation, and finite (*) or infinite ($^\omega$) iteration is quite standard. We just recall these definitions.

Let L be a subset of A^* we define $L^+, L^* \subseteq A^*$ and $L^\omega \subseteq A^\infty$ by

- $u \in L^+$ if and only if there exist $u_1, \ldots, u_n \in L$, with $n > 0$, such that $u = u_1 u_2 \cdots u_n$.
  In other words, $L^+ = \bigcup_{n \geq 1} L^n$ where $L^n = \{u_1 \cdots u_n \in A^* \mid u_i \in L\}$, and we have $L^+ = L \cup LL^+$.
- $L^* = \{\varepsilon\} \cup L^+$. Thus, $LL^* = L^+$.
- $u \in L^\omega$ if and only if there exists an infinite sequence $u_1, \ldots, u_n, \ldots$, with $u_n \in L$, such that $u = u_1 \cdots u_n \cdots$

Let us remark that if $u = u_0 u_1 \cdots$, the length $|u|$ of u is equal to $\sum_{n \geq 0} |u_i|$. If u_i is ε for almost all n, then u is a finite word. Therefore, if ε belongs to L then $L^* \subseteq L^\omega$. Conversely if L^ω contains finite words then $\varepsilon \in L$. More precisely, $L^\omega \cap A^*$ is empty if $\varepsilon \notin L$ and is equal to L^* if $\varepsilon \in L$.

The rational languages of A^* are defined by induction as follows

- $\emptyset$ is a rational language; for each letter a, $\{a\}$ is a rational language,
- if L and L' are rational languages, so are $L \cup L'$ and LL'.
- if L is a rational language, so is L^*.

Let $\mathrm{Rat}(A^*)$ be the set of all rational languages of A^*.

If we also use the infinite iteration L^ω in the above inductive definition, we get the set $\mathrm{Rat}(A^\infty)$ of rational languages of A^∞:

- if L is in $\mathrm{Rat}(A^*)$, it is also in $\mathrm{Rat}(A^\infty)$,
- if L and L' are in $\mathrm{Rat}(A^\infty)$, so are $L \cup L'$ and LL',
- if L is in $\mathrm{Rat}(A^\infty)$, so are L^* and L^ω.

5.1.3 Arden's lemma

We are going to prove that L^* and L^ω can be defined as the least and greatest fixed points of some monotonic mapping over $\mathcal{P}(A^\infty)$.

Lemma 5.1.4 (Arden's Lemma).

Let $L \subseteq A^*$ and $L' \subseteq A^\infty$. The extremal fixed points of the monotonic mapping that associates $L' \cup LK$ with $K \subseteq A^\infty$ are

$$\begin{aligned}\mu x.(L' \cup Lx) &= L^*L',\\ \nu x.(L' \cup Lx) &= \begin{cases} A^\infty & \text{if } \varepsilon \in L,\\ L^*L' \cup L^\omega & \text{if } \varepsilon \notin L.\end{cases}\end{aligned}$$

Proof. Let $K = \mu x.(Lx \cup L')$ and $K' = \nu x.(Lx \cup L')$.

Since $L' \cup LL^*L' = L^*L'$ we have $K \subseteq L^*L'$. Since $K = L' \cup LK$ we have $LK \subseteq K$ and $L' \subseteq K$. It follows that $LL' \subseteq LK \subseteq K$. More generally, if $L^nL' \subseteq K$ then $L^{n+1}L' = LL^nL' \subseteq LK \subseteq K$. Hence $L^*L' \subseteq K$.

If $\varepsilon \in L$, we have $A^\infty = \{\varepsilon\}A^\infty \subseteq LA^\infty \subseteq L' \cup LA^\infty$. Hence $A^\infty \subseteq K'$, thus $K' = A^\infty$.

Let us assume $\varepsilon \notin L$. Since, by definition of $L^\omega, LL^\omega = L^\omega$, we have $L' \cup L(L^*L' \cup L^\omega) = L' \cup LL^*L' \cup LL^\omega = L^*L' \cup L^\omega$. It follows that $L^*L' \cup L^\omega \subseteq K'$.

Let us show that, if $\varepsilon \notin L$, any set K'' such that $K'' \subseteq LK''$ is included in L^ω. Let $w_0 \in K''$, then $w_0 \in LK''$ and $w_0 = u_0w_1$ with $\varepsilon \neq u_0 \in L$ and $w_1 \in K''$. By induction, it can be shown that there exist two sequences $u_0, u_1, \ldots \in L$, $w_0, w_1, \ldots \in K''$ such that, for any $n \geq 0$, $w_0 = u_0u_1 \cdots u_nw_{n+1}$. It follows that the word $w = u_0u_1 \cdots \in L^\omega$ is a prefix of w_0 and since none of the u_i is the empty word, w is an infinite word, that implies $w = w_0$.

Let $K'' = K' - K$. Since $K \subseteq K'$, we have $K' = K \cup K''$. Hence, $K' = LK' \cup L' = LK \cup LK'' \cup L' = K \cup LK''$. It follows that $K'' = K' - K$ is included in LK'', and, by the previous result, $K'' \subseteq L^\omega$. Hence, $K' \subseteq K \cup L^\omega = L^*L' \cup L^\omega$. □

It should be observed that $\nu x.(L' \cup Lx) = \mu x.(L' \cup Lx) \cup \nu x.Lx$. This is a special case of a more general result relating least and greatest fixed points in some specific cases (see [90]). It shoud be also observed that $L^\omega A^\infty$ is equal to A^∞ if $\varepsilon \in L$ and to L^ω otherwise, so that $\nu x.(L' \cup Lx)$ is always equal to $L^*L' \cup L^\omega A^\infty$.

Corollary 5.1.5. *Let $L \subseteq A^\infty$ and let $L_* = L \cap A^*$, $L_\omega = L \cap A^\omega$. Since $LK = L_\omega \cup L_*K$, we have*

$$\begin{aligned}\mu x.Lx &= L_*^*L_\omega,\\ \nu x.Lx &= \begin{cases} A^\infty & \text{if } \varepsilon \in L,\\ L_*^*L_\omega \cup L_*^\omega & \text{if } \varepsilon \notin L.\end{cases}\end{aligned}$$

There is a *matricial* version of this lemma.

Let $M^{n\times m}$ be the set of matrices $\boldsymbol{L} = \begin{bmatrix} L_{11} & \cdots & L_{1j} & \cdots & L_{1m} \\ \vdots & & & & \vdots \\ L_{i1} & \cdots & L_{ij} & \cdots & L_{im} \\ \vdots & & & & \vdots \\ L_{n1} & \cdots & L_{nj} & \cdots & L_{nm} \end{bmatrix}$

where the L_{ij} are subsets of A^∞.

Obviously, we make $M^{n\times m}$ a complete lattice by using the product ordering (see Section 1.1.7, page 7). We denote by $+$ the least upper bound in this lattice. We also define a matrix product from $M^{n\times m} \times M^{m\times p}$ into $M^{n\times p}$ by $(\boldsymbol{L}\boldsymbol{L}')_{ij} = \sum_{k=1}^{m} \boldsymbol{L}_{ik}\boldsymbol{L}'_{kj}$. This product is monotonic in its two arguments, so that for $\boldsymbol{L} \in M^{n\times n}, \boldsymbol{L}' \in M^{n\times 1}$, the mapping denoted $\boldsymbol{L}\boldsymbol{x} + \boldsymbol{L}' : M^{n\times 1} \to M^{n\times 1}$ has extremal fixed points.

Let $\boldsymbol{L} \in M^{n\times n}$. We define $\boldsymbol{L}^+, \boldsymbol{L}^* \in M^{n\times n}$ and $\boldsymbol{L}^\omega \in M^{n\times 1}$ as follows.

- $\boldsymbol{L}^+ = \sum_{n\geq 1} \boldsymbol{L}^n$.
- $\boldsymbol{L}^* = [\varepsilon_n] + \boldsymbol{L}^+$, where $[\varepsilon_n]$ is the matrix defined by $[\varepsilon_n]_{ij} = \varepsilon$ if $i = j$, $\emptyset$ otherwise.
- $u \in (\boldsymbol{L}^\omega)_i$ if and only if for every $j \geq 0$, there exists $i_j \in \{1, \dots, n\}, u_j \in A^\infty$ such that $i_0 = i$, $u_j \in \boldsymbol{L}_{i_j i_{j+1}}$, $u = u_0 u_1 \cdots$

It is easy to see that for any $\boldsymbol{K} \in M^{n\times m}$, $[\varepsilon_n]\boldsymbol{K} = \boldsymbol{K} = \boldsymbol{K}[\varepsilon_m]$, and that for any $\boldsymbol{L} \in M^{n\times n}$, $\boldsymbol{L}^+ = \boldsymbol{L} + \boldsymbol{L}\boldsymbol{L}^+ = \boldsymbol{L}\boldsymbol{L}^*$.

Let us also remark that $u \in (\boldsymbol{L}^p)_{ij}$ if and only if there are $i_0, i_1, \dots, i_p \in \{1, \dots, n\}, u_1, \dots, u_p \in A^\infty$ such that $i_0 = i, i_p = j$, $u_j \in \boldsymbol{L}_{i_{j-1}i_j}$, $u = u_0 u_1 \cdots u_p$.

Lemma 5.1.6.

$$\begin{aligned} \mu\boldsymbol{x}.(\boldsymbol{L}\boldsymbol{x} + \boldsymbol{L}') &= \boldsymbol{L}^*\boldsymbol{L}', \\ \nu\boldsymbol{x}.(\boldsymbol{L}\boldsymbol{x} + \boldsymbol{L}') &= \boldsymbol{L}^*\boldsymbol{L}' + \boldsymbol{L}^\omega A^\infty \end{aligned}$$

Proof. The proof is similar to the *scalar* case.

Let $\boldsymbol{K} = \mu\boldsymbol{x}.(\boldsymbol{L}\boldsymbol{x} + \boldsymbol{L}')$ and $\boldsymbol{K}' = \nu\boldsymbol{x}.(\boldsymbol{L}\boldsymbol{x} + \boldsymbol{L}')$.

Since $\boldsymbol{L}^*\boldsymbol{L}' = \boldsymbol{L}' + \boldsymbol{L}\boldsymbol{L}^*\boldsymbol{L}'$, $\boldsymbol{K} \subseteq \boldsymbol{L}^*\boldsymbol{L}'$. Since $\boldsymbol{K} = \boldsymbol{L}' + \boldsymbol{L}\boldsymbol{K}$ we have $\boldsymbol{L}\boldsymbol{K} \subseteq \boldsymbol{K}$ and $\boldsymbol{L}' \subseteq \boldsymbol{K}$. It follows that $\boldsymbol{L}\boldsymbol{L}' \subseteq \boldsymbol{L}\boldsymbol{K} \subseteq \boldsymbol{K}$. More generally, if $\boldsymbol{L}^n\boldsymbol{L}' \subseteq \boldsymbol{K}$ then $\boldsymbol{L}^{n+1}\boldsymbol{L}' = \boldsymbol{L}\boldsymbol{L}^n\boldsymbol{L}' \subseteq \boldsymbol{L}\boldsymbol{K} \subseteq \boldsymbol{K}$. Hence $\boldsymbol{L}^*\boldsymbol{L}' \subseteq \boldsymbol{K}$.

It is easy to see that $\boldsymbol{L}_{ij}(\boldsymbol{L}^\omega)_j \subseteq (\boldsymbol{L}^\omega)_i$, hence $\boldsymbol{L}(\boldsymbol{L}^\omega) \subseteq \boldsymbol{L}^\omega$. On the other hand, if $u \in (\boldsymbol{L}^\omega)_i$, then $u = u_0 v$ with $u_0 \in \boldsymbol{L}_{ij}$ and $v \in (\boldsymbol{L}^\omega)_j$, hence $\boldsymbol{L}(\boldsymbol{L}^\omega) = \boldsymbol{L}^\omega$. It follows that $\boldsymbol{L}(\boldsymbol{L}^*\boldsymbol{L}' + \boldsymbol{L}^\omega A^\infty) + \boldsymbol{L}' = \boldsymbol{L}^*\boldsymbol{L}' + \boldsymbol{L}^\omega A^\infty$, hence $\boldsymbol{L}^*\boldsymbol{L}' + \boldsymbol{L}^\omega A^\infty \subseteq \boldsymbol{K}'$. To prove the converse inclusion, it is enough to prove $\boldsymbol{K}' - \boldsymbol{K} \subseteq \boldsymbol{L}^\omega A^\infty$. Let $w_0 \in (\boldsymbol{K}' - \boldsymbol{K})_{i_0}$, then there is i_1 and $u_0 \in \boldsymbol{L}_{i_0 i_1}$ such that $w_0 = u_0 w_1$ with $w_1 \in (\boldsymbol{K}' - \boldsymbol{K})_{i_1}$. By induction, it can be shown that there exist three sequences $i_0, i_1, \dots$, $u_0, u_1, \dots$, with

$u_j \in \boldsymbol{L}_{i_j i_{j+1}}$, $w_0, w_1, \ldots$, with $w_j \in (\boldsymbol{K}' - \boldsymbol{K})_{i_j}$, such that, for any $j \geq 0$, $w_0 = u_0 u_1 \cdots u_j w_{j+1}$. It follows that the word $w = u_0 u_1 \cdots \in (\boldsymbol{L}^\omega)_{i_0}$ is a prefix of w_0, that implies $w_0 \in (\boldsymbol{L}^\omega A^\infty)_{i_0}$. □

5.1.4 The μ-calculus of extended languages

Let t be a term with $ar(t) = \{x\}$. Let us assume that there exists $L \subseteq A^*$ such that for any $v : Var \to \mathcal{P}(A^\infty)$, $[\![t]\!][v] = Lv(x)$. By analogy with $[\![ax]\!][v] = av(x)$, we find it convenient to write $[\![t]\!] = Lx$.

Then $[\![\mu x.(y \vee t)]\!][v]$ (with $y \neq x$) is the least fixed point of the mapping that associates with K the set $[\![y \vee t]\!][v\{K/x\}] = [\![y]\!][v\{K/x\}] \cup [\![t]\!][v\{K/x\}] = v(y) \cup LK$. By the Arden's Lemma (Lemma 5.1.4, page 107), $[\![\mu x.(y \vee t)]\!][v] = L^* v(y)$ and, by the same analogy as above, we could write $[\![\mu x.(y \vee t)]\!] = L^* y$.

We are going to formalize this way of denoting the interpretation $[\![t]\!]$ of an intersection-free term by introducing the following μ-calculus whose objects are called *extended languages*.

Definition 5.1.7. An *extended language* is a pair $E = \langle X, L \rangle$ where X is a subset of Var and L is a subset of $A^\infty \cup A^* X$.

If $E = \langle X, L \rangle$ is an extended language, we denote by $E_\sharp$ the set $L \cap A^\infty$, and by E_x for $x \in X$, the set $\{u \in A^* \mid ux \in L\}$, so that L can be uniquely written $E_\sharp \cup \bigcup_{x \in X} E_x x$. Moreover since E_x is empty for any x not in X, we can also write $E_\sharp \cup \bigcup_{x \in Var} E_x x$.

Note that if $E = \langle X, L \rangle$ is an extended language, and $Y \supset X$ then $E' = \langle Y, L \rangle$ is still an extended language, not equal to E, where for any $y \in Y - X$, $E'_y = \emptyset$. In particular, if $L \subseteq A^\infty$, for any $X \subseteq Var$, $\langle X, L \rangle$ is an extended language.

We denote by $\mathcal{EL}$ the set of all extended languages. We are showing that $\mathcal{EL}$, together with the following operations, is a μ-calculus.

- $ar(\langle X, L \rangle) = X$,
- For each variable x, the extended language $\hat{x}$ is the pair $\langle \{x\}, x \rangle$ i.e., $\hat{x}_\sharp = \emptyset$, $\hat{x}_x = \{\varepsilon\}$.
- Let us define $E' = \langle X', L' \rangle = \langle X, L \rangle[\rho]$ for $\langle X, L \rangle \in \mathcal{EL}$ and $\rho : Var \to \mathcal{EL}$.
  - $X' = \bigcup_{x \in X} ar(\rho(x))$,
  - $E'_\sharp = E_\sharp \cup \bigcup_{x \in Var} E_x \rho(x)_\sharp$,
  - $E'_y = \bigcup_{x \in Var} E_x \rho(x)_y$.

  It is easy to see that $E' = \langle X, L \rangle[\rho]$ is still an extended language: If $E'_y \neq \emptyset$ then there exists x such that $E_x \neq \emptyset$ and $\rho(x)_y \neq \emptyset$, which implies $x \in X$ and $y \in ar(\rho(x))$, hence $y \in X'$.
- If $E = \langle X, L \rangle$ is an extended language, then
  - $\mu x.E$ is the extended language E' of arity $X' = X - \{x\}$ defined by
    - $E'_\sharp = E_x^* E_\sharp$,

  - $E'_y = E_x^* E_y$, for $y \in X'$,
- $\nu x.E$ is equal to $\langle X', A^\infty \cup A^* X' \rangle$ if $\varepsilon \in E_x$, otherwise it is equal to the extended language $\langle X', E' \rangle$ where E' is defined by
  - $E'_\sharp = E_x^* E_\sharp \cup E_x^\omega$,
  - $E'_y = E_x^* E_y$, for $y \in X'$.

Note that if $x \notin X$, then $E'_x = \emptyset$. Since $\emptyset^* = \{\varepsilon\}$ and $\emptyset^\omega = \emptyset$, we get $\mu x.E = \nu x.E = E$.

Proposition 5.1.8. *The set of extended languages is a μ-calculus. Moreover it satisfies a strengthened version of Axiom 7, namely: For any z not in $ar((\theta x.E)[\rho])$, $(\theta x.E)[\rho] = \theta z.(E[\rho\{\hat{z}/x\}])$.*

Proof.

1. Obviously $ar(\hat{x}) = \{x\}$.
2. By definition, $ar(E[\rho])$ is always equal to $ar'(E, \rho)$.
3. Obviously, by construction, $ar(\theta x.E) = ar(E) - \{x\}$.
4. if $E' = \hat{x}[\rho]$ then $ar(E') = ar(\rho(x))$ and
   - $E'_\sharp = \rho(x)_\sharp$,
   - $E'_y = \rho(x)_y$, for any variable $y \in ar(E')$.

   It follows that $E' = \rho(x)$.
5. Only the sets $\rho(x)$ for $x \in ar(E)$ are needed to define $E[\rho]$.
6. Let us show that $E[\rho][\pi] = E[\rho \star \pi]$. It is easy to see that

$$\begin{aligned} E[\rho][\pi]_\sharp &= E[\rho]_\sharp \cup \bigcup_{y \in Var} E[\rho]_y \pi(y)_\sharp \\ &= E_\sharp \cup \bigcup_{x \in Var} E_x \rho(x)_\sharp \cup \bigcup_{x \in Var} \bigcup_{y \in Var} E_x \rho(x)_y \pi(y)_\sharp \end{aligned}$$

On the other hand

$$\begin{aligned} E[\rho \star \pi]_\sharp &= E_\sharp \cup \bigcup_{x \in Var} E_x (\rho(x)[\pi])_\sharp \\ &= E_\sharp \cup \bigcup_{x \in Var} E_x \rho(x)_\sharp \cup \bigcup_{y \in Var} E_x \rho(x)_y \pi(y)_\sharp . \end{aligned}$$

We also have, for $z \in ar(E[\rho][\pi]) = ar(E[\rho \star \pi])$,

$$\begin{aligned} E[\rho][\pi]_z &= \bigcup_{y \in Var} E[\rho]_y \pi(y)_z \\ &= \bigcup_{x \in Var} \bigcup_{y \in Var} E_x \rho(x)_y \pi(y)_z . \end{aligned}$$

On the other hand,

$$\begin{aligned} E[\rho \star \pi]_z &= \bigcup_{x \in Var} E_x(\rho(x)[\pi])_z \\ &= \bigcup_{x \in Var} E_x(\bigcup_{y \in Var} \rho(x)_y \pi(y)_z. \end{aligned}$$

7. Let $E' = \mu x.E$, $X = ar(E)$, $Y = X - \{x\}$, and $Z = ar(E'[\rho])$. Then $(E'[\rho])_\sharp = E'_\sharp \cup \bigcup_{y \in Y} E'_y \rho(y)_\sharp = E^*_x E_\sharp \cup \bigcup_{y \in Y} E^*_x E_y \rho(y)_\sharp$ and, for any $z \in Z$, $(E'[\rho])_z = \bigcup_{y \in Y} E'_y \rho(y)_z = \bigcup_{y \in X'} E^*_x E_y \rho(y)_z$.
   Let v be any variable not in Z, let $F = E[\rho\{\hat{v}/x\}]$ and let $F' = \mu v.F$. First, we have $ar(F') = ar(E'[\rho]) = Z$. This is because $ar(F') = ar'(E, \rho\{\hat{v}/x\}) - \{v\}$ where

$$ar'(E, \rho\{\hat{v}/x\}) = \bigcup_{y \in Y} ar(\rho(y)) \cup \{v \mid x \in X\} = Z \cup \{v \mid x \in X\}.$$

   Since $v \notin Z$, $ar(F') = Z$.
   Now, $F'_\sharp = F^*_v F_\sharp$ and, for $z \in Z$, $F'_z = F^*_v F_z$. But it is easy to see that $F_v = E_x$, that $F_\sharp = \bigcup_{y \in Y} E_y \rho(y)_\sharp$, and that $F_z = \bigcup_{y \in Y} E_y \rho(y)_z$. Hence, it follows that $E'[\rho] = F'$.
   The proof is similar for $E' = \nu x.E$.

□

If E is an extended language, we denote by aE the unique extended language E' such that

- $ar(E') = ar(E)$,
- $E'_\sharp = aE_\sharp$,
- $\forall x \in ar(E)$, $E'_x = aE_x$.

If E' and E'' are two extended languages, we define their union $E \cup E'$ as the unique extended language E of arity $ar(E) \cup ar(E')$ such that

- $E_\sharp = E'_\sharp \cup E''_\sharp$,
- for any $x \in ar(E)$, $E_x = E'_x \cup E''_x$. Note that if $E_x \neq \emptyset$ then $E'_x \neq \emptyset$ or $E''_x \neq \emptyset$, which implies $x \in ar(E')$ or $x \in ar(E'')$.

Then we can define the mapping $\sigma : \mathit{fix}\mathcal{T}(A_{\{\vee\}}) \to \mathcal{EL}$ by $\sigma(x) = \hat{x}$, $\sigma(at) = a\sigma(t)$, $\sigma(t \vee t') = \sigma(t) \cup \sigma(t')$, $\sigma(\theta x.t) = \theta x.\sigma(t)$.

Proposition 5.1.9. *The mapping σ is a homomorphism of μ-calculi.*

Proof. We have just to prove $ar(t) = ar(\sigma(t))$ and $\sigma(t[\rho]) = \sigma(t)[\sigma(\rho)]$ by induction on t. The first point is obvious. Let us show the second one.

- $\sigma(x[\rho]) = \sigma(\rho(x)) = \sigma(\rho)(x) = \hat{x}[\sigma(\rho))]$.
- $\sigma(at)[\sigma(\rho)] = a\sigma(t)[\sigma(\rho)] = a\sigma(t[\rho]) = \sigma(at[\rho])$, since, in $\mathcal{EL}$, $(aE)[\rho] = a(E[\rho])$.

- $\sigma(t \vee t')[\sigma(\rho)] = (\sigma(t) \cup \sigma(t'))[\sigma[\rho)] = \sigma(t)[\sigma(\rho)] \cup \sigma(t')[\sigma(\rho)] = \sigma(t[\rho] \vee t'[\rho])$ $= \sigma((t \vee t')[\rho])$, since in $\mathcal{EL}$, $(E \cup E')[\rho] = E[\rho] \cup E'[\rho]$.
- Since $(\theta x.t)[\rho] = \theta z.(t[\rho\{\hat{z}/x\}])$ with $z \notin ar((\theta x.t)[\rho])$, we have

$$\sigma((\theta x.t)[\rho]) = \theta z.\sigma(t[\rho\{\hat{z}/x\}]) = \theta z.\sigma(t)[\sigma(\rho)\{\hat{z}/x\}].$$

On the other hand $\sigma(\theta x.t)[\sigma(\rho)] = (\theta x.\sigma(t))[\sigma(\rho)]$. Since $z \notin ar(\theta x.t)[\rho)] \subseteq ar(t[\rho]) = ar(\sigma(t)[\sigma(\rho)])$, z is not in $ar((\theta x.\sigma(t))[\sigma(\rho)])$ and, by Proposition 5.1.8 (page 111), $(\theta x.\sigma(t))[\sigma(\rho)] = \theta z.\sigma(t)[\sigma(\rho)\{\hat{z}/x\}]$.

□

Identifying any subset of A^∞ with an extended language of arity $\emptyset$, we get:

Proposition 5.1.10. *For any term $t \in fix\mathcal{T}(A_{\{\vee\}})$ and any $v : Var \to \mathcal{P}(A^\infty)$, $[t][v] = \sigma(t)[v]$.*

If t is a closed term, $[t][v] = \sigma(t)_\sharp$.

Proof. The proof is by induction on t, applying Proposition 5.1.9. □

As a direct consequence of the definition of the μ-calculus of extended languages, we get:

Proposition 5.1.11. *For every term t, the set $\sigma(t)_\sharp$ is in $Rat(A^\infty)$, and for each $x \in Var$, the set $\sigma(t)_x$ is in $Rat(A^*)$.*

Before proving the converse we need two results.

Proposition 5.1.12. *For any rational language L in $Rat(A^\infty)$ the language $L - \{\varepsilon\}$ is also in $Rat(A^\infty)$.*

Proof. The proof is by induction on L.

- If $L = \emptyset$ or $L = \{a\}$, then $L - \{\varepsilon\} = L$.
- $(L \cup L') - \{\varepsilon\} = (L - \{\varepsilon\}) \cup (L' - \{\varepsilon\}$.
- Note that $\varepsilon \in LL'$ if and only if $\varepsilon \in L \cap L'$. In this case

$$(LL') - \{\varepsilon\} = (L - \{\varepsilon\})(L' - \{\varepsilon\});$$

otherwise, $(LL') - \{\varepsilon\} = LL'$.
- $L^* - \{\varepsilon\} = (L - \{\varepsilon\})^+$.
- Note that $\varepsilon \in L^\omega$ if and only if $\varepsilon \in L$. In this case $L^\omega = L^* \cup (L - \{\varepsilon\})^\omega$, hence, $L^\omega - \{\varepsilon\} = (L - \{\varepsilon\})^+ \cup (L - \{\varepsilon\})^\omega$. □

Definition 5.1.13. A language is ε-free if it does not contain ε.

Proposition 5.1.14. *Every language in Rat(A^∞) is equal to a finite union*

$$L_0 \cup \bigcup_{i=1}^{m} L_i K_i^\omega$$

where the L_i are rational languages in Rat(A^*) and the K_i are ε-free rational languages in Rat(A^*).

Proof. The proof is by induction on L.

- The result is true for $L = \emptyset$, $\{a\}$, and the induction step for $L' \cup L''$ is obvious.
- Assume that $L = L_0 \cup \bigcup_{i=1}^m L_i K_i^\omega$ where each K_i is ε-free. Then $LL' = L_0 L' \cup \bigcup_{i=1}^m L_i K_i^\omega$. If, moreover, $L' = L'_0 \cup \bigcup_{i=1}^{m'} L'_i K_i'^\omega$ where each K'_i is ε-free, then $L_0 L' = L_0 L'_0 \cup \bigcup_{i=1}^{m'} L_0 L'_i K_i'^\omega$ and the result is proved.
- If $L = L_0 \cup \bigcup_{i=1}^m L_i K_i^\omega$ where each K_i is ε-free, then $L^* = L_0^* \cup \bigcup_{i=1}^m L_0^* L_i K_i^\omega$ and $L^\omega = L_0^\omega \cup \bigcup_{i=1}^m L_0^* L_i K_i^\omega$. But $L_0^\omega = L_0^* \cup (L_0 - \{\varepsilon\})^\omega$. □

Corollary 5.1.15. *If L is in Rat(A^∞) then $L \cap A^*$ is in Rat(A^*), and $L \cap A^\omega$ is in Rat(A^∞).*

Proposition 5.1.16. *If E is an extended language such that $E_\sharp$ and E_x, for any variable $x \in ar(E)$, are rational, then there exists a term t of arity $ar(E)$ such that $\sigma(t) = E$.*

If $E_\sharp$ is a subset of A^* then t is in Σ_1, otherwise it is in Π_2.

Proof. The proof is in several steps.

In a first step we prove by induction on L that for every language L in Rat(A^*) and any variable x, there exists a term $t \in \Sigma_1$ of arity $\{x\}$ such that $\sigma(t) = \langle\{x\}, Lx\rangle$.

- If $L = \emptyset$ then we take $t = \mu x.x$. Thus $\sigma(\mu x.x) = \mu x.\sigma(x)$. But $\sigma(x)_\sharp = \emptyset$ and $\sigma(x)_x = \varepsilon$, thus, $\mu x.\sigma(x) = \{\varepsilon\}^*\emptyset = \emptyset$.
- If $L = \{a\}$ then we take $t = ax$, and $\sigma(t) = \langle\{x\}, \{ax\}\rangle$.
- If $L = L_1 \cup L_2$ with $\langle\{x\}, L_i x\rangle = \sigma(t_i)$ then we take $t = t_1 \vee t_2$. Then, $\sigma(t) = \sigma(t_1) \cup \sigma(t_2)$.
- If $L = L_1 L_2$ with $\langle\{x\}, L_i x\rangle = \sigma(t_i)$ then we take $t = t_1[id\{t_2/x\}]$. Then $\sigma(t) = \sigma(t_1)[id\{\sigma(t_2)/x\}] = \langle\{x\}, L_1 L_2 x\rangle$.
- If $L = L'^*$ with $\langle\{x\}, L'x\rangle = \sigma(t')$ then we take $t = \mu y.(t[id\{y/x\}] \vee x)$, where $y \neq x$. In this case $\sigma(t) = \mu y.(\sigma(t[id\{y/x\}])t \cup \sigma(x)) = \mu y.\langle\{x, y\}, L'y \cup \{x\}\rangle = \langle\{x\}, L'^* x\rangle$.

Next, we prove that if $L \subseteq A^\infty$ is rational, there exists a term t in Σ_1 or in Π_2 of arity $\emptyset$ such that $\sigma(t) = \langle \emptyset, L\rangle$. It is easy to see that if $\sigma(t) = \langle \{x\}, Kx\rangle$ where K is ε-free, and where $t \in \Sigma_1$, then $\sigma(\nu x.t) = \langle \emptyset, K^\omega\rangle$. It follows that if $\sigma(t) = \langle \{x\}, Lx\rangle$ and $\sigma(t') = K^\omega$ then $\sigma(t[id\{t'/x\}]) = \langle \emptyset, LK^\omega\rangle$. Since $t \in \Sigma_1$ and $t' \in \Pi_2$, the term $t[id\{t'/x\}]$ is in Π_2. Finally, if $\sigma(t) = \langle \{x\}, Lx\rangle$ then $\sigma(t[id\{E_\varepsilon/x\}] = \langle \emptyset, L\rangle$, where E_ε is the extended language $\langle \emptyset, \{\varepsilon\}\rangle$.

Thus, by using Proposition 5.1.14, we get the result.

Now let $t_\sharp$ of arity $\emptyset$ be such that $\sigma(t_\sharp) = \langle \emptyset, E_\sharp\rangle$, and, for $x \in ar(E)$, t_x of arity $\{x\}$ be such that $\sigma(t_x) = \langle \{x\}, E_x\rangle$. Then $E = \sigma(t_\sharp \vee \bigvee_{x \in ar(E)} t_x)$. □

Theorem 5.1.17. *Every rational language over an alphabet A is a component of the standard interpretation of some vectorial fixed-point term*

$$\nu \boldsymbol{x}.\mu \boldsymbol{y}.\boldsymbol{t}(\boldsymbol{x}, \boldsymbol{y})$$

where $\boldsymbol{t}$ is built up only with ε, $\vee$, and the unary symbols a for a in A. Moreover, if this language is included in A^*, it is the standard interpretation of a simpler term $\mu \boldsymbol{x}.\boldsymbol{t}(\boldsymbol{x})$.

5.2 Nondeterministic automata

This section is devoted to the presentation of nondeterministic automata on words. In a first part, we recall the definition of an automaton over finite words. Then we will focus on automata for infinite words emphasizing the notion of *chain automaton* or *parity automaton* because of their close relationship with the μ-calculus.

In contrast to the previous section, we shall not consider automata on both finite and infinite words: by Corollary 5.1.15 (page 114), we can deal separately with languages of finite words and languages of infinite words.

5.2.1 Automata on finite words

An *automaton* is a tuple $\mathcal{A} = \langle S, T, I, F\rangle$ where

- S is a nonempty finite set of states,
- T is a subset of $S \times A \times S$, the set of *transitions*,
- I is a nonempty subset of S, the *initial states*,
- F is a subset of S, elements of which are called the *accepting states*.

An automaton $\mathcal{A}$ is said to be *deterministic* if I is a singleton and if for all $(s, a) \in S \times A$, there exists at most one s' in S such that $(s, a, s') \in T$.

A *run* of $\mathcal{A}$ on a word $u = a_1 \cdots a_n$ in A^* is a sequence of states $\rho = s_0 s_1 \cdots s_n s_n$ such that $s_0 \in I$ and $\forall i : 1 \le i \le n, \langle s_{i-1}, a_i, s_i \rangle \in T$. (In case $\mathcal{A}$ is deterministic, there is at most one such run). A run ρ is *accepting* if its last state s_n is an accepting state. A word u is accepted by $\mathcal{A}$ if there exists an accepting run of $\mathcal{A}$ on u. We denote by $L(\mathcal{A})$ the set of words accepted by $\mathcal{A}$ and we say that this set is *recognized* by $\mathcal{A}$. A language is *recognizable* if it is recognized by some automaton.

We assume that the reader is already familiar with the following closure properties of recognizable languages, which we state without proof. We refer, for instance, to [46].

Theorem 5.2.1. *For every automaton $\mathcal{A}$ there exists a deterministic automaton $\mathcal{A}'$ such that $L(\mathcal{A}) = L(\mathcal{A}')$.*

If $L \subseteq A^*$ is recognizable, then the complement $A^* - L$ of L is recognizable.

If L and L' are recognizable, then so are $L \cup L'$ and $L \cap L'$.

Definition 5.2.2. Given an automaton $\mathcal{A}$, we define the relation $s \xrightarrow{u}_{\mathcal{A}} s'$ in $S \times A^* \times S$, which we write $s \xrightarrow{u} s'$ when $\mathcal{A}$ is clear from the context, by

$s \xrightarrow{\varepsilon} s' \Leftrightarrow s = s'$,
$s \xrightarrow{a_1 \cdots a_n} s'$ if and only if there exist $s_0, \ldots, s_n \in S$, with $s = s_0$, $s' = s_n$, such that $(s_i, a_{i+1}, s_{i+1}) \in T$.

Let us assume that $S = \{1, \ldots, n\}$ and let $L_{ss'} = \{u \in A^* \mid s \xrightarrow{u} s'\}$. Let $\boldsymbol{K}$ be the vector defined by $\forall s \in S, \boldsymbol{K}_s = \bigcup_{s' \in F} L_{ss'}$. It is easy to see that $\boldsymbol{K} = \boldsymbol{T}^* \boldsymbol{U}$ where $\boldsymbol{T}$ is the $n \times n$ matrix defined by $\boldsymbol{T}_{ss'} = \{a \in A \mid s \xrightarrow{a} s'\}$ and $\boldsymbol{U}$ is the vector defined by $\boldsymbol{U}_s = \varepsilon$ if $s \in F$, $\emptyset$ otherwise.

Now we prove the following result which can be viewed as a formulation of the celebrated Kleene's theorem.

Theorem 5.2.3. *For any language L included in A^*, the following statements are equivalent.*

1. *L is recognizable.*
2. *There exists a term t in $\Sigma_1(A_{\{\vee\}})$, of arity $\emptyset$ such that $\|t\| = L$.*
3. *L is rational.*

Proof. 1 $\Rightarrow$ 2. The language recognized by $\mathcal{A}$ is equal to $\bigcup_{s \in I, s' \in F} L_{ss'} = \bigcup_{s \in I} \boldsymbol{L}_s$ where $\boldsymbol{L} = \boldsymbol{T}^* \boldsymbol{U}$. By Lemma 5.1.6 (page 109) $\boldsymbol{L} = \mu \boldsymbol{x}.(\boldsymbol{T}\boldsymbol{x} \vee \boldsymbol{U})$.

Then $\boldsymbol{L}$ is the interpretation of a vectorial term in $\Sigma_1^{\rightarrow}(A_{\{\vee\}})$, and, by Proposition 2.7.8 (page 63), each $\boldsymbol{L}_s$ is the interpretation of a term t_s in $\Sigma_1(A_{\{\vee\}})$. Since L is the union of some of the $\boldsymbol{L}_s$, it is the interpretation of the union of some of the t_s, which is also in $\Sigma_1(A_{\{\vee\}})$.

2 $\Rightarrow$ 3. By Proposition 5.1.11 (page 113) and Corollary 5.1.15 (page 114).

3 $\Rightarrow$ 1. This is a classical result of the elementary theory of languages and automata. □.

5.2.2 Automata on infinite words

An automaton for infinite words is almost the same as an automaton for finite words. The main difference lies in the criterion to accept a word or not.

An *automaton* is a tuple $\mathcal{A} = \langle S, T, I, C \rangle$ where

- S is a nonempty finite set of states,
- T is a subset of $S \times A \times S$, the set of *transitions*,
- I is a nonempty subset of S, the *initial states*,
- C is a subset of S^ω, called the *acceptance criterion.*

A *run* of $\mathcal{A}$ on the word $u = a_0 a_1 \cdots a_n \cdots$ in A^ω is a sequence of states $\rho = s_0 s_1 \cdots s_n \cdots$ such that $s_0 \in I$ and $\forall i \geq 0, \langle s_i, a_i, s_{i+1} \rangle \in T$. A run ρ is *accepting* if it belongs to the acceptance criterion C. A word u is accepted by $\mathcal{A}$ if there exists an accepting run of $\mathcal{A}$ on u. It follows that the set of words accepted by an automaton $\mathcal{A} = \langle S, T, I, C \rangle$ does not change if S is restricted to its subset of *reachable states* defined inductively by

- every initial state is reachable,
- s' is reachable if there exist a letter a and a reachable state s such that $\langle s, a, s' \rangle \in T$.

We denote by $L(\mathcal{A})$ the set of words accepted by $\mathcal{A}$ and we say that this set is *recognized* by $\mathcal{A}$.

The following closure properties of recognizable languages are also given without proofs. We refer to [95].

Theorem 5.2.4. *Let $L, L' \subseteq A^\omega, K \subseteq B^\omega$ and let $h : A \to B$ which is extended into a mapping from A^ω to B^ω.*

If L an L' are recognizable, so are $L \cup L'$, $L \cap L'$, and $h(L) \subseteq B^\omega$.

If K is recognizable, so is $h^{-1}(K) \subseteq A^\omega$.

An automaton is *deterministic* if (i) there is only one initial state and (ii) for any s in S and any a in A, there is at most one s' in S such that $\langle s, a, s' \rangle$. In this case, for any word u, there is at most one run of $\mathcal{A}$ on u.

An automaton is *complete* if for any s in S and any a in A, there is at least one s' in S such that $\langle s, a, s' \rangle$. In this case, for any word u, there is at least one run of $\mathcal{A}$ on u.

Now we define several kinds of automata according to how the acceptance criterion C is defined. Firstly, for any element $\rho \in S^\omega$ we denote by $\mathrm{Inf}(\rho)$ the set of states that occur infinitely often in ρ:

$$\mathrm{Inf}(s_0 s_1 \cdots s_n \cdots) = \{s \in S \mid \forall m \geq 0, \exists n \geq m : s = s_n\}.$$

An acceptance criterion C is a *Muller criterion* if there is a subset $\mathcal{F}$ of $\mathcal{P}(S)$ ($\mathcal{F}$ is called a *Muller condition*) such that $C = \{\rho \in S^\omega \mid \mathrm{Inf}(\rho) \in \mathcal{F}\}$. In this case we write $M(\mathcal{F})$ instead of C.

It is a *Rabin criterion* if there is a finite set of pairs $\langle L_i, U_i\rangle (i = 1, \dots, k)$ (called a *Rabin condition*) such that $\rho \in C$ if and only $L_i \cap \mathrm{Inf}(\rho) = \emptyset$ and $U_i \cap \mathrm{Inf}(\rho) \neq \emptyset$ for some $i : 1 \leq i \leq k$. We write $R(\langle L_i, U_i\rangle_{i=1,\dots,k})$ for C.

It is a *chain Rabin criterion* if it is a Rabin criterion $R(\langle L_i, U_i\rangle_{i=1,\dots,k})$ such that $L_i \subseteq U_i$ for $i = 1, \dots, k$ and $U_{i+1} \subseteq L_i$ for $i = 1, \dots, k-1$. In other words, C is a chain Rabin criterion if there is an increasing sequence $E_{2k-1} \subseteq E_{2k-2} \subseteq \cdots, \subseteq E_1 \subseteq E_0$ with $k \geq 1$ (called a *chain Rabin condition*) such that $C = R(\langle E_{2i-1}, E_{2i-2}\rangle_{i=1,\dots,k})$.

It is a *Büchi criterion* if there is a subset F of S (called a *Büchi condition*)such that $C = \{\rho \in S^\omega \mid \mathrm{Inf}(\rho) \cap F \neq \emptyset\}$, and we write $B(F)$ for C. We recall the trivial observation that $B(F \cup F') = B(F) \cup B(F')$.

We will refer to the automata with Muller criterion as to Muller automata, and similarly for other criteria.

Obviously, a Büchi criterion is a special case of a chain Rabin criterion since $B(F) = R(\langle \emptyset, F\rangle)$. On the other hand, a Rabin criterion is a special case of a Muller criterion since

$$R(\langle L_i, U_i\rangle_{i=0,\dots,k}) = M(\{X \subseteq S \mid \exists i, 1 \leq i \leq k : L_i \cap X = \emptyset, U_i \cap X \neq \emptyset\}.$$

Also, it is possible to express Rabin and Muller criteria as Boolean combination of Büchi criteria.

Proposition 5.2.5.

$$\begin{aligned} R(\langle L_i, U_i\rangle_{i=1,\dots,k}) &= \bigcup_{i=1}^{k} (B(U_i) - B(L_i)), \\ M(\mathcal{F}) &= \bigcup_{F \in \mathcal{F}} (\bigcap_{s \in F} B(\{s\}) - B(S - F)). \end{aligned}$$

Proof. Obvious. □

It follows that a non empty chain Rabin criterion can always be defined by a strictly increasing sequence $E_{2k-1} \subset E_{2k-2} \subset \cdots \subset E_1 \subset E_0$. If $E_{2i-1} = E_{2i-2}$, we can withdraw this pair because $B(E_{2i-2}) - B(E_{2i-1}) = \emptyset$. If $E_{2i} = E_{2i-1}$, we can replace the two pairs $\langle E_{2i+1}, E_{2i}\rangle$ and $\langle E_{2i-1}, E_{2i-2}\rangle$ by the single pair $\langle E_{2i+1}, E_{2i-2}\rangle$ because $(B(E_{2i}) - B(E_{2i+1})) \cup (B(E_{2i-2}) - B(E_{2i-1})) = B(E_{2i-21}) - B(E_{2i+1})$.

5.2.3 Parity automata

Another way of expressing the chain Rabin criterion is the *parity criterion*, similar to the one used in the definition of games in Chapter 4. Let r be a mapping from S into $\mathbb{N}$ (called a *parity condition*). For any subset E of

S, let $r(E)$ be $\{r(s) \mid s \in E\}$. Then the parity criterion $P(r)$ is the set $\{\rho \in S^\omega \mid \max(r(\mathrm{Inf}(\rho)))$ is even$\}$. Equivalently, $\rho = s_0 s_1 \cdots s_n \cdots \in P(r)$ if and only if $\limsup_{n\to\infty} r(s_n)$ is even.

Proposition 5.2.6. *A criterion $C \subseteq S^\omega$ is a chain Rabin criterion if and only if it is a parity criterion.*

Proof. Let $C = R(\langle E_{2i+1}, E_{2i}\rangle_{i=1,\dots,k})$, with $E_j \subseteq E_{j-1}$ be a chain Rabin criterion. Let $E_1 = S$ and $E_{2k+2} = \emptyset$. For any subset $s \in S$, let $r(s)$ be the least $i \in \{1, \dots 2k\}$ such that $s \in E_i - E_{i+1}$. If follows that $\rho \in C$ if and only if $\max(r(\mathrm{Inf}(\rho)))$ is even.

Conversely, let $r : S \to \mathbb{N}$. Note that if r' is defined by $r'(s) = r(s) + 2$, we have $P(r) = P(r')$. Thus, we may assume that $r(S) \subseteq \{1, \dots, 2k\}$ and we define $E_i = \{s \in S \mid r(s) \geq i\}$. Obviously $E_i \subseteq E_{i-1}$. It follows that the least i such that $s \in E_i$ is $r(s)$ and $P(r) = R(\langle E_{2i+1}, E_{2i}\rangle_{i=1,\dots,k})$ for the same reason as above.

□

The fact that a chain Rabin criterion can always be defined by a strictly increasing sequence is equivalent to the fact that a parity criterion can always be defined by some r such that $r(S)$ is always one of the following four sets: $\{1, \dots, 2k\}$, $\{1, \dots, 2k+1\}$, $\{0, \dots, 2k-1\}$,$\{0, \dots, 2k-2\}$. In this case the associated chain Rabin condition has k pairs, so we say that this parity condition has k pairs. However, the four sets above provide a finer classification of parity criteria, called the Mostowski index (see Definition 7.1.7, page 160).

In particular, the Büchi criterion is a special case of chain Rabin criterion and it is easy to see that Büchi automata are the parity automata where the rank function has its range equal to $\{1, 2\}$.

We shall see at the end of this section that parity automata are tightly related to the μ-calculus on words without intersection. We first show that any language recognized by an automaton can be recognized by a parity automaton.

First we are going to transform a Muller or Rabin automaton into a chain Rabin one, using the technique of *LAR* (later appearance record) *with hit* introduced by Büchi (see [96] for references).

Let S be any finite set of cardinality n. We denote by $\hat{S}$ the set, of cardinality $n!$, of all permutations of S, i.e., the set of words $s_1 \cdots s_n$ in S^n such that $i \neq j \Rightarrow s_i \neq s_j$. For any w in $\hat{S}$ and any s in S, w can be written in a unique way as usv with $u, v \in S^*$. We define an (infixed) binary operation $*$ from $\hat{S} \times S$ into $\hat{S}$ by $w * s = uvs$ where $w = usv$.

Let us choose an arbitrary element w_* of $\hat{S}$. Given this element, we define inductively a mapping $\lambda : S^* \to \hat{S}$ by $\lambda(\varepsilon) = w_*$, $\lambda(us) = \lambda(u) * s$.

Now let $\rho = s_0 s_1 \cdots s_m \cdots$ be an element of S^ω. We define the following sequence of elements of $\hat{S}$: $w_0 = w_*$, $w_{m+1} = \lambda(s_0 \cdots s_m)$, so that $w_{m+1} =$

uvs_m where $w_m = us_mv$. For any letter $s \in S$, let $\text{rank}_m(s)$ be the position of s in w_m. Since w_{m+1} is obtained from w_m by moving s_m at the end, we have

(i) $\text{rank}_{m+1}(s) = n$ if and only if $s = s_m$,
(ii) if $s_m \neq s$ then $\text{rank}_{m+1}(s) \leq \text{rank}_m(s)$, and
(iii) if $\text{rank}_{m+1}(s) < \text{rank}_m(s)$, then $\text{rank}_m(s_m) < \text{rank}_m(s)$,
(iv) if $\text{rank}_m(s') < \text{rank}_m(s) = \text{rank}_{m+1}(s)$, then $\text{rank}_m(s') = \text{rank}_{m+1}(s')$,
(v) if $\text{rank}_m(s) = \text{rank}_{m+1}(s) \neq n$, then $\text{rank}_m(s_m) > \text{rank}_m(s)$.

We also define the sequences x_m and y_m by: $x_m y_m$ is the unique decomposition of w_m such that the first letter of y_m is s_m. Finally, let X_m (resp.Y_m) be the set of letters of x_m (resp. y_m). Since (X_m, Y_m) is a partition of S, $X_i = X_j \Leftrightarrow Y_i = Y_j$.

Proposition 5.2.7. *Let $Z = Inf(\rho)$ be the set of letters that occur infinitely often in ρ. Then*
(1) for almost all m, $Y_m \subseteq Z$,
(2) there exists infinitely many m such that $Y_m = Z$.

Proof. Let $Z' = S - Z$ be the set of letters that occurs only finitely many times in ρ. For $s \in Z'$, let s_p be its last occurrence in ρ. Then we have, by property (ii) mentioned above, $\text{rank}_{p+1}(s) \geq \text{rank}_{p+2}(s) \geq \cdots \geq \text{rank}_{p+i}(s) \geq \cdots$, thus $\text{rank}_m(s)$ is ultimately constant: there exists m_s such that $\forall m \geq m_s, \text{rank}_m(s) = \text{rank}_{m_s}(s)$, and by property (i), this constant rank cannot be n. By property (iv), we get

(vi) for any s' such that $\text{rank}_{m_s}(s') < \text{rank}_{m_s}(s)$, we have
$\forall m \geq m_s, \text{rank}_m(s') = \text{rank}_{m_s}(s')$,

which implies, by property (i), $\forall m \geq m_s, s' \neq s_m$, thus $s' \in Z'$.

Let $m' = \max\{m_s \mid s \in Z'\}$ which is not equal to n and let ℓ be equal to $\max\{\text{rank}_{m'}(s) \mid s \in Z'\}$ (if Z' is empty, ℓ is equal to 0). By property (vi) we get $s \in Z' \Leftrightarrow 1 \leq \text{rank}_{m'}(s) \leq \ell$. Therefore there exists a word x of length ℓ consisting of all the letters in Z' which is a prefix of all w_m for $m \geq m'$. Moreover, for $m \geq m'$, $s_m \notin Z'$, so that x is also a prefix of x_m. It follows that $Z' \subseteq X_m$ hence $Y_m = S - X_m \subseteq S - Z' = Z$. This proves the point (1).

Now assume that point (2) is false, i.e., there is an m' such that for any $m \geq m'$, Y_m is strictly included in Z, or equivalently, Z' is strictly included in X_m. We can choose m' large enough so that z is a prefix of all x_m for $m \geq m'$ and indeed a strict prefix since Z' is strictly included in X_m. Therefore, for all $m \geq m'$ there exists $s'_m \in Z$ such that zs'_m is a prefix of x_m. It is easy to see that $s'_m \neq s_{m+1} \Rightarrow s'_{m+1} = s'_m$, for no letter in zs'_m can be equal to s_{m+1}. Since $s'_{m'}$ is in Z there exists a $p > m'$ such that $s_p = s'_{m'}$ and let us consider the least p that has this property. Then for $m' \leq j \leq p$, $s'_j = s'_{m'}$, thus

$zs'_p = zs'_{m'} = zs_p$ is a prefix of w_p. The canonical decomposition (x_p, y_p) of w_p is such that the first letter of y_p has to be s_p, hence $x_p = z$, a contradiction. □

Theorem 5.2.8. *Let $\mathcal{A} = \langle S, T, I, C\rangle)$ be an automaton where the set S has n states, and C is either a Muller criterion $M(\mathcal{F})$ where $\mathcal{F}$ is a family of k subsets of S, or a Rabin criterion $R(\langle L_i, U_i\rangle_{i=1,\dots,k})$ of k pairs. Then $L(\mathcal{A})$ is recognized by a parity automaton $\mathcal{A}'$ with $n \times n!$ states whose ranks are included in $\{0, \dots, 2k\}$ or in $\{1, \dots, 2k+1\}$. Moreover, if $\mathcal{A}$ is deterministic, so is $\mathcal{A}'$.*

Proof. Let us first define the parity automaton $\mathcal{A}' = \langle S', T', I', r\rangle$ by

- $S' = S \times \hat{S}$, which is of cardinality $n \times n!$,
- $I' = \{\langle s, w_*\rangle \mid s \in I\}$ where w_* is the distinguished elements of $\hat{S}$,
- $(\langle s, w\rangle, a, \langle s', w'\rangle) \in T'$ if and only if $(s, a, s') \in T$ and $w' = w * s$,
- the parity condition r will be defined later on.

Obviously, if $\mathcal{A}$ is deterministic, so is $\mathcal{A}'$.

A state $s' \in S'$ is a pair $\langle s, w\rangle$ where w can be uniquely decomposed into (x, y) such that $w = xy$ and s is the first letter of y, and we denote by $|s'|$ the length of y and by $\mathrm{alph}(s')$ the set of letters in y.

A run of $\mathcal{A}'$ on any word $u \in A^\omega$ is a sequence $\rho' = s'_0 s'_1 \cdots = \langle s_0, w_0\rangle\langle s_1, w_1\rangle \cdots$ where $\rho = s_0 s_1 \cdots$ is a run of $\mathcal{A}$ on u and the sequence $w_0, w_1, \dots$ is defined by $w_0 = w_* = \lambda(\varepsilon)$ and $w_{m+1} = \lambda(s_0 s_1 \cdots s_m)$.

Let $Z' = \mathrm{Inf}(\rho')$ and $Z = \mathrm{Inf}(\rho)$.

As a consequence of the previous proposition, we have;
(1) $s' \in Z' \Rightarrow \mathrm{alph}(s') \subseteq Z$, (2) there are infinitely many m such that $\mathrm{alph}(s'_m) = Z$. Since there are only finitely many $s' \in S'$ such that $\mathrm{alph}(s') = Z$, we get (3) Z' contains some s' with $\mathrm{alph}(s') = Z$.

Now assume that we can define $r : S' \to \mathbb{N}$ so that the condition (4) $\mathrm{alph}(s'_1) \subset \mathrm{alph}(s'_2) \Rightarrow r(s'_1) \le r(s'_2)$ holds. By (1) and (3), we have (5) $\max(r(Z')) = \max\{r(s') \mid \mathrm{alph}(s') = Z\}$.

Now assume that C is $M(\mathcal{F})$. We define r as follows:

$$r(s') = \begin{cases} 2|s'| - 2 & \text{if } \mathrm{alph}(s') \in \mathcal{F}, \\ 2|s'| - 1 & \text{otherwise.} \end{cases}$$

The above condition (4) is satisfied since $\mathrm{alph}(s'_1) \subset \mathrm{alph}(s'_2) \Rightarrow |s'_1| \le |s'_2| - 1$ and $r(s'_1) \le 2|s'_1| - 1 \le 2|s'_2| - 3 \le r(s'_2)$. And it is easy to see, by (5) that $Z \in \mathcal{F}$ if and only if $\max(r(Z'))$ is even.

Since the value of $|s'|$ is always in $\{1, \dots, n\}$, the range of r is included in $\{0, \dots, 2n-1\}$, and the associated chain Rabin criterion $\langle E_{2i+1}, E_{2i}\rangle_{i=1,\dots,n}$ has n pairs. A closer look shows that if for a given length ℓ, there is no s' of length ℓ such that $\mathrm{alph}(s') \in \mathcal{F}$, (i.e., $\mathcal{F}$ does not contain a subset

F of S of cardinality ℓ), then $2l-1$ is not in the range of r, that makes it possible to get a chain Rabin criterion with one less pair. Therefore, we can reduce the number of pairs to $n-p$, where p is the cardinal of the set $\{1,\dots,n\} - \{|F| \mid F \in \mathcal{F}\}$. But $\mathcal{F}$ has k elements, thus $\{|F| \mid F \in \mathcal{F}\}$ has at most k elements and $p \geq n-k$, hence $n-p \leq k$.

In case C is the Rabin criterion $R(\langle L_i, U_i\rangle_{i=1,\dots,k})$, we define r as follows. First for any subset X of S, we define $L(X) = \{i \mid X \cap L_i \neq \emptyset\}$ and $U(X) = \{i \mid X \cap (L_i \cup U_i) \neq \emptyset\}$. Obviously, $L(X) \subseteq U(X)$. Let $\ell(X)$ (resp. $u(X)$) be the number of indices in $L(X)$ (resp. $U(X)$). Obviously, $1 \leq \ell(X) \leq u(X) \leq k$. Moreover, $\ell(X) < u(X)$ if and only if there exists an i such that $i \in U(X) - L(X)$, i.e., $X \cap L_i = \emptyset$ and $X \cap U_i \neq \emptyset$. Let us define $r' : \mathcal{P}(S) \to \{1,\dots,2k+1\}$ by

$$r'(X) = \begin{cases} \ell(X) + u(X) + 1 & \text{if } \ell(X) = u(X), \\ \ell(X) + u(X) + 1 & \text{if } \ell(X) < u(X) \text{ and } \ell(X) + u(X) \text{ is odd}, \\ \ell(X) + u(X) & \text{if } \ell(X) < u(X) \text{ and } \ell(X) + u(X) \text{ is even}. \end{cases}$$

If $X \subset X'$ we have $L(X) \subseteq L(X')$ and $U(X) \subseteq U(X')$. If $L(X) = L(X')$ and $U(X) = U(X')$ then $r'(X) = r'(X')$. Otherwise, $r'(X) \leq \ell(X) + u(X) + 1 \leq \ell(X') + u(X') \leq r'(X')$.

Now let $r(s') = r'(\text{alph}(s'))$. The above condition (4) is satisfied, so that, by (5), $\max(r(Z'))$ is even if and only if $r'(Z)$ is even. This is the case only when there exists i such that $Z \cap L_i = \emptyset$ and $Z \cap U_i \neq \emptyset$. □

The interest of the constructions given in the previous theorem is that they preserves determinism. Without this constraint, we have a better complexity bound for $\mathcal{A}'$ as shown in Definition 5.2.14, page 125.

5.2.4 Recognizable languages

By Theorem 5.2.8, we know that any recognizable language is recognizable by an automaton with the parity acceptance criterion. Now we are going to prove that such parity automata are in some sense equivalent to vectorial fixed-point terms without intersection.

Let $\mathcal{A}$ be a parity automaton with $S = \{1,\dots,n\}$ and $r(S) \subseteq \{1,\dots,k\}$.

Let $S_{\leq j} = \{s \in S \mid r(s) \leq j\}$, and $S_j = \{s \in S \mid r(s) = j\}$. Let $\{x_{i,j} \mid 1 \leq i \leq n, 1 \leq j \leq k\}$ be nk distinct variables. For $j : 1 \leq j \leq k$, we set $\boldsymbol{x}_j = \langle x_{1,j},\dots,x_{n,j}\rangle$. For $i : 1 \leq i \leq n$, we set t_i equal to the *extended language* $\bigcup_{(i,a,j)\in T} a x_{j,r(j)}$ and $\boldsymbol{t} = \langle t_1,\dots,t_n\rangle$. Indeed, we can write $\boldsymbol{t}$, using the matrix notation introduced in Section 5.1.3 (page 107), in the form $\boldsymbol{T}_1\boldsymbol{x}_1 + \cdots \boldsymbol{T}_k\boldsymbol{x}_k$ where $(\boldsymbol{T}_m)_{ss'} = \{a \mid (s,a,s') \in T, r(s') = m\}$.

Finally, let $\theta_i = \mu$ if i is odd and $\theta_i = \nu$ if i is even.

For each j between 1 and k, and for each $m > j$ we define the matrix $\boldsymbol{L}_m^{(j)}$ by: $(\boldsymbol{L}_m^{(j)})_{ss'}$ equal to $\emptyset$ if $r(s') \neq m$, and otherwise it is the set of all

non empty words $u = a_0 \cdots a_p \in A^+$ such that $\exists s_1, \ldots, s_p \in S_{\leq j} : s \xrightarrow{a_0} s_0 \cdots s_i \xrightarrow{a_i} s_{i+1} \cdots s_p \xrightarrow{a_p} s'$. Note that none of the entries of $\boldsymbol{L}_m^{(j)}$ contains the empty word. We define also the vectors $\boldsymbol{K}^{(j)}$ by: $\boldsymbol{K}_s^{(j)}$ is the set of all infinite words $u = a_0 \cdots a_p \cdots$ such that
$\exists s_1, \ldots, s_p \ldots \in S_{\leq j} : s \xrightarrow{a_0} s_0 \cdots s_p \xrightarrow{a_p} s' \cdots$ and
$\limsup_{n\to\infty} r(s_n)$ is even.

Proposition 5.2.9. *For $j \in \{1, ..., k\}$,*

$$\boldsymbol{K}^{(j)} + \sum_{m=j+1}^{k} \boldsymbol{L}_m^{(j)} \boldsymbol{x}_m = \theta_j \boldsymbol{x}_j . \cdots . \mu \boldsymbol{x}_1 . (\boldsymbol{T}_1 \boldsymbol{x}_1 + \cdots \boldsymbol{T}_k \boldsymbol{x}_k).$$

Proof. The proof is by induction on j.

First it is easy to see that $\boldsymbol{L}_m^{(1)} = (\boldsymbol{T}_1)^* \boldsymbol{T}_m$ and that $\boldsymbol{K}^{(1)}$ is the empty set.

By Lemma 5.1.6 (page 109), $\mu \boldsymbol{x}_1 . (\boldsymbol{T}_1 \boldsymbol{x}_1 + \cdots \boldsymbol{T}_k \boldsymbol{x}_k)$ is equal to

$$\sum_{m=2}^{k} (\boldsymbol{T}_1)^* \boldsymbol{T}_m \boldsymbol{x}_m = \boldsymbol{K}^{(1)} + \sum_{m=2}^{k} \boldsymbol{L}_m^{(k)} \boldsymbol{x}_m .$$

By the induction hypothesis,

$$\theta_j \boldsymbol{x}_j . \cdots . \mu \boldsymbol{x}_1 . (\boldsymbol{T}_1 \boldsymbol{x}_1 + \cdots \boldsymbol{T}_k \boldsymbol{x}_k) = \boldsymbol{K}^{(j)} + \sum_{m=j+1}^{k} \boldsymbol{L}_m^{(j)} \boldsymbol{x}_m ,$$

hence, $\theta_{j+1} \boldsymbol{x}_{j+1} . \cdots . \mu \boldsymbol{x}_1 . (\boldsymbol{T}_1 \boldsymbol{x}_1 + \cdots \boldsymbol{T}_k \boldsymbol{x}_k) =$

$$\theta_{j+1} \boldsymbol{x}_{j+1} . (\boldsymbol{K}^{(j)} + \sum_{m=j+2}^{k} \boldsymbol{L}_m^{(j)} \boldsymbol{x}_m + \boldsymbol{L}_{j+1}^{(j)} \boldsymbol{x}_{j+1})$$

which is equal, by Proposition 5.1.6, to $(\boldsymbol{L}_{j+1}^{(j)})^* (\boldsymbol{K}^{(j)} + \sum_{m=j+2}^{k} \boldsymbol{L}_m^{(j)} \boldsymbol{x}_m)$, if j is odd, and to $(\boldsymbol{L}_{j+1}^{(j)})^\omega A^\infty + (\boldsymbol{L}_{j+1}^{(j)})^* (\boldsymbol{K}^{(j+1)} + \sum_{m=j+2}^{k} \boldsymbol{L}_m^{(j)} \boldsymbol{x}_m)$, if j is even.

But, clearly, $\boldsymbol{L}_m^{(j)} = (\boldsymbol{L}_{j+1}^{(j)})^* \boldsymbol{L}_m^{(j)}$, and $\boldsymbol{K}^{(j+1)}$ is equal to $(\boldsymbol{L}_{j+1}^{(j)})^* \boldsymbol{K}^{(j)}$, if j is odd, and to $(\boldsymbol{L}_{j+1}^{(j)})^\omega + (\boldsymbol{L}_{j+1}^{(j)})^* \boldsymbol{K}^{(j)}$ if j is even. Moreover, since ε does not occur in any of the $\boldsymbol{L}_m^{(j)}$, it does not occur in any of the $\boldsymbol{L}_m^{(j+1)}$, and $(\boldsymbol{L}_{j+1}^{(j)})^\omega = (\boldsymbol{L}_{j+1}^{(j)})^\omega A^\infty$. □

In particular, for $j = k$ we get $\theta_k \boldsymbol{x}_k . \cdots . \mu \boldsymbol{x}_1 . (\boldsymbol{T}_1 \boldsymbol{x}_1 + \cdots \boldsymbol{T}_k \boldsymbol{x}_k) = \boldsymbol{K}^{(k)}$.

Corollary 5.2.10. *Let $\mathcal{A}$ be a parity automaton. Then the language recognized by $\mathcal{A}$ is a union of components of $\boldsymbol{K}^{(k)} = \theta_k \boldsymbol{x}_k . \cdots . \mu \boldsymbol{x}_1 . (\boldsymbol{T}_1 \boldsymbol{x}_1 + \cdots + \boldsymbol{T}_k \boldsymbol{x}_k)$.*

Proof. It is easy to see, from the definition of $\boldsymbol{K}_s^{(k)}$, that this set is the set of all words that have an accepting run of $\mathcal{A}$ starting in state s. Therefore the language recognized by $\mathcal{A}$ is the union of all $\boldsymbol{K}_s^{(k)}$ such that s is an initial state of $\mathcal{A}$. □

Theorem 5.2.11. *Every language recognized by a parity automaton is rational.*

Proof. This is a direct consequence of the previous corollary and of Propositions 5.1.10 (page 113) and 5.1.11 (page 113). □

Now, we prove the converse of this theorem.

Theorem 5.2.12. *Every component of the standard interpretation of a closed intersection-free vectorial fixed-point term is recognizable.*

Proof. Let $\{x_{i,j} \mid 1 \leq i \leq n, 1 \leq j \leq k\}$ be nk distinct variables. For $j : 1 \leq j \leq k$, we set $\boldsymbol{x}_j = \langle x_{1,j}, \dots, x_{n,j}\rangle$. For $i : 1 \leq j \leq n$, let t_i be a term that has the form $\bigvee_{m\in M_i} a_m z_m$ where M_i is a finite set of indices, $a_m \in A$ and $z_m \in \{x_{i,j} \mid 1 \leq i \leq n, 1 \leq j \leq k\}$.

Let us consider the fixed-point term $\theta_k.\boldsymbol{x}_k \cdots \theta_1.\boldsymbol{x}_1.\langle t_1, \dots, t_n\rangle$.

Obviously we can rewrite this term as

$$\theta_k.\boldsymbol{x}_k \cdots \theta_1.\boldsymbol{x}_1.(\boldsymbol{T}_1\boldsymbol{x}_1 + \cdots + \boldsymbol{T}_k\boldsymbol{x}_k)$$

where $(\boldsymbol{T}_j)_{ii'} = \{a \in A \mid \exists m \in M_i : a = a_m, x_{i',j} = z_m\}$.

We define the parity automaton by $S = \{\langle i,j\rangle \mid 1 \leq i \leq n, 1 \leq j \leq k\}$, $T = \{(\langle i,j\rangle, a, \langle i',j'\rangle) \mid \exists m \in M_i : a = a_m, x_{i',j'} = z_m\}$, and
$$r(\langle i,j\rangle) = \begin{cases} 2j+1 & \text{if } \theta_j = \mu \\ 2j & \text{if } \theta_j = \nu. \end{cases}$$
By a method similar to that used in the proof of the above proposition, one can show that the language recognized by this automaton with $\langle i,j\rangle$ as initial state, is the i-th component of $\theta_k.\boldsymbol{x}_k \cdots \theta_1.\boldsymbol{x}_1.\langle t_1, \dots, t_n\rangle$. □

Proposition 5.2.13. *Any recognizable language can be recognized by a Büchi automaton.*

Proof. Every recognizable language is rational. We also know, by Proposition 5.1.17 (page 115) that each rational language is the component of some vectorial term $\nu\boldsymbol{x}.\mu\boldsymbol{y}.\boldsymbol{t}(\boldsymbol{x},\boldsymbol{y})$. The associated automaton has its ranks in the set $\{1, 2\}$, then it is a Büchi automaton. □

In particular, any parity automaton is equivalent to a Büchi automaton, in the sense that they recognize the same language. We give a direct construction of this Büchi automaton.

Definition 5.2.14. Let $\mathcal{A} = \langle S, T, I, r\rangle$ with $r : S \to \{1, \dots, k\}$ be a parity automaton. We define the Büchi automaton $\mathcal{B} = \langle S', T', I', F\rangle$ by:

- $S' = S \cup S \times \{2i \mid 1 \le 2i \le k\}$, $I' = I$,
- $T' = T \cup T_0 \cup \bigcup_{1\le 2i\le k} T_{2i}$,
- $F = \{\langle s, 2i\rangle \mid 1 \le r(s) = 2i \le k\}$,

where $T_0 = \{(s, a, \langle s', 2i\rangle) \mid (s, a, s') \in T\}$ and

$$T_{2i} = \{(\langle s, 2i\rangle, a, \langle s', 2i\rangle) \mid (s, a, s') \in T, r(s) \le 2i, r(s') \le 2i\}.$$

The intuitive idea behind this construction is the following. At some point the automaton guesses the maximal even rank $2i$ of the states of a run (rules of T_0). Then, it is obliged to use only states of rank less than or equal to $2i$ (rules of T_{2i}). A run of $\mathcal{B}$ is accepting if and only if there is a state of rank $2i$ that occurs infinitely often, i.e., the maximal rank of states occurring infinitely often in the run is even.

5.2.5 McNaughton's determinization theorem

McNaughton [61] has proved the following determinization theorem. Several different proofs can be found in the litterature, among them the Safra's one [86] is the most algorithmic.

Theorem 5.2.15 (McNaughton). *Every recognizable language is recognizable by a deterministic automaton with a Rabin acceptance criterion.*

From McNaughton's Theorem it is easy to infer that the complement of any recognizable language is recognizable. We give below a proof which is not the simplest one but uses a lemma which has its own interest.

Lemma 5.2.16. *Let $\mathcal{A} = \langle S, T, I, r\rangle$ be a complete deterministic parity automaton. Let $r' : S \to \mathbb{N}$ be defined by $r'(s) = r(s) + 1$, and let $\mathcal{A}' = \langle S, T, I, r'\rangle$. Then $L(\mathcal{A}')$ is the complement of $L(\mathcal{A})$ in A^ω.*

Proof. Obvious since with each word one can associate one and only one run of $\mathcal{A}$. □

Proposition 5.2.17. *The complement of a recognizable language is recognizable.*

Proof. By McNaughton'theorem and by Theorem 5.2.8, every recognizable language is recognized by a deterministic parity automaton. It is not hard to make this automaton complete by adding sink states of odd rank. Then the result follows from the previous lemma. □

Since, by Proposition 5.2.5 (page 118) every acceptance criterion is a Boolean combination of Büchi criteria, another immediate consequence of McNaughton's theorem is the following proposition.

Proposition 5.2.18. *Any recognizable language is a Boolean combination of languages recognized by deterministic Büchi automata.*

Proof. let $\mathcal{A}$ be a deterministic complete Rabin automaton recognizing L, with acceptance criterion $R(\langle L_i, U_i\rangle_{i=1,\dots,k}) = \bigcup_{i=1}^{k}(B(U_i) - B(L_i))$. Let K_i (resp. K_i') be the language accepted by $\mathcal{A}$ with the acceptance criterion $B(U_i)$ (resp. $B(L_i)$). Then $L = \bigcup_{i=1}^{k}(K_i - K_i')$. If ρ is the unique run of $\mathcal{A}$ on the word u, this run is accepting if and only there is an i such that $\mathrm{Inf}(\rho) \cap U_i \neq \emptyset$ and $\mathrm{Inf}(\rho) \cap L_i = \emptyset$. But $\mathrm{Inf}(\rho) \cap U_i \neq \emptyset \Leftrightarrow u \in K_i$ and $\mathrm{Inf}(\rho) \cap L_i = \emptyset \Leftrightarrow u \notin K_i'$. □

5.3 Terms with intersection

We have seen that, without intersection, every recognizable language can be defined by a fixed-point term in Π_2. Using intersection allows us to define such a language by a yet simpler fixed-point term in the alternation-depth hierarchy of Definition 2.6.3, page 56.

Theorem 5.3.1. *Every recognizable language is defined by a term in*

$$\mathit{Comp}(\Sigma_1(A_{\{\vee,\wedge\}}) \cup \Pi_1(A_{\{\vee,\wedge\}})).$$

Before proving this theorem we consider a simple example which explains the result. Moreover, the proof of the general case is a simple generalization of the construction done in this simple case.

Example 5.3.2. Let $A = \{a, b\}$ and consider the language $L = (a^*b)^\omega$ consisting of all words having an infinite number of b's.

We have $L = \nu x.\mu y.(ay \cup bx)$. Indeed, by Arden's lemma, $\mu y.(ay \cup bx) = a^*bx$ and $\nu x.\mu y.(ay \cup bx) = \nu x.a^*bx = (a^*b)^\omega$.

Let $f(x) = \mu y.(ay \cup bx)$. Then $f(\nu x.x) = f(A^\omega) = a^*bA^\omega$ is the set of all words that contain at least one b.

Now, consider $g(x) = \nu y.(x \cap (ay \cup by))$. Let us show that for any language K, $g(K)$ is equal to the set $G(K)$ of all words that have all their suffixes in

K. Since, for $c = a, b$, a suffix of cu is either cu or a suffix of u, we get $cu \in G(K)$ implies $cu \in K$ and $u \in G(K)$. It follows that $G(K)$ is included in $K \cap (aG(K) \cup bG(K))$, thus, $G(K) \subseteq g(K)$. Conversely, let $u_0 u_1 \cdots u_n \cdots$ be in $g(K) = K \cap (ag(K) \cup bg(K))$. Obviously, $u_0 u_1 \cdots u_n \cdots$ is in K and $u_1 \cdots u_n \cdots$ is in $g(K)$. By iterating this argument, one can show that for every i, the word $u_i u_{i+1} \cdots$ is in K, hence $u_0 u_1 \cdots u_n \cdots$, which has all its suffixes in K, is in $G(K)$.

Finally, $g(f(\nu x.x))$ is the set of words that have all their suffixes in $f(\nu x.x) = a^* b A^\omega$, i.e., the set of words whose all suffixes contain at least one b. That is, obviously, the set of words that have an infinite number of b's.

Let us borrow another lesson from this example. We have previously seen that $\nu x.\mu y.(ay \cup bx) = (a^*b)^\omega$. Let us evaluate $\mu y.\nu x.(ay \cup bx)$: $\nu x.(ay \cup bx) = b^* ay \cup b^\omega$ and $\mu y.\nu x.(ay \cup bx) = \mu y.(b^* ay \cup b^\omega) = (b^*a)^* b^\omega = A^* b^\omega$ which is strictly included in L. This is an illustration of Proposition 1.3.4 (page 20). □

Proof of Theorem 5.3.1 Let $\mathcal{A} = \langle S, T, I, F \rangle$ be a deterministic Büchi automaton with $S = \{s_1, \ldots, s_n\}$.

For each state $s_i \in S$, let x_i and y_i be variables, let $t_i = \bigcup_{(s_i,a,s_j)\in T} a x_j$. Let $\langle g_1(y_1, \ldots, y_n), \ldots, g_n(y_1, \ldots, y_n) \rangle =$

$$\nu\langle x_1, \ldots, x_n \rangle.\langle y_1 \cap t_1, \ldots, y_n \cap t_n \rangle.$$

For any $L_1, L_2, \ldots L_n \subseteq A^\omega$ and for any j, $g_j(L_1, \ldots, L_n)$ is the set of all infinite words $u = a_1 a_2 \cdots a_n \cdots$ that have the following property: if $s_{i_1} = s_j, s_{i_2}, \ldots, s_{i_k}, \ldots$ is the unique run of $\mathcal{A}$ on u starting in state s_j, then any suffix $a_k a_{k+1} \cdots$ of u belongs to L_{i_k}.

Let z be a new variable. For any i and j, let $t_i^{(j)} = \begin{cases} t_i & \text{if } i \neq j, \\ t_i \cup z & \text{if } i = j. \end{cases}$

Let $\langle f_1^{(j)}(z), \ldots, f_n^{(j)}(z) \rangle = \mu\langle x_1, \ldots, x_n \rangle.\langle t_1^{(j)}, \ldots, t_n^{(j)} \rangle$. Then $f_i^{(j)}(A^\omega)$ is the set of all words u such that the unique run of $\mathcal{A}$ on u starting in state s_i contains the state s_j.

It follows that $g_i(f_1^{(j)}(A^\omega), \ldots, f_n^{(j)}(A^\omega))$ is the set of words u such that the unique run of $\mathcal{A}$ on u starting in state s_i contains infinitely often the state s_j.

Therefore, the language recognized by $\mathcal{A}$ is equal to

$$\bigcup_{s_i \in I, s_j \in F} g_i(f_1^{(j)}(A^\omega), \ldots, f_n^{(j)}(A^\omega)),$$

and the result holds for languages recognized by deterministic Büchi automata.

Since the dual of ax is $ax \cup \bigcup_{b \neq a} bA^\omega$, and by Proposition 1.2.25 (page 17) the result also holds for complements of deterministic Büchi languages. By

Proposition 5.2.18, every recognizable language is a Boolean combination of deterministic Büchi ones and the theorem is proved. □

5.3.1 The μ-calculus of constrained languages

By Theorem 5.2.11 (page 124) and Theorem 5.1.17 (page 115), we know that rational languages are exactly those in $\Sigma_2(A_{\{\vee\}})$ and that they are all in $Comp(\Sigma_1(A_{\{\vee,\wedge\}}) \cup \Pi_1(A_{\{\vee,\wedge\}}))$. We are going to show that fixed-point terms with intersection which are higher in the alternation-depth hierarchy do not increase the class of languages that can be defined, i.e., the μ-calculus on infinite words defines only rational languages.

Let us assume now that the intersection operator is allowed in the fixed-point terms. Therefore, it is no longer possible to represent the interpretation of a term by an extended language. For instance, let us assume that $\sigma(t) = \{ax\}$ and $\sigma(t') = \{aby\}$. We know that $[\![t \wedge t']\!][v\{K/x, K'/y\}] = aK \cap abK'$ which is the set of all words w that can be written abv with $bv \in K$ and $v \in K'$. If $[\![t \wedge t']\!]$ could be described by an extended language $L = L_\sharp \cup L_x x \cup L_y y$, then $[\![t \wedge t']\!][v\{b/x, \varepsilon/y\}] = \{ab\} \cap \{ab\} = \{ab\}$ should be equal to $L_\sharp \cup L_x b \cup L_y$, while $[\![t \wedge t']\!][v\{a/x, \varepsilon/y\}] = \{aa\} \cap \{ab\} = \emptyset$ should be equal to $L_\sharp \cup L_x a \cup L_y$, and this is impossible.

However one can describe $[\![t \wedge t']\!]$ by a word like $a\{x\}b\{y\}$ with the intuitive meaning that a substitution $[K/x, K'/y]$ yields all the words abv such that $bv \in K$ and $v \in K'$ as shown by the following picture

$a\{x\}b\{y\}\cdots\cdots$
| | → a suffix in K' substituted for y
| → a suffix in K substituted for x

In this section, we are going to define a μ-calculus on the sets of such encodings.

Definition 5.3.3. Let $\xi : \mathbb{N} \to \mathcal{P}(Var)$. For $i \in \mathbb{N}$, $\xi[i] : \mathbb{N} \to \mathcal{P}(Var)$ is defined by $\xi[i](j) = \xi(i+j)$ so that $\xi[0] = \xi$ and $\xi[i][j) = \xi[i+j)$. For any $X \subseteq Var$, let $\xi \upharpoonright X : \mathbb{N} \to \mathcal{P}(Var)$ be defined by $(\xi \upharpoonright X)(i) = \xi(i) \cap X$.

As usual $\xi \subseteq \xi'$ means $\forall i,\ \xi(i) \subseteq \xi'(i)$. For example, we have $\xi \upharpoonright X \subseteq \xi$.

For a set X of variables, we denote by X^ω the mapping ξ such that $\xi(i) = X$. (Indeed, it is the same notation as for the set of all infinite words over X, but the context will make the difference clear. It is for instance the case when we write $\xi \upharpoonright X \subseteq X^\omega$.) Finally, we define $ar(\xi) = \bigcup_{i \geq 0} \xi(i)$.

Definition 5.3.4. A *constrained language* is a pair $C = \langle X, L \rangle$ where X is a subset of Var and L is a subset of $\mathcal{P}(A^\infty \times \mathcal{P}(X)^\omega)$, i.e., a set of pairs $\langle u, \xi \rangle$ with $u \in A^\infty$ and $\xi : \mathbb{N} \to \mathcal{P}(X)$ which is *closed* in the following sense:

$$\forall u \in A^\infty, \forall \xi, \xi' : \mathbb{N} \to \mathcal{P}(X), \quad \xi \subseteq \xi' \ \&\ \langle u, \xi \rangle \in L \Rightarrow \langle u, \xi' \rangle \in L.$$

For instance, the expression $a\{x\}b\{y\}$ we have considered above will be identified to the pair $\langle ab, \{x\}\{y\}\emptyset\emptyset\cdots\rangle$.

The arity of the constrained language $C = \langle X, L\rangle$ is equal to X.

Let $\mathcal{CL}$ be the set of all constrained languages. We make this set a μ-calculus as follows.

Definition 5.3.5. If $C = \langle X, L\rangle$ is a constrained language and if $\rho : Var \to \mathcal{CL}$ is a substitution, where $\rho(x) = \langle Y_x, L_x\rangle$, we define $C[\rho]$ as the pair $\langle Y, L'\rangle$ where $Y = ar'(C, \rho)$ and L' is defined by: $\langle u, \xi\rangle \in L'$ if and only if

- $\forall i \in \mathbb{N}, \xi(i) \subseteq Y$,
- there exists $\langle u, \zeta\rangle \in L$ such that $\forall i \in \mathbb{N}, \forall x \in \zeta(i)$, $\langle u[i], \xi[i] \upharpoonright Y_x\rangle \in L_x$.

If $\xi \subseteq \xi' \subseteq X^\omega$, then $\langle u[i], \xi[i] \upharpoonright Y_x\rangle \in L_x \Rightarrow \langle u[i], \xi'[i] \upharpoonright Y_x\rangle \in L_x$. It follows that $\langle u, \xi\rangle \in L' \Rightarrow \langle u, \xi'\rangle \in L'$. That implies that $\langle Y, L'\rangle$ is a constrained language.

The following proposition can be seen as an alternative definition of $C[\rho]$, which is sometimes more useful.

Proposition 5.3.6. *Let $C[\rho] = \langle Y, L'\rangle$. Given $\langle u, \xi\rangle$ with $\xi : \mathbb{N} \to \mathcal{P}(X)$, we define $\zeta : \mathbb{N} \to \mathcal{P}(X)$ by $\zeta(i) = \{x \in X \mid \langle u[i], \xi[i] \upharpoonright Y_x\rangle \in L_x\}$. Then $\langle u, \xi\rangle \in L' \Leftrightarrow \langle u, \zeta\rangle \in L$.*

Proof. Let us assume $\langle u, \zeta\rangle \in L$. For any $i \in \mathbb{N}$ and any $x \in \zeta(i)$, we have $\langle u[i], \xi[i] \upharpoonright Y_x\rangle \in L_x$, hence, $\langle u, \xi\rangle \in L'$.

Conversely, let us assume $\langle u, \xi\rangle \in L'$. There exists $\langle u, \zeta'\rangle \in L$ such that $\forall i \in \mathbb{N}, \forall x \in \zeta'(i)$, $\langle u[i], \xi[i] \upharpoonright Y_x\rangle \in L_x$). It follows that $\zeta' \subseteq \zeta$, and since L is closed, $\langle u, \zeta\rangle \in L$. □

Definition 5.3.7. For any variable x we define $\hat{x} = \langle \{x\}, K(x)\rangle$ where $K(x)$ is the constrained language defined by $\langle u, \xi\rangle \in K(x)$ if and only if $u \in A^\infty$, $\forall i \in \mathbb{N}, \xi(i) \subseteq \{x\}$ and $x \in \xi(0)$.

Finally we have to define $\mu x.C$ and $\nu x.C$.

Definition 5.3.8. Let $C = \langle X, L\rangle \in \mathcal{CL}$. If $x \notin X$ then $\mu x.C = C$. Otherwise, $\mu x.C = \langle Y, L'\rangle$ where $Y = X - \{x\}$ and $\langle u, \xi\rangle \in L'$ if and only if ξ is a mapping from $\mathbb{N}$ to $\mathcal{P}(Y)$ and $\forall \psi : \mathbb{N} \to \mathcal{P}(\{x\})$,

$$(\forall i \in \mathbb{N},\ \langle u[i], \xi[i] \cup \psi[i]\rangle \in L \Rightarrow x \in \psi(i)) \Rightarrow \langle u, \xi \cup \psi\rangle \in L.$$

Let us show that L' is closed. Let $\xi \subseteq \xi'$ and let ψ be such that

$$\forall i \in \mathbb{N},\ \langle u[i], \xi'[i] \cup \psi[i]\rangle \in L \Rightarrow x \in \psi(i).$$

We have to show that $\langle u, \xi\rangle \in L' \Rightarrow \langle u, \xi' \cup \psi\rangle \in L$.

Let i be such that $\langle u[i], \xi[i] \cup \psi[i]\rangle \in L$. We also have $\langle u[i], \xi'[i] \cup \psi[i]\rangle \in L$, hence, $x \in \psi(i)$. Thus, since $\langle u, \xi\rangle \in L'$, we have $\langle u, \xi \cup \psi\rangle \in L$, and also $\langle u, \xi' \cup \psi\rangle \in L$.

Definition 5.3.9. Let $C = \langle X, L\rangle \in \mathcal{CL}$. If $x \notin X$ then $\nu x.C = C$. Otherwise, $\nu x.C = \langle Y, L'\rangle$ where $Y = X - \{x\}$ and $\langle u, \xi\rangle \in L'$ if and only if ξ is a mapping from $\mathbb{N}$ to $\mathcal{P}(Y)$ and there exists $\psi : \mathbb{N} \to \mathcal{P}(\{x\})$ such that

- $\langle u, \xi \cup \psi\rangle \in L$,
- $\forall i \in \mathbb{N}$, $x \in \psi(i) \Rightarrow \langle u[i], \xi[i] \cup \psi[i]\rangle \in L$.

It is easy to see that L' is closed, and thus is a constrained language: Let $\langle u, \xi\rangle \in L'$ and let $\xi \subseteq \xi'$. Since $\langle u, \xi\rangle \in L'$, there exists ψ such that $\xi \cup \psi$ has the required properties. Since $\xi[i] \cup \psi[i] \subseteq \xi'[i] \cup \psi[i]$, $\xi' \cup \psi$ has also the required properties. It follows that $\langle u, \xi'\rangle \in \nu x.L$.

Proposition 5.3.10. *$\mathcal{CL}$ is a μ-calculus.*

Proof. 1. By definition, $ar(\hat{x}) = \{x\}$. *2.* By definition, $ar(C[\rho]) = ar'(C, \rho)$. *3.* By definition, $ar(\theta x.C) = ar(C) - \{x\}$.

4. Let $\rho(x) = \langle X, L\rangle$, let $\hat{x}[\rho] = \langle X', L'\rangle$. By definition, $X' = ar'(\hat{x}, \rho) = ar(\rho(x)) = X$. Let us show that $L = L'$.

Let $u \in A^\infty$ and $\xi : \mathbb{N} \to \mathcal{P}(X)$. By Proposition 5.3.6, $\langle u, \xi\rangle \in L' \Leftrightarrow \langle u, \zeta\rangle \in K(x)$, where $\zeta : \mathbb{N} \to \mathcal{P}(\{x\})$ is defined by $\zeta(i) = \{x\} \Leftrightarrow \langle u[i], \xi[i]\rangle \in L$. But $\langle u, \zeta\rangle \in K(x) \Leftrightarrow x \in \zeta(0) \Leftrightarrow \langle u[0], \xi[0]\rangle = \langle u, \xi\rangle \in L$.

5. The value of $C[\rho]$ depends only on the value of $\rho(x)$ for $x \in ar(C)$.

6. Let us show that $C[\rho][\pi] = C[\rho \star \pi]$.

NB. For the sake of avoiding heavy notation, we identify any constrained language $C = \langle X, L\rangle$ with its second component L.

Let $X = ar(C)$, $Y_x = ar(\rho(x))$, $Z_y = ar(\pi(y))$. Let $Y = ar(C[\rho])$, $U_x = ar(\rho(x)[\pi]$, and $V = ar(C[\rho][\pi])$. By definition we have $Y = \bigcup_{x\in X} Y_x$, $U_x = \bigcup_{y\in Y_x} Z_y$, and $V = \bigcup_{y\in Y} Z_y = \bigcup_{x\in X}\bigcup_{y\in Y_x} Z_y$. We also have $ar(C[\rho\star\pi]) = \bigcup_{x\in X} U_x = \bigcup_{x\in X}\bigcup_{y\in Y_x} Z_y$. Hence $ar(C[\rho][\pi]) = ar(C[\rho \star \pi]) = V$.

Let $u \in A^\infty$ and $\xi : \mathbb{N} \to \mathcal{P}(V)$. Let $\zeta : \mathbb{N} \to \mathcal{P}(Y)$ be defined by $\zeta(i) = \{y \in Y \mid \langle u[i], \xi[i] \upharpoonright Z_y\rangle \in \pi(y)\}$ and $\eta : \mathbb{N} \to \mathcal{P}(X)$ be defined by $\eta(i) = \{x \in X \mid \langle u[i], \zeta[i] \upharpoonright Y_x\rangle \in \rho(x)\}$. By Proposition 5.3.6 (page 129),

$$\langle u, \xi\rangle \in C[\rho][\pi] \Leftrightarrow \langle u, \zeta\rangle \in C[\rho] \Leftrightarrow \langle u, \eta\rangle \in C.$$

Let $\eta' : \mathbb{N} \to \mathcal{P}(X)$ be defined by

$$\eta'(i) = \{x \in X \mid \langle u[i], \xi[i] \upharpoonright U_x\rangle \in \rho(x)[\pi]\}.$$

Again by Proposition 5.3.6 $\langle u, \xi\rangle \in C[\rho \star \pi] \Leftrightarrow \langle u, \eta'\rangle \in C$.

Thus, to prove that $C[\rho][\pi] = C[\rho \star \pi]$, it suffices to prove that $\eta = \eta'$, i.e., for any i and x,

$$\langle u[i], \zeta[i] \upharpoonright Y_x\rangle \in \rho(x) \Leftrightarrow \langle u[i], \xi[i] \upharpoonright U_x\rangle \in \rho(x)[\pi].$$

Again by Proposition 5.3.6,

$$\langle u[i), \xi[i) \upharpoonright U_x \rangle \in \rho(x)[\pi] \Leftrightarrow \langle u[i), \zeta_{i,x} \rangle \in \rho(x)$$

where $\zeta_{i,x} : \mathbb{N} \to \mathcal{P}(Y_x)$ is defined by

$$\zeta_{i,x}(j) = \{y \in Y_x \mid \langle u[i)[j), \xi[i)[j) \upharpoonright U_x \upharpoonright Z_y \rangle \in \pi(y)\}.$$

But $y \in Y_x \Rightarrow Z_y \subseteq U_x$, hence

$$\zeta_{i,x}(j) = \{y \in Y_x \mid \langle u[i+j), \xi[i+j) \upharpoonright Z_y \rangle \in \pi(y)\}.$$

It follows that $y \in \zeta_{i,x}(j)$ if and only if $y \in Y_x$ and $y \in \zeta(i+j)$, that is, $\zeta_{i,x} = \zeta[i) \upharpoonright Y_x$.

7. Let $X = ar(\theta x.C)$ and $Y = ar'(\theta x.C, \rho)$. Let z be any variable not in Y and let us show that $(\theta x.C)[\rho] = \theta z.C[\rho\{\hat{z}/x\}]$.

Obviously $(\theta x.C)[\rho]$ and $\theta z.C[\rho\{\hat{z}/x\}]$ have the same arity Y.

If $x \notin ar(C)$, then $(\theta x.C)[\rho] = C[\rho]$, and $C[\rho\{\hat{z}/x\}] = C[\rho]$. Hence, $\theta z.C[\rho\{\hat{z}/x\}] = \theta z.C[\rho]$. Since $z \notin Y = ar'(\theta x.C, \rho) = ar'(C, \rho) = ar(C[\rho])$, we get $\theta z.C[\rho] = C[\rho]$.

Thus, we assume that $x \in ar(C)$. In this case we have to prove that for any $u \in A^\infty$ and for any $\xi : \mathbb{N} \to \mathcal{P}(Y)$,

$$\langle u, \xi \rangle \in (\theta x.C)[\rho] \Leftrightarrow \langle u, \xi \rangle \in \theta z.C[\rho\{\hat{z}/x\}].$$

Let $\zeta : \mathbb{N} \to \mathcal{P}(X)$ be defined by

$$\zeta(i) = \{y \in X \mid \langle u[i), \xi[i) \upharpoonright ar(\rho(y)) \rangle \in \rho(y)\}.$$

Then $\forall i \in \mathbb{N}$, $\langle u[i), \xi[i) \rangle \in (\theta x.C)[\rho] \Leftrightarrow \langle u[i), \zeta[i) \rangle \in \theta x.C$.

Let $\rho' = \rho\{\hat{z}/x\}$. For any $i \in \mathbb{N}$, and any $\psi' : \mathbb{N} \to \mathcal{P}(\{z\})$, we have $\langle u, \xi \cup \psi' \rangle \in C[\rho'] \Leftrightarrow \langle u, \delta \rangle \in C$, where $\delta : \mathbb{N} \to \mathcal{P}(ar(C))$ is defined by

$$\delta(i) = \{y \in ar(C) \mid \langle u[i), \xi[i) \upharpoonright ar(\rho'(y)) \cup \psi'[i) \upharpoonright ar(\rho'(y)) \rangle \in \rho'(y)\}.$$

Now, let us remark that if $y \in X$, then $\rho'(y) = \rho(y)$ and $\psi'[i) \upharpoonright ar(\rho(y)) = \emptyset^\omega$, since $z \notin ar(\rho(y))$. Thus, $y \in \delta(i) \Leftrightarrow \langle u[i), \xi[i) \upharpoonright ar(\rho(y)) \rangle \in \rho(y) \Leftrightarrow y \in \zeta(i)$, hence, $\delta \upharpoonright X = \zeta$.

Finally,

$$x \in \delta(i) \Leftrightarrow \langle u[i), \xi[i) \upharpoonright \{z\} \cup \psi'[i) \upharpoonright \{z\} \rangle \in \hat{z}.$$

But, by construction $\xi[i) \upharpoonright \{z\} \cup \psi'[i) \upharpoonright \{z\} = \psi'[i)$. Hence $x \in \delta(i) \Leftrightarrow z \in \psi'(i)$. It follows that $\delta = \zeta \cup \psi$ where $\psi : \mathbb{N} \to \mathcal{P}(\{x\})$ is defined by $x \in \psi(i) \Leftrightarrow z \in \psi'(i)$.

Therefore,

$$\forall i \in \mathbb{N},\ \langle u[i], \xi[i] \cup \psi'[i]\rangle \in C[\rho\{\hat{z}/x\}] \Leftrightarrow \langle u[i], \zeta[i] \cup \psi[i]\rangle \in C$$

and the end of the proof is straightforward. □

Let $\mathcal{CL}(X)$ be the set of constrained languages of arity X, ordered by inclusion. Obviously, it is a complete lattice.

Let C be any constrained language, let x be a variable, and let $X = ar(C) - \{x\}$ We consider the mapping $g : \mathcal{CL}(X) \to \mathcal{CL}(X)$ defined by $g(D) = C[id\{D/x\}]$. Obviously, this mapping is monotonic, thus it has a least and a greatest fixed point. We claim that $\mu x.L$ and $\nu x.L$ are these fixed points.

We prove here that $\nu x.C$ is the greatest fixed point of g. The fact that $\mu x.L$ is the least fixed point of this mapping will be shown later on (Proposition 5.3.16, page 134), by a duality argument, after the introduction of a notion of duality for constrained languages.

Proposition 5.3.11. *$\nu x.C$ is the greatest fixed point of the mapping $g : \mathcal{CL}(X) \to \mathcal{CL}(X)$ defined by $g(D) = C[id\{D/x\}]$, where $X = ar(C) - \{x\}$.*

Proof. If $x \notin ar(C)$ then, for any D, $g(D) = C$ and the result is proved.

Let us assume $x \in ar(C)$, To prove that $\nu x.C$ is the greatest fixed point of g, we have to show:

1. For any $D \in \mathcal{CL}(X)$, $D \subseteq g(D) \Rightarrow D \subseteq \nu x.C$,
2. $\nu x.C \subseteq g(\nu x.C)$.

Proof of $D \subseteq g(D) \Rightarrow D \subseteq \nu x.C$.

With $\langle u, \xi\rangle \in \mathcal{CL}(X)$ we associate $\psi : \mathbb{N} \to \mathcal{P}(\{x\})$ defined by $\psi(i) = \{x\} \Leftrightarrow \langle u[i], \xi[i]\rangle \in D$. We show that if $D \subseteq g(D)$ then for any i such that $\langle u[i], \xi[i]\rangle \in D$ we have $\langle u[i], \xi[i] \cup \psi[i]\rangle \in C$. By Definition 5.3.9, we immediately deduce that if $\langle u, \xi\rangle$ is in D, it is also in $\nu x.C$.

Hence, let $u' = u[i]$, $\xi' = \xi[i]$, and $\psi' = \psi[i]$. Let us assume that $\langle u', \xi'\rangle \in D$ and that $D \subseteq g(D) = C[id\{D/x\}]$. By definition, there exists $\langle u', \zeta'\rangle \in C$ such that for any $j \in N$,

- $x \in \zeta'(j) \Rightarrow \langle u'[j], \xi'[j] \upharpoonright X\rangle \in D$.
- for all $y \in X$, $y \in \zeta'(j) \Rightarrow \langle u'[j], \xi'[j] \upharpoonright \{y\}\rangle \in \hat{y}$.

But $\langle u'[j], \xi'[j] \upharpoonright \{y\}\rangle \in \hat{y} \Rightarrow y \in \xi'(j)$, hence we get $\zeta' \upharpoonright X \subseteq \xi'$.

On the other hand, since $\xi' \upharpoonright X = \xi'$, we have

$$x \in \zeta'[j] \Rightarrow \langle u'[j], \xi'[j] \upharpoonright X\rangle = \langle u[i+j), \xi[i+j)\rangle \in D$$

and, by definition of ψ, $\psi'(j) = \psi(i+j) = \{x\}$. It follows that $\zeta' \upharpoonright \{x\} \subseteq \psi'$.

Hence, $\zeta' = \zeta' \upharpoonright X \cup \zeta' \upharpoonright \{x\} \subseteq \xi' \cup \psi'$, and since $\langle u', \zeta'\rangle \in C$ we have $\langle u', \xi' \cup \psi'\rangle \in C$.

Proof of $\nu x.C \subseteq g(\nu x.C)$.
Assume $\langle u, \xi \rangle \in \nu x.C$. By definition, there exists $\psi : \mathbb{N} \to \mathcal{P}(\{x\})$ such that $\langle u, \xi \cup \psi \rangle \in C$ and $\forall i, x \in \psi(i) \Rightarrow \langle u[i], \xi[i] \cup \psi[i] \rangle \in C$.

Let i be such that $x \in \psi(i)$. We have $\langle u[i], \xi[i] \cup \psi[i] \rangle \in C$. We also have, for any j such that $x \in \psi[i](j) = \psi(i+j)$,

$$\langle u[i][j], \xi[i][j] \cup \psi[i][j] \rangle = \langle u[i+j), \xi[i+j) \cup \psi[i+j) \rangle \in C.$$

Hence $\langle u[i), \xi[i) \rangle \in \nu x.C$.

In other words, there exists $\zeta = \xi \cup \psi$ such that $\langle u, \zeta \rangle \in C$ and $\forall i : x \in \zeta(i), \langle u[i), \zeta[i) \upharpoonright X \rangle \in \nu x.C$. It follows that $\langle u, \xi \rangle \in g(\nu x.C)$ □

5.3.2 Duality

It is easy to define the *dual* $\tilde{C}$ of a constrained language $C \in \mathcal{CL}$: Let $C = \langle X, L \rangle$. Then $\tilde{C} = \langle X, \tilde{L} \rangle$ where $\tilde{L}$ is defined by

$$\langle u, \xi \rangle \in \tilde{L} \Leftrightarrow \langle u, \overline{\xi} \rangle \notin L$$

where $\overline{\xi}$ is the mapping defined by $\overline{\xi}(i) = X - \xi(i)$. In particular, if $ar(C) = \emptyset$ (i.e. if C is identified with a subset L of A^∞), then $\tilde{C}$ is identified with the complement $\overline{L}$ of L in A^∞, i.e., $\overline{L} = \{\langle u, \emptyset^\omega \rangle \mid \langle u, \emptyset^\omega \rangle \notin L\}$.

Obviously we have $\tilde{\tilde{C}} = C$. We also have:

Proposition 5.3.12. *Let C, D be of arity X. If $C \subseteq D$ then $\tilde{D} \subseteq \tilde{C}$.*

Proof. $\langle u, \xi \rangle \in \tilde{D} \Rightarrow \langle u, \overline{\xi} \rangle \notin D \Rightarrow \langle u, \overline{\xi} \rangle \notin C \Rightarrow \langle u, \xi \rangle \in \tilde{C}$. □

Moreover, we have:

Proposition 5.3.13. *Let $C' = C[\rho]$. Then $\widetilde{C'} = \tilde{C}[\tilde{\rho}]$, where $\tilde{\rho}$ is defined by $\tilde{\rho}(x) = \widetilde{\rho(x)}$.*

Proof. Obviously $ar(\widetilde{C'}) = ar(C') = ar'(C, \rho) = ar'(\tilde{C}, \tilde{\rho})$, hence, here again, we identify an element of $\mathcal{CL}$ with its second component.

Given $u \in A^\infty$, $\xi : \mathbb{N} \to \mathcal{P}(ar(C'))$, we define $\zeta, \zeta' : \mathbb{N} \to \mathcal{P}(ar(C))$ by
$\zeta(i) = \{x \in ar(L) \mid \langle u[i), \xi[i) \upharpoonright ar(\rho(x)) \rangle \in \widetilde{\rho(x)}\}$ and
$\zeta'(i) = \{x \in ar(L) \mid \langle u[i), \overline{\xi}[i) \upharpoonright ar(\rho(x)) \rangle \in \rho(x)\}$.

By Proposition 5.3.6 (page 129)

$$\langle u, \xi \rangle \in \tilde{C}[\tilde{\rho}] \Leftrightarrow \langle u, \zeta \rangle \in \tilde{C},$$
$$\langle u, \overline{\xi} \rangle \notin C' = C[\rho] \Leftrightarrow \langle u, \zeta' \rangle \notin C.$$

But $\langle u, \zeta\rangle \in \tilde{C} \Leftrightarrow \langle u, \overline{\zeta}\rangle \notin C$, (the complement of ζ being taken in $ar(C)$), and

$$x \in \overline{\zeta}(i) \Leftrightarrow x \notin \zeta(i) \quad \Leftrightarrow \quad \langle u[i], \xi[i] \upharpoonright ar(\rho(x))\rangle \notin \widetilde{\rho(x)}$$
$$\Leftrightarrow \quad \langle u[i], \overline{\xi}[i] \upharpoonright ar(\rho(x))\rangle \in \rho(x).$$

It follows that $\overline{\zeta} = \zeta'$ and we get

$$\langle u, \xi\rangle \in \tilde{C}[\tilde{\rho}] \quad \Leftrightarrow \quad \langle u, \zeta\rangle \in \tilde{C} \Leftrightarrow \langle u, \zeta'\rangle \notin C$$
$$\Leftrightarrow \quad \langle u, \overline{\xi}\rangle \notin C' \Leftrightarrow \langle u, \xi\rangle \in \widetilde{C'}.$$

□

Proposition 5.3.14. $\widetilde{id} = id$.

Proof. For any x, $\langle u, \psi\rangle$ (with $\psi : \mathbb{N} \to \mathcal{P}(\{x\})$) is in the dual of $\hat{x}$ if and only if $\langle u, \overline{\psi}\rangle \notin \hat{x}$ if and only if $x \notin \overline{\psi}(0)$ if and only if $x \in \psi(0)$ if and only if $\langle u, \xi\rangle \in \hat{x}$. □

Proposition 5.3.15. $\widetilde{\mu x.C} = \nu x.\tilde{C}$, $\widetilde{\nu x.C} = \mu x.\tilde{C}$.

Proof. The second equality is a consequence of the first one and of the fact that $\tilde{\tilde{C}} = C$.

If $x \notin ar(C)$ then $\widetilde{\mu x.C} = \nu x.\tilde{C} = \tilde{C}$. let us consider the case where $x \in ar(C)$.

First we remark that for $\xi : \mathbb{N} \to \mathcal{P}(X - \{x\})$ and $\psi : \mathbb{N} \to \mathcal{P}(\{x\})$, $\overline{\xi \cup \psi} = \overline{\xi} \cup \overline{\psi}$ where the complements are implicitly taken in the adequate sets of variables.

Using the definitions of $\mu x.C$ and $\nu x.C$ we get $\langle u, \xi\rangle \in \widetilde{\mu x.C}$

$$\begin{aligned}
&\Leftrightarrow \quad \langle u, \overline{\xi}\rangle \notin \mu x.C\\
&\Leftrightarrow \quad \exists\psi : \langle u, \overline{\xi} \cup \psi\rangle \notin C \,\&\, \forall i, \langle u[i], \overline{\xi}[i] \cup \psi[i]\rangle \in C \Rightarrow x \in \psi(i)\\
&\Leftrightarrow \quad \exists\psi : \langle u, \xi \cup \overline{\psi}\rangle \in \tilde{C} \,\&\, \forall i, \langle u[i], \xi[i] \cup \overline{\psi}[i]\rangle \notin \tilde{C} \Rightarrow x \in \psi(i)\\
&\Leftrightarrow \quad \exists\psi : \langle u, \xi \cup \overline{\psi}\rangle \in \tilde{C} \,\&\, \forall i, x \in \overline{\psi}(i) \Rightarrow \langle u[i], \xi[i] \cup \overline{\psi}[i]\rangle \in \tilde{C}\\
&\Leftrightarrow \quad \exists\psi : \langle u, \xi \cup \psi\rangle \in \tilde{C} \,\&\, \forall i, x \in \psi(i) \Rightarrow \langle u[i], \xi[i] \cup \psi[i]\rangle \in \tilde{C}\\
&\Leftrightarrow \quad \langle u, \xi\rangle \in \nu x.\tilde{C}.
\end{aligned}$$

□

Now, we can extend Proposition 5.3.11 (page 132).

Proposition 5.3.16. *The constrained language $\mu x.C$ is the least fixed point of the mapping $g : \mathcal{CL}(X) \to \mathcal{CL}(X)$ defined by $g(D) = C[id\{D/x\}]$, where $X = ar(C) - \{x\}$.*

Proof. We know that $\mu x.C$ is the dual of $\nu x.\tilde{C}$. By Proposition 5.3.11 (132), $\nu x.\tilde{C}$ is the greatest fixed point of h defined by $h(D) = \tilde{C}[id\{D/x\}]$. Using the properties of duals mentioned above, we get $\nu x.\tilde{C} = \tilde{C}[id\{\nu x.\tilde{C}/x\}]$. It follows that $\mu x.C = C[id\{\mu x.C/x] = g(\mu x.C)$. Now, let D be such that $D = g(D) = C[id\{D/x\}]$. We get $\tilde{D} = \tilde{C}[\{\tilde{D}/x\}] = h(\tilde{D})$, hence $\nu x.\tilde{C} \supseteq \tilde{D}$. It follows that the dual of $\nu x.\tilde{C}$ is included in D. □

Finally, with each term t we associate its *syntactic dual* $\tilde{t}$ defined by induction as follows.

- If $t = x$ then $\tilde{t} = x$.
- If $t = \varepsilon$ then $\tilde{t} = \nu x. \bigvee_{a \in A}(ax \vee a\varepsilon)$.
- If $t = at'$ then $\tilde{t} = \varepsilon \vee a\tilde{t'} \vee \bigvee_{b \neq a} b(\nu x.x)$.
- If $t = t' \vee t''$ (resp. $t' \wedge t''$) then $\tilde{t} = \tilde{t'} \wedge \tilde{t''}$ (resp. $\tilde{t'} \vee \tilde{t''}$).
- If $t = \mu x.t'$ (resp $\nu x.t'$) then $\tilde{t} = \nu x.\tilde{t'}$ (resp. $\mu x.\tilde{t'}$).

It is easy to see that $ar(\tilde{t}) = ar(t)$.

5.3.3 Interpretation of terms by constrained languages

Let x and y be two fixed variables, and let us define the following constrained languages (still identifying a constrained language with its second component when its arity is defined).

- For any letter a, let C_a be the constrained language of arity $\{x\}$ such that $\langle u, \xi\rangle \in C_a \Leftrightarrow u(0) = a$ and $\xi(1) = \{x\}$.
- $C_\vee$ is the constrained language of arity $\{x, y\}$ such that $\langle u, \xi\rangle \in C_\vee \Leftrightarrow \xi(0) \neq \emptyset$.
- $C_\wedge$ is the constrained language of arity $\{x, y\}$ such that $\langle u, \xi\rangle \in C_\wedge \Leftrightarrow \xi(0) = \{x, y\}$.

With each term t of $\mathit{fix}\mathcal{T}(A_{\{\vee,\wedge\}})$, we associate the constrained language $\sigma(t)$ of arity $ar(t)$ defined by induction as follows.

Definition 5.3.17.

- $\sigma(x) = \{\hat{x}\}$,
- $\sigma(\varepsilon) = \langle\emptyset, \{\langle\varepsilon, \emptyset^\omega\rangle\}\rangle$.
- $\sigma(at) = C_a[id\{\sigma(t)/x\}]$,
- $\sigma(t \vee t') = C_\vee[id\{\sigma(t)/x, \sigma(t')/y\}]$,
- $\sigma(t \wedge t') = C_\wedge[id\{\sigma(t)/x, \sigma(t')/y\}]$,
- $\sigma(\theta x.t) = \theta x.\sigma(t)$.

It is easy to check that $\sigma(t)$ is a constrained language and that

- $\sigma(at) = \{\langle au, \xi\rangle \mid \langle u, \xi[1)\rangle \in \sigma(t)\}$,
- $\sigma(t \wedge t') = \{\langle u, \xi\rangle \mid \langle u, \xi\upharpoonright ar(t)\rangle \in \sigma(t)$ or $\langle u, \xi\upharpoonright ar(t')\rangle \in \sigma(t')\}$,

- $\sigma(t \wedge t') = \{\langle u, \xi\rangle \mid \langle u, \xi \upharpoonright ar(t)\rangle \in \sigma(t)$ and $\langle u, \xi \upharpoonright ar(t')\rangle \in \sigma(t')\}$.

Proposition 5.3.18. *σ is a homomorphism from the μ-calculus of terms to the μ-calculus of constrained languages.*

Proof. It is sufficient to show by induction on t, that $\sigma(t[\rho]) = \sigma(t)[\sigma(\rho)]$.

- Since $x[\rho] = \rho(x)$, $\sigma(x[\rho]) = \sigma(\rho)(x)$, while $\sigma(x)[\sigma(\rho)] = \hat{x}[\sigma(\rho)] = \sigma(\rho)(x)$.
- $\sigma(\varepsilon[\rho]) = \sigma(\varepsilon) = \sigma(\varepsilon)[\sigma(\rho)]$,
- For $t = at'$, $t' \vee t''$, $t' \wedge t''$, this is a consequence of the associativity of composition.
- Since $(\theta x.t)[\rho] = \theta z.(t[\rho\{\hat{z}/x\}])$ for some $z \notin ar(\theta x.t)$, we get $\sigma((\theta x.t)[\rho]) = \theta z.\sigma(t[\rho\{\hat{z}/x\}])$.
  By the induction hypothesis, $\theta z.\sigma(t[\rho\{\hat{z}/x\}]) = \theta z.(\sigma(t)[\sigma(\rho)\{\hat{z}/x\}])$. On the other hand, $\sigma(\theta x.t)[\sigma(\rho)] = (\theta x.\sigma(t))[\sigma(\rho)]$, and since $z \notin ar(\theta x.\sigma(t)) = ar(\theta x.t)$, we get $(\theta x.\sigma(t))[\sigma(\rho)] = \theta z.\sigma(t[\rho\{\hat{z}/x\}])$. □

We can also easily prove the following by induction on t.

Proposition 5.3.19. $\sigma(\tilde{t}) = \widetilde{\sigma(t)}$

Finally let us define the bijection $\sigma' : \mathcal{P}(A^\infty) \to \mathcal{CL}(\emptyset)$ defined by $\sigma'(L) = \langle \emptyset, \{\langle u, \phi\rangle \mid u \in L\}\rangle$ where ϕ is defined by $\phi(i) = \emptyset$. With a valuation $v : Var \to \mathcal{P}(A^\infty)$, we associate the valuation $\sigma'(v) \to \mathcal{CL}(\emptyset)$ defined by $\sigma'(v)(x) = \sigma'(v(x))$.

The following result can be easily proved by induction on t.

Proposition 5.3.20. *For any valuation $v : Var \to \mathcal{P}(A^\infty)$, and for any term $t \in fix\mathcal{T}(A_{\{\vee,\wedge\}})$,*

$$\sigma'(\llbracket t \rrbracket[v]) = \sigma(t)[\sigma'(v)].$$

In particular, if t is closed, $\sigma'(\llbracket t \rrbracket) = \sigma(t)$.

5.3.4 Recognizable constrained languages

In th[is s]ection we show that the constrained language $\sigma(t)$ associated with a term $t \in fix\mathcal{T}(A_{\{\vee,\wedge\}})$ (see Definition 5.3.17, page 135) is recognizable. Then we [d]educe that for any closed term t, $\llbracket t \rrbracket$ is a rational subset of A^∞. For any finite set X of variables, let M_X be the set

$$(A \times \mathcal{P}(X))^*\mathcal{P}(X)^\omega \cup (A \times \mathcal{P}(X))^\omega$$

which can be considered as a recognizable language over the alphabet $(A \times \mathcal{P}(X)) \cup \mathcal{P}(X)$. An element of M_X is written as a pair of words $\langle u, \xi\rangle$ with

$u \in A^\infty$, $\xi \in \mathcal{P}(X)^\omega$, i.e., ξ is a mapping from $\mathbb{N}$ to $\mathcal{P}(X)$, so that any constrained language C of arity X can be identified with a subset of M_X. We say that a constrained language C of arity X is recognizable if it is identified with a recognizable subset of words over the alphabet $((A \times \mathcal{P}(X)) \cup \mathcal{P}(X)$.

For any subset Y of X we define

$$h_Y : (A \times \mathcal{P}(X)) \cup \mathcal{P}(X) \to (A \times \mathcal{P}(Y)) \cup \mathcal{P}(Y)$$

by $h_Y(\langle a, Z\rangle) = \langle a, Z \cap Y\rangle$ and $h_Y(Z) = Z \cap Y$. This mapping is extended into a homomorphism $h_Y : M_X \to M_Y$, as usual.

Proposition 5.3.21. *For any term $t \in \mathit{fix}\mathcal{T}(A_{\{\vee,\wedge\}})$, $\sigma(t)$ is a recognizable language included in $M_{ar(t)}$.*

For any closed term $t \in \mathit{fix}\mathcal{T}(A_{\{\vee,\wedge\}})$, $|t|$ is a recognizable language included in A^∞.

Proof. The proof is by induction on t.

If $t = x$ then $\sigma(t) = \hat{x}$ which is obviously recognizable (just check that x occurs in the first position of an element of $M_{\{x\}}$). If $t = \varepsilon$, $\sigma(t)$ is identified with $\emptyset^\omega$.

If $t = at'$ with $ar(t') = X$, then $\sigma(t) = \{\langle au, Y\xi\rangle \mid \langle u, \xi\rangle \in \sigma(t'), Y \subseteq X\}$.

If $t = t' \vee t''$, with $X' = ar(t')$, $X'' = ar(t'')$, then

$$\begin{aligned}\sigma(t) &= \{\langle u, \xi\rangle \mid \langle u, \xi \upharpoonright X'\rangle \in \sigma(t') \text{ or } \langle u, \xi \upharpoonright X''\rangle \in \sigma(t'')\}\\ &= h_{X'}^{-1}(\sigma(t')) \cup h_{X''}^{-1}(\sigma(t'')),\end{aligned}$$

which is recognizable. Similarly, $\sigma(t' \wedge t'') = h_{X'}^{-1}(\sigma(t')) \cap h_{X''}^{-1}(\sigma(t''))$.

By definition of $\nu x.\sigma(t)$ (page 130), $\langle u, \xi\rangle \in \sigma(\nu x.t) = \nu x.\sigma(t)$ if and only if there exists $\psi : \mathbb{N} \to \mathcal{P}(\{x\})$ such that

- $\langle u, \xi \cup \psi\rangle \in \sigma(t)$,
- $\forall i \in \mathbb{N}$, $x \in \psi(i) \Rightarrow \langle u[i), \xi[i) \cup \psi[i)\rangle \in \sigma(t)$.

Let $Y = ar(t)$ and $X = ar(\nu x.t) = Y - \{x\}$. Since $\xi = (\xi \cup \psi) \upharpoonright X$ and $\psi = (\xi \cup \psi) \upharpoonright \{x\}$, we have $\nu x.\sigma(t) = h_X(K)$ where

$$K = \{\langle u, \zeta\rangle \in \sigma(t) \mid \forall i,\ x \in \zeta(i) \Rightarrow \langle u[i), \zeta[i)\rangle \in \sigma(t)\}.$$

But $K = \sigma(t) - K'$ where

$$\begin{aligned}K' &= \{\langle u, \zeta\rangle \in \sigma(t) \mid \exists i : x \in \zeta(i) \,\&\, \langle u[i), \zeta[i)\rangle \notin \sigma(t)\}\\ &= \{\langle u, \zeta\rangle \in \sigma(t) \mid \exists i : x \in \zeta(i) \,\&\, \langle u[i), \zeta[i)\rangle \in M_Y - \sigma(t)\}.\end{aligned}$$

Since, by the induction hypothesis, $\sigma(t)$ is recognizable by an automaton $\mathcal{A}$, $M_Y - \sigma(t)$ is recognizable by an automaton $\mathcal{B}$. It follows that K' is recognized by an automaton $\mathcal{C}$ obtained as a product of $\mathcal{A}$ and a variant of $\mathcal{B}$: $\mathcal{A}$

reads the input word while $\mathcal{B}$ stays in its initial state. At some nondeterministically chosen position i such that $x \in \zeta(i)$, $\mathcal{A}$ goes on reading and $\mathcal{B}$ starts reading $\langle u[i], \zeta[i] \rangle$.

Since K' is recognizable, $K = \sigma(t) - K'$ and $\nu x.\sigma(t) = h_X(K)$ are also recognizable.

Since $\sigma(\mu x.t) = \mu x.\sigma(t) = \widetilde{\nu x.\widetilde{\sigma(t)}}$ it is sufficient to show that if D is a recognizable constrained language, so is $\tilde{D}$. Indeed, if D is recognizable, then $M_{ar(D)} - D$ is recognized by an automaton $\mathcal{A} = \langle S, T, I, C \rangle$. Then $\tilde{D}$ is recognized by $\mathcal{A}' = \langle S, T', I, C \rangle$ where

$$T' = \{\langle s, \langle a, Z \rangle, s' \rangle \mid \langle s, \langle a, \overline{Z} \rangle, s' \rangle \in T\} \cup \{\langle s, Z, s' \rangle \mid \langle s, \overline{Z}, s' \rangle \in T\}.$$

Finally, if t is closed, we have $\sigma'(\llbracket t \rrbracket) = \sigma(t)[\sigma'(v)] = \sigma(t)$ which is recognizable. But $\sigma'(\llbracket t \rrbracket) = \{\langle u, \phi \rangle \mid u \in \llbracket t \rrbracket\}$. It follows that $\llbracket t \rrbracket$ is the image of $\sigma(t)$ under the mapping that sends $\langle a, \emptyset \rangle$ on A, and is recognizable by Theorem 5.2.4 (page 117). □

5.4 Bibliographic notes and sources

We refer the reader to a comprehensive presentation of the subject of automata on infinite words by Thomas [95, 97]. The idea of finite–state recognizability of infinite objects is due to Büchi who used finite automata running over infinite words as a tool for proving decidability of the monadic second-order theory of the natural numbers with successor [21]. Further variations of this concept by different acceptance criteria were introduced by Muller [70] and McNaughton in connection with the search of deterministic counterpart of Büchi automata. This was accomplished by the McNaughton's theorem [61] which we have presented in Section 5.2.5 (page 125).

The parity acceptance criterion appeared, to our knowledge, for the first time essentially in the work by K. Wagner [100] (see also [101]), and has been later on reinvented by several authors [65, 33, 62]; the actual appealing formulation in terms of a rank function is due to Emerson and Jutla [33].

It is worth to note that an application to infinite games shed a new light on different acceptance criteria, by revealing their connection with the issue of the memory of strategies (see bibliographic notes in Chapter 4). While the Muller criterion always guarantees the existence of a finite–state strategy (*viz* forgetful, in terms of Gurevich and Harrington) for the winner, the parity criterion guarantees the existence of a positional (*viz* memoryless) strategy. For a game with Rabin criterion, Klarlund [52] showed that if a winner plays for the criterion (but not for its complement), her strategy can also be made positional. Moreover, viewing the Rabin and parity criteria as special cases of Muller criterion, the above conditions are also necessary for the respective properties of the strategies to hold in any game (see [105]).

The idea of characterizing finite–state recognizability by algebraic fixed-point expressions can be traced back to the fundamental work by Kleene [53], introducing regular expressions. Indeed, the star operator is definable by a least fixed point, as demonstrated by the Arden's Lemma [2]. The operator L^{ω} is implicit in the very concept of Büchi automaton; more complex expressions defining ω-regular languages by means of greatest fixed points appeared in [7]. A complete characterization of ω-regular languages by fixed point terms without intersection (summarized in Theorems 5.2.11 and 5.2.13 above) was given by Park [81], who also showed that the alternation hierarchy collapses to the level Π_2 (Proposition 5.1.16 above; as reported in [81], the results were obtained in collaboration with J. Tiuryn). The characterization of ω-regular languages by means of fixed-point terms with intersection of the alternation–free level $Comp\,(\Sigma_1 \cup \Pi_1)$ was given in [10].

6. The μ-calculus over powerset algebras

In Chapter 2, Section 2.3.3 (page 49), we have discussed the interplay of syntax and semantics of the μ-calculus on an abstract level, using the concept of a μ-interpretation, the universe of which can be an arbitrary complete lattice. We have subsequently observed that any μ-interpretation is in some sense equivalent to a powerset interpretation, i.e., one whose universe consists of a powerset (see Section 2.5). In Chapters 3 and 5 we have studied two important powerset interpretations: the Boolean algebra of logical values (the universe of $\mathbb{B}$ can be viewed as the powerset of a singleton set), and the lattice of languages of (finite or infinite) words. Now we are going to concentrate on a wide subclass of powerset interpretations, subsuming those two and many others (in particular, the modal μ-calculus of [55]). The main feature of these interpretations, is that the basic operations are not only monotonic, but also additive; in other words, they can be constructed "from below", from some underlying simpler structure. Our interest in such interpretations is motivated mainly by the considerations of the next chapter, where we will establish a connection between the μ-calculus, and a very general concept of automata. But we will show some applications of the new concept already in this chapter, by generalizing the notion of a bisimulation between transition systems.

6.1 Powerset algebras

6.1.1 Semi-algebras

A *signature* (sometimes called *type*) is any finite set of function symbols $Sig \subseteq Fun$. Recall (see Section 2.3.1, page 47) that each symbol f in *Fun* has a fixed finite arity; we will denote it by $\rho(f)$. According to the common sense of the word, an *algebra* over *Sig* consists of a set (universe), and an interpretation of each symbol in *Sig* by a function (operation) of corresponding arity over this universe. In the more general case of a *partial algebra*, operations need not to be everywhere defined. We will consider a yet more general situation where the operations may be also over-defined.

Definition 6.1.1. A semi-algebra over a signature *Sig* is presented by $\mathcal{B} = \langle B, \{f^{\mathcal{B}} \mid f \in Sig\}\rangle$, where B is a set called the *universe* of $\mathcal{B}$, and, for each $f \in Sig$ of arity $\rho(f)$, $f^{\mathcal{B}}$ is a relation over B of arity $\rho(f) + 1$, that is, $f^{\mathcal{B}} \subseteq B^{\rho(f)+1}$. We call $f^{\mathcal{B}}$ a $\rho(f)$-ary *basic operation* from $B^{\rho(f)}$ to B, and write $b \doteq f(b_1, \dots, b_{\rho(f)})$ to mean $(b_1, \dots, b_{\rho(f)}, b) \in f^{\mathcal{B}}$. Note that a basic operation need not be a total function from $B^{\rho(f)}$ to B; if it is the case for each $f \in Sig$, a semi-algebra is an *algebra* over *Sig* (briefly: *Sig*-algebra), and then clearly $\doteq$ coincides with $=$.

Example 6.1.2. A directed graph $G = \langle V, E\rangle$ with a set of vertices V and a set of edges $E \subseteq V \times V$ can be viewed as a semi-algebra over a signature consisting of one unary symbol, say $Sig = \{f\}$. The universe of this semi-algebra is V and $f^G = E$. Note that $f^G(v) \doteq w$ if and only if there is an edge from v to w.

Example 6.1.3. A *syntactic tree* over an arbitrary signature *Sig* can be presented as a mapping $t : \mathrm{dom}\ t \to Sig$, where dom t is a set of finite sequences (i.e., words) over natural numbers, $\mathrm{dom}\ t \subseteq \mathbb{N}^*$, and moreover the following conditions are satisfied.

1. dom t is closed under prefixes: if $xy \in \mathrm{dom}\ t$ then $x \in \mathrm{dom}\ t$. In particular, the empty word ε is always in dom t.
2. If $w \in \mathrm{dom}\ t$ and $t(w)$ has arity k then $wi \in \mathrm{dom}\ t$ if and only if $1 \le i \le k$.

Note that $w \in \mathrm{dom}\ t$ is a *leaf* of the tree (a node without successors) if and only if $t(w)$ has arity 0, i.e. is a constant symbol.

Finite syntactic trees coincide with closed terms over *Sig* in an obvious manner, for example a term $f(c, f(c, d))$ can be presented as the tree t with $\mathrm{dom}\ t = \{\varepsilon, 1, 2, 2\cdot 1, 2\cdot 2\}$, such that $t(\varepsilon) = t(2) = f$, $t(1) = t(2\cdot 1) = c$, and $t(2\cdot 2) = d$. Therefore, infinite syntactic trees can be viewed as infinite terms.

A syntactic tree t can be organized into a semi-algebra

$$\mathbf{t} = \langle \mathrm{dom}\ t, \{f^{\mathbf{t}} \mid f \in Sig\}\rangle$$

of universe dom t, where, for each $f \in Sig$,

$$f^{\mathbf{t}} = \{(w1, \dots, w\rho(f), w) \mid t(w) = f\}.$$

Note that $w \doteq f^{\mathbf{t}}(v_1, \dots, v_k)$ only if $t(w) = f$ and $v_1, \dots, v_k$ are the successors of w. This $\mathbf{t}$ is in general not an algebra since no operation is defined at the root ε, unless t is a constant. For $f \in Sig$ of arity $\rho(f) > 0$, the basic operation $f^{\mathbf{t}}$ is a partial function from $(\mathrm{dom}\ t)^{\rho(f)}$ to dom t, but if $\rho(f) = 0$, $f^{\mathbf{t}} \subseteq \mathrm{dom}\ t$ is not a partial function of arity 0 whenever it contains at least two elements.

Let T_{Sig} denote the set of all syntactic trees over *Sig*. This set can be also organized into a semi-algebra (which is actually an algebra), $\mathcal{T}_{Sig} = \langle T_{Sig}, \{f^{\mathcal{T}_{Sig}} \mid f \in Sig\}\rangle$, where, for $f \in Sig$ and $t_1, \ldots, t_{\rho(f)} \in T_{Sig}$, $f^{\mathcal{T}_{Sig}}(t_1, \ldots, t_{\rho(f)})$ is the syntactic tree of domain

$$\{\varepsilon\} \cup \bigcup_{i=1,\ldots,\rho(f)} \{iw \mid w \in \mathrm{dom}\ t_i\}$$

defined by

$$\begin{aligned} t(\varepsilon) &= f \\ t(iw) &= t_i(w) \text{ for } w \in \mathrm{dom}\ t_i \end{aligned}$$

The finite syntactic trees form a subalgebra of this algebra which coincides with the classical free algebra of closed terms. Therefore, $\mathcal{T}_{Sig}$ can be viewed as an algebra of infinite terms.

6.1.2 Powerset algebras

For each signature *Sig* we fix a signature $Sig_{\sim}$ which can be presented by $Sig_{\sim} = Sig \cup \{\tilde{f} \mid f \in Sig\}$, where, for each $f \in Sig$, $\tilde{f}$ is a symbol not in *Sig* of arity $\rho(\tilde{f}) = \rho(f)$.

Definition 6.1.4. Let $\mathcal{B} = \langle B, \{f^{\mathcal{B}} \mid f \in Sig\}\rangle$ be a semi-algebra over signature *Sig*. The *powerset algebra* over $\mathcal{B}$ is an algebra over the signature $Sig_{\sim}$ of the form

$$\wp\mathcal{B} = \langle \mathcal{P}(B), \{f^{\wp\mathcal{B}} \mid f \in Sig\} \cup \{\tilde{f}^{\wp\mathcal{B}} \mid f \in Sig\}\rangle$$

where $\mathcal{P}(B)$ is the set of all subsets of B, and, for each $f \in Sig$, and for $L_1, \ldots, L_{\rho(f)} \subseteq B$,

$$f^{\wp\mathcal{B}}(L_1, \ldots, L_{\rho(f)}) =$$
$$\{b \mid \exists a_1 \in L_1, \ldots, \exists a_{\rho(f)} \in L_{\rho(f)} : f^{\mathcal{B}}(a_1, \ldots, a_{\rho(f)}) \doteq b\},$$

$$\tilde{f}^{\wp\mathcal{B}}(L_1, \ldots, L_{\rho(f)}) =$$
$$\{b \mid \forall a_1, \ldots, a_{\rho(f)} \in B,\ f^{\mathcal{B}}(a_1, \ldots, a_{\rho(f)}) \doteq b \Rightarrow a_i \in L_i \text{ for some } i\}.$$

Clearly, $\langle \mathcal{P}(B), \subseteq\rangle$ is a complete lattice and the operations $f^{\wp\mathcal{B}}$ and $\tilde{f}^{\wp\mathcal{B}}$, for $f \in Sig$, are monotonic with respect to $\subseteq$. Therefore the powerset algebra of $\mathcal{B}$ constitutes a μ-interpretation of $Sig_{\sim}$ in the sense of Definition 2.3.7 (page 49). It induces an interpretation of the fixed-point terms over $Sig_{\sim}$, according to Definition 2.3.8 (page 49). We shall denote this interpretation by $[\![t]\!]_{\wp\mathcal{B}}$.

Considering $\langle\mathcal{P}(B),\subseteq\rangle$ as a Boolean algebra, we can further observe that the operation $\tilde{f}^{\wp\mathcal{B}}$ is dual to $f^{\wp\mathcal{B}}$ in the sense of Definition 1.2.27, (page 18) since

$$\tilde{f}^{\wp\mathcal{B}}(L_1,\ldots,L_{\rho(f)}) \quad = \quad \overline{f^{\wp\mathcal{B}}(\overline{L_1},\ldots,\overline{L_{\rho(f)}})}$$

for all $L_1,\ldots,L_{\rho(f)}\subseteq B$.

This remark allows us to syntactically define the dual $\tilde{t}$ of a fixed-point term t: it is a term obtained by exchanging in t, μ and ν as well as f and $\tilde{f}$. More precisely, we define $\tilde{t}$ by induction of t as follows.

Definition 6.1.5. The *dual* $\tilde{t}$ of a term t is defined inductively by

- if t is a variable x, , $\tilde{t}=t=x$,
- if $t=f(t_1,\ldots,t_{\rho(f)})$ then $\tilde{t}=\tilde{f}(\widetilde{t_1},\ldots,\widetilde{t_{\rho(f)}})$,
- if $t=\tilde{f}(t_1,\ldots,t_{\rho(f)})$ then $\tilde{t}=f(\widetilde{t_1},\ldots,\widetilde{t_{\rho(f)}})$,
- if $t=\mu x.t'$ (resp. $\nu x.t'$) then $\tilde{t}=\nu x.\widetilde{t'}$ (resp. $\mu x.\widetilde{t'}$).

As an immediate consequence, we get:

Proposition 6.1.6. *The mapping* $[\![\tilde{t}]\!]_{\wp\mathcal{B}}:\mathcal{P}(B)^{ar(t)}\to\mathcal{P}(B)$ *is the dual of* $[\![t]\!]_{\wp\mathcal{B}}$. *In particular, if t is closed, then* $[\![\tilde{t}]\!]_{\wp\mathcal{B}}=\overline{[\![t]\!]_{\wp\mathcal{B}}}$.

6.1.3 Logical operations

The lattice operations $\vee$ and $\wedge$ in $\langle\mathcal{P}(B),\subseteq\rangle$ coincide with the usual set-theoretical operations of union and intersection. We will see that they can also be "constructed from below" if the underlying semi-algebra is enriched by one particular operation.

Let $\mathcal{B}$ be a semi-algebra over signature *Sig*, and let $eq\in \mathit{Fun}$ be a symbol of arity 2 not in *Sig*. We let $\mathcal{B}_{eq}$ be a semi-algebra over signature $\mathit{Sig}\cup\{eq\}$, such that, for $f\in \mathit{Sig}$, the interpretation $f^{\mathcal{B}_{eq}}$ coincides with $f^{\mathcal{B}}$, and $eq^{\mathcal{B}_{eq}}=\{(b,b,b)\mid b\in B\}$. In other words, $eq^{\mathcal{B}_{eq}}(a,b)\doteq c$ if and only if $a=b=c$. Then, in the powerset algebra of $\mathcal{B}_{eq}$, we have

$$eq^{\wp\mathcal{B}_{eq}}(L_1,L_2) \quad = \quad L_1\cap L_2$$

and

$$\widetilde{eq}^{\wp\mathcal{B}_{eq}}(L_1,L_2) \quad = \quad L_1\cup L_2$$

for any $L_1,L_2\subseteq B$.

Example 6.1.7. Let *Sig* be a signature whose only symbol is *eq*. Consider a semi-algebra $\mathcal{B}$ over this signature, whose universe consists of a single element, say $B = \{*\}$, and *eq* is interpreted as above (which, in this case is the only nontrivial interpretation). Then it is plain to see that the powerset algebra $\wp\mathcal{B}$ is isomorphic to the Boolean algebra $\mathbb{B}$ of Chapter 3 (via the mapping $\emptyset \mapsto 0$, $\{*\} \mapsto 1$), up to the identification of the symbols *eq* with $\wedge$, and $\tilde{eq}$ with $\vee$.

6.2 Modal μ-calculus

The propositional modal μ-calculus introduced by Kozen [55] is perhaps the best known realization of the μ-calculus. It extends the propositional modal logic, possibly with many modalities (sometimes referred to as system K) by the least fixed-point operator; the greatest fixed point is definable by the duality law : $\nu X.p(X) = \neg\mu X.\neg p(\neg X)$ (see Section 1.2.4,page 17). In our setting, the Kozen's logic can be presented as a calculus of fixed points interpreted in a powerset algebra, under the only restriction that all symbols in the signature except for *eq* have arity at most one.

To be more precise, let us first recall the classical definition. Let *Prop* ("propositions") and *Act* ("actions") be two finite sets of symbols. Let *Var*, as always, be a countably infinite set of variables. The *formulas* of the modal μ-calculus are defined inductively by the following clauses.

- Any variable $x \in \mathit{Var}$ is a formula (called, in this context, a propositional variable).
- For a proposition $p \in \mathit{Prop}$, p, as well as $\bar{p}$, are formulas.
- If φ, ψ are formulas, so are $(\varphi \vee \psi)$ and $(\varphi \wedge \psi)$.
- If φ is a formula and $a \in \mathit{Act}$ an action then $\langle a \rangle \varphi$, as well as $[a]\varphi$, are formulas.
- If φ is a formula and $x \in \mathit{Var}$ a variable, then $\mu x.\varphi$, as well as $\nu x.\varphi$, are formulas.

The *semantics* of the calculus is provided by *Kripke structures*. Such a structure can be presented as a tuple

$$M = \langle S^M, \{p^M \subseteq S^M \mid p \in \mathit{Prop}\}, \{a^M \subseteq S^M \times S^M \mid a \in \mathit{Act}\}\rangle$$

where S^M is an underlying set of *states* (or worlds), and the p^M's and a^M's are interpretations of propositions and actions, respectively. Note that M can be viewed as a labeled graph (or a transition system) whose nodes (S^M) are labeled by sets of propositions, and edges by actions.

The *interpretation* of formulas is defined relatively to a valuation $\mathit{val} : \mathit{Var} \to \mathcal{P}(S^M)$. The interpretation $[\varphi]_M[\mathit{val}]$ of a formula φ is a set of states; the relation $s \in [\![\varphi]\!]_M[val]$ is also written by $M, val, s \models \varphi$, and read: φ is true in M at the state s, under the valuation val. It is defined inductively by the following clauses.

- $M, val, s \models p$ if and only if $s \in val(p)$,
- $M, val, s \models \bar{p}$ if and only if $s \notin val(p)$,
- $M, val, s \models (\varphi \vee \psi)$ if and only if $M, val, s \models \varphi$ or $M, val, s \models \psi$,
- $M, val, s \models (\varphi \wedge \psi)$ if and only if $M, val, s \models \varphi$ and $M, val, s \models \psi$,
- $M, val, s \models \langle a \rangle \varphi$ if and only if there exists s' such that $(s, s') \in a^M$ and $M, val, s' \models \varphi$,
- $M, val, s \models [a]\varphi$ if and only if for all s', $(s, s') \in a^M$ implies $M, val, s' \models \varphi$,
- $M, val, s \models \mu x.\varphi$ if and only if for all $W \subseteq S^M$, $[\![\varphi]\!]_M[val\{W/x\}] \subseteq W$ implies $s \in W$,
- $M, val, s \models \nu x.\varphi$ if and only if there exists W, such that $s \in W$ and $W \subseteq [\![\varphi]\!]_M[val\{W/x\}]$.

Note that, by the Knaster–Tarski theorem (Theorem 1.2.8, page 10), the last two conditions mean that the state s belongs to the least (respectively, the greatest) fixed point (in $\mathcal{P}(S^M)$) of the operator $W \mapsto [\![\varphi]\!]_M[val\{W/x\}]$ induced by the formula φ and the variable x.

Remark. In the above presentation, the negation operator is absent, except for that $\bar{p}$ can be viewed as the negation of p. Some authors include the negation operator in the language, by letting $(\neg\varphi)$ be a formula whenever φ is, and setting, of course, $M, val, s \models \neg\varphi$ if and only if $M, val, s \not\models \varphi$. But then the clauses concerning fixed-point formulas need to be modified, in order to guarantee the existence of the fixed points, e.g., by forcing monotonicity of the induced operators. To this end, it is usually assumed that $\mu x.\varphi$ (as well $\nu x.\varphi$) is a formula only if x occurs under an even number of negations. Note that the new set of formulas is an extension of the previous one (up to identifying $\bar{p}$ with $(\neg p)$). In the extended logic, one of the modalities, as well as one of the fixed-point operators, become redundant, because a formula $[a]\varphi$ is semantically equivalent to $\neg\langle a \rangle(\neg\varphi)$, and $\nu x.\varphi(x)$ is equivalent to $\neg\mu x.\varphi(\neg x)$. On the other hand, any *sentence* (i.e., a formula without free variables) of the extended logic can be transformed to a sentence in which the negation is applied only to propositional variables, i.e., to a sentence of the μ-calculus that we have presented above. It can be achieved by using the aforementioned equivalences, as well as the De Morgan laws, and the size of the resulted formula increases at most by a constant factor. Therefore, since we are ultimately interested in sentences, the absence of the explicit negation is not an essential restriction.

The propositional modal μ-calculus can be presented in our setting as follows. Let $Sig = Prop \cup Act \cup \{eq\}$ be a signature in which the propositions are considered of arity 0, and actions of arity 1. Then, there is a straightforward translation *f2t* of the formulas of the modal μ-calculus into the fixed-point terms over signature $Sig_\sim$, as well as a converse translation $t2f$ (terms to formulas). The $f2t$ mapping is given by the following rules.

$$
\begin{array}{lll}
f2t & : x \mapsto x & \\
 & : p \mapsto p & : \tilde{p} \mapsto \tilde{p} \\
 & : (\varphi \wedge \psi) \mapsto eq(f2t(\varphi), f2t(\psi)) & : (\varphi \vee \psi) \mapsto \tilde{eq}(f2t(\varphi), f2t(\psi)) \\
 & : \langle a \rangle \varphi \mapsto a(f2t(\varphi)) & : [a]\varphi \mapsto \tilde{a}(f2t(\varphi)) \\
 & : \mu x.\varphi \mapsto \mu x.f2t(\varphi) & : \nu x.\varphi \mapsto \nu x.f2t(\varphi)
\end{array}
$$

The converse transformation is analogous:

$$
\begin{array}{lll}
t2f & : x \mapsto x & \\
 & : p \mapsto p & : \tilde{p} \mapsto \tilde{p} \\
 & : eq(t_1, t_2) \mapsto (t2f(t_1) \wedge t2f(t_2)) & : \tilde{eq}(t_1, t_2) \mapsto (t2f(t_1) \vee t2f(t_2)) \\
 & : a(t) \mapsto \langle a \rangle t2f(t) & : \tilde{a}(t) \mapsto [a]t2f(t) \\
 & : \mu x.t \mapsto \mu x.t2f(t) & : \nu x.t \mapsto \nu x.t2f(t)
\end{array}
$$

It is plain to see that $t2f \circ f2t$ and $f2t \circ t2f$ are identity mappings on their respective domains.

Now, a Kripke structure M as above can be identified with a semi-algebra over Sig, let us denote it by $\mathcal{K}_M$, as follows. The universe of $\mathcal{K}_M$ is S^M; for each $p \in Prop$, and $s \in S^M$, we let $p^{\mathcal{K}_M} \doteq s$ if and only if $s \in p^M$; for each $a \in Act$, and $s_1, s_2 \in S^M$, we let $s_1 \doteq a^{\mathcal{K}_M}(s_2)$ if and only if $(s_1, s_2) \in a^M$. In other words, the binary relation $a^{\mathcal{K}_M}$ amounts precisely to the converse of a^M (recall that $s_1 \doteq a^{\mathcal{K}_M}(s_2)$ if and only if $(s_2, s_1) \in a^{\mathcal{K}_M}$), while the relations $p^{\mathcal{K}_M}$ and p^M coincide. Passing to the powerset algebra $\wp\mathcal{K}_M$, we have the following.

Proposition 6.2.1. *Let Sig be a signature constructed as above from the sets* Prop *and* Act. *Let φ be a formula of the modal μ-calculus, and t a fixed-point term over $Sig_\sim$. Then, for any Kripke structure M, and any valuation $val : Var \to \mathcal{P}(S^M)$,*

$$[\![\varphi]\!]_M[val] \quad = \quad [\![f2t(\varphi)]\!]_{\wp\mathcal{K}_M}[val]$$

and

$$[\![t]\!]_{\wp\mathcal{K}_M}[val] \quad = \quad [\![t2f(t)]\!]_M[val].$$

Proof. The claims follow easily from the definitions by induction on φ and on t respectively. Let us check the induction step for $\langle a \rangle \varphi$. We have

$$
\begin{array}{rcl}
[\![\langle a \rangle \varphi]\!]_M[val] & = & \{s \mid \exists s', (s, s') \in a^M \text{ and } s' \in [\![\varphi]\!]_M[val]\} \\
 & = & a^{\mathcal{K}_M}([\![\varphi]\!]_M[val]) \\
 & = & a^{\mathcal{K}_M}([\![f2t(\varphi)]\!]_{\wp\mathcal{K}_M}[val]) \text{ (by the induction hypothesis)} \\
 & = & [\![a(f2t(\varphi))]\!]_{\wp\mathcal{K}_M}[val] \\
 & = & [\![f2t(\langle a \rangle \varphi)]\!]_{\wp\mathcal{K}_M}[val]
\end{array}
$$

$\square$

The above considerations show that Kripke structures can be identified with certain semi-algebras, and the modal μ-calculus over the Kripke structures coincides with the calculus of fixed-point terms over the corresponding powerset algebras. But the converse interpretation is also easy. Let *Sig* be a signature such that all symbols except for *eq* are of arity at most 1. Viewing the symbols of arity 0 as propositions, and those of arity 1 as actions, we can identify any semi-algebra $\mathcal{B}$ with a Kripke structure, say, $M_{\mathcal{B}} = \langle B, \{p^{\mathcal{B}} \mid p \in Sig,\ ar(p) = 0\}, \{(f^{\mathcal{B}})^{-1} \mid f \in Sig,\ ar(f) = 1\}$. It is obvious that, by applying the previous transformation $M \mapsto \mathcal{K}_M$ to $M_{\mathcal{B}}$, we shall obtain again $\mathcal{K}_{M_{\mathcal{B}}} = \mathcal{B}$. Thus, from Proposition 6.2.1, we can also get the equalities $[t]_{\wp\mathcal{B}}[val] = [t2f(t)]_{M_{\mathcal{B}}}[val]$, and $[\varphi]_{M_{\mathcal{B}}}[val] = [f2t(\varphi)]_{\wp\mathcal{B}}[val]$. This shows that the modal μ-calculus can indeed be derived from the μ-calculus over powerset algebras, by merely restricting the arity of the function symbols in the signatures.

6.3 Homomorphisms, μ-homomorphisms, and bisimulations

The classical notion of a homomorphism for algebras extends easily to semi-algebras.

Definition 6.3.1. Let

$$\mathcal{B} = \langle B, \{f^{\mathcal{B}} \mid f \in Sig\}\rangle \qquad \text{and} \qquad \mathcal{B}' = \langle B', \{f^{\mathcal{B}'} \mid f \in Sig\}\rangle$$

be two semi-algebras over the same signature *Sig*.

A mapping $h : B \to B'$ is a homomorphism if for any $f \in Sig$, and for any $b, b_1, \ldots, b_{\rho(f)} \in B$, $b \doteq f^{\mathcal{B}}(b_1, \ldots, b_{\rho(f)})$ implies $h(b) \doteq f^{\mathcal{B}'}(h(b_1), \ldots, h(b_{\rho(f)}))$.

Note that if both $\mathcal{B}$ and $\mathcal{B}'$ are algebras, h is a homomorphism in the usual sense.

Definition 6.3.2. We call a homomorphism of semi-algebras $h : B \to B'$ *reflective* if whenever $h(b) \doteq f^{\mathcal{B}'}(b'_1, \ldots, b'_{\rho(f)})$, for some $b \in B$ and $b'_1, \ldots, b'_{\rho(f)} \in B'$, there exist $b_1, \ldots, b_{\rho(f)} \in B$ such that

- $b \doteq f^{\mathcal{B}}(b_1, \ldots, b_{\rho(f)})$,
- $b'_i = h(b_i)$, for $i = 1, \ldots, \rho(f)$.

Example 6.3.3. Let $t \in T_{Sig}$ be a syntactic tree over *Sig* as defined in Example 6.1.3, page 142. For $w \in \text{dom}\ t$, the *subtree* of t induced by w is the mapping $t.w$ over the domain $\{v : wv \in \text{dom}\ t\}$ defined by $t.w(v) = t(wv)$. It is plain to see that $t.w$ is a syntactic tree.

Now, it follows easily by our definitions that the mapping $h : \text{dom}\, t \to T_{Sig}$ defined by $h(w) = t.w$ is a homomorphism from the semi-algebra $\mathbf{t}$ to the algebra $\mathcal{T}_{Sig}$. If $w \doteq f^{\mathbf{t}}(v_1, \ldots, v_k)$ then $t(w) = f$ and $v_i = wi$. Then $h(w) = t.w = f^{\mathcal{T}_{Sig}}(t.w1, \ldots, t.wk) = f^{\mathcal{T}_{Sig}}(h(v_1), \ldots, h(v_k))$.

Let us observe that this homomorphism is reflective. Indeed, suppose $t.w = f^{\mathcal{T}_{Sig}}(t_1, \ldots, t_k)$. Then $t(w) = f$, and w has k successors in dom t : $w1, \ldots, wk$; moreover an easy calculation shows that $t.wi = t_i$, for $i = 1, \ldots, k$. Hence, we have $w \doteq f^{\mathbf{t}}(w1, \ldots, wk)$, and $h(wi) = t_i$, for $i = 1, \ldots, k$, as required.

Adding the special symbol *eq* to a signature *Sig*, interpreted as explained in Section 6.1.3 (page 144) does not change the nature of a homomorphism h. The following proposition shows that if we want to check whether or not a mapping is a (reflective) homomorphism, we do not have to care about the special symbol *eq*.

Proposition 6.3.4. *Let $\mathcal{B}$ and $\mathcal{B}'$ be two semi-algebras over Sig and let $\mathcal{B}_{eq}$ and $\mathcal{B}'_{eq}$ be the semi-algebras over $Sig \cup \{eq\}$ obtained by extending $\mathcal{B}$ and $\mathcal{B}'$ by $eq^{\mathcal{B}_{eq}}$ and $eq^{\mathcal{B}'_{eq}}$.*

If $h : B \to B'$ is a (reflective) homomorphism from $\mathcal{B}$ to $\mathcal{B}'$, it is also a (reflective) homomorphism from $\mathcal{B}_{eq}$ to $\mathcal{B}'_{eq}$.

Proof. Any mapping h is homomorphic with respect to *eq*: If $b \doteq eq^{\mathcal{B}_{eq}}(b', b'')$ then $b = b' = b''$ which implies $h(b) \doteq eq^{\mathcal{B}'_{eq}}(h(b'), h(b''))$.

Any mapping h is reflective with respect to *eq*: If $h(b) \doteq eq^{\mathcal{B}'_{eq}}(b', b'')$ then $b \doteq eq^{\mathcal{B}_{eq}}(b, b)$ with $h(b) = b' = b''$. □

Our first observation is that a reflective homomorphism induces a homomorphism of the respective powerset algebras.

Proposition 6.3.5. *Let $\mathcal{B}$ and $\mathcal{B}'$ be semi-algebras over Sig, and let $h : \mathcal{B} \to \mathcal{B}'$ be a reflective homomorphism. Then the mapping $h^{-1} : \mathcal{P}(B') \to \mathcal{P}(B)$, where $h^{-1}(L) = \{b \in B \mid h(b) \in L\}$, is a homomorphism, whenever $\mathcal{P}(\mathcal{B})$ and $\mathcal{P}(\mathcal{B}')$ are considered as algebras over $Sig_{\sim}$.*

Proof. Let $f \in Sig$. We first verify that, for any $L_1, \ldots, L_{\rho(f)} \subseteq B'$,

$$h^{-1}(f^{\wp\mathcal{B}'}(L_1, \ldots, L_{\rho(f)})) = f^{\wp\mathcal{B}}(h^{-1}(L_1), \ldots, h^{-1}(L_{\rho(f)}))$$

Indeed, the inclusion $\supseteq$ is true for any homomorphism, and the inclusion $\subseteq$ is a straightforward consequence of the reflectiveness of h. It remains to verify that

$$h^{-1}(\tilde{f}^{\wp\mathcal{B}'}(L_1, \ldots, L_{\rho(f)})) = \tilde{f}^{\wp\mathcal{B}}(h^{-1}(L_1), \ldots, h^{-1}(L_{\rho(f)}))$$

Using the above equality , the characterization of $\tilde{f}$ in terms of complementation (page 143), and the fact that $\overline{h^{-1}(L)} = h^{-1}(\overline{L})$, for any $L \subseteq B'$, we have

$$
\begin{aligned}
h^{-1}(\tilde{f}^{\wp\mathcal{B}'}(L_1, \ldots, L_{\rho(f)})) &= h^{-1}(\overline{f^{\wp\mathcal{B}'}(\overline{L_1}, \ldots, \overline{L_{\rho(f)}})}) \\
&= \overline{h^{-1}(f^{\wp\mathcal{B}'}(\overline{L_1}, \ldots, \overline{L_{\rho(f)}}))} \\
&= \overline{f^{\wp\mathcal{B}}(h^{-1}(\overline{L_1}), \ldots, h^{-1}(\overline{L_{\rho(f)}}))} \\
&= \overline{f^{\wp\mathcal{B}}(\overline{h^{-1}(L_1)}, \ldots, \overline{h^{-1}(L_{\rho(f)})})} \\
&= \tilde{f}^{\wp\mathcal{B}}(h^{-1}(L_1), \ldots, h^{-1}(L_{\rho(f)}))
\end{aligned}
$$

□

As we have said at the beginning of this chapter, we wish to consider powerset algebras as interpretations for the μ-calculus. Therefore, we are interested in morphisms that preserve the whole clone of fixed-point definable operations, and not only the basic algebraic structure.

Recall that for any semi-algebra over *Sig*, and a fixed-point term t over $Sig_{\sim}$, the interpretation $[\![t]\!]_{\wp\mathcal{B}}$ is well defined; that is, for any valuation $v : ar(t) \to \mathcal{P}(B)$, $[\![t]\!]_{\wp\mathcal{B}}(v)$ is a subset of B.

Definition 6.3.6. Let $\mathcal{C}$ and $\mathcal{D}$ be two powerset algebras over the signature $Sig_{\sim}$. A homomorphism $h : \mathcal{C} \to \mathcal{D}$ is a μ-homomorphism if, for every fixed-point term t and for every valuation $v : Var \to C$,

$$h([\![t]\!]_{\mathcal{C}}[v]) = [\![t]\!]_{\mathcal{D}}[h \circ v]$$

where $h \circ v : Var \to \mathcal{D}$ is defined by $(h \circ v)(x) = h(v(x))$, for $x \in Var$. Since t can be a base term (see Definition 2.3.1, page 47), a μ-homomorphism is always a homomorphism.

Remark. In the general perspective of Chapter 2, we could organize the set of all interpretations $[\![t]\!]_{\wp\mathcal{B}}$ into a μ-calculus, actually a sub-calculus of the functional calculus over $\langle \mathcal{P}(B), \subseteq \rangle$. In this setting, the property defining a μ-homomorphism means precisely that h is a homomorphism of μ-calculi.

Proposition 6.3.7. *The homomorphism $h^{-1} : \mathcal{P}(B') \to \mathcal{P}(B)$ considered in Proposition 6.3.5 is a μ-homomorphism.*

Proof. We show by induction on a term t that, for any valuation $v : Var \to \mathcal{P}(B')$, $h^{-1}([\![t]\!]_{\wp\mathcal{B}'}[v]) = [\![t]\!]_{\wp\mathcal{B}}[h^{-1} \circ v]$.

If t is a variable, the claim is obvious.

If $t = f(t_1, \ldots, t_{\rho(f)})$, or $t = \tilde{f}(t_1, \ldots, t_{\rho(f)})$, the result follows by the induction hypothesis about $t_1, \ldots, t_{\rho(f)}$, and the fact h^{-1} that is a homomorphism (Proposition 6.3.5).

Finally assume that $t = \theta x.t'$. Then $[t]_{\wp\mathcal{B}'}[v]$ is the extremal fixed point of the mapping $f : \mathcal{P}(B') \to \mathcal{P}(B')$ defined by $f(L') = [t']_{\wp\mathcal{B}'}[v\{L'/x\}]$ and $[t]_{\wp\mathcal{B}}[h^{-1} \circ v]$ is the extremal fixed point of the mapping $g : \mathcal{P}(B) \to \mathcal{P}(B)$ defined by $g(L) = [t']_{\wp\mathcal{B}}[(h^{-1} \circ v)\{L/x\}]$. By the induction hypothesis, $h^{-1}(f(L')) = g(h^{-1}(L'))$, and since $h^{-1} : \mathcal{P}(B') \to \mathcal{P}(B)$ is obviously β-inf- and β-sup-continuous for any ordinal β, the result is a consequence of Lemma 1.2.15 (page 13). □

From Propositions 6.3.5 and 6.3.7 we get immediately:

Corollary 6.3.8. *Let $\mathcal{B}$ and $\mathcal{B}'$ be semi-algebras over Sig, and let $h : \mathcal{B} \to \mathcal{B}'$ be a reflective homomorphism.*

For any closed fixed-point term τ and any $b \in B$, $b \in [\tau]_{\wp\mathcal{B}} \Leftrightarrow h(b) \in [\tau]_{\wp\mathcal{B}'}$.

As an example of use of the previous proposition, let us consider the reflective homomorphism $h : \mathbf{t} \to \mathcal{T}_{Sig}$ defined in the Example 6.3.3 by $h(w) = t.w$. Let τ be a closed fixed point term. We have $w \in [\tau]_{\wp\mathbf{t}}$ if and only if $t.w \in [\tau]_{\wp\mathcal{T}_{Sig}}$, and we get, in particular, the following characterization of $[\tau]_{\wp\mathcal{T}_{Sig}}$.

Proposition 6.3.9. $t \in [\tau]_{\wp\mathcal{T}_{Sig}}$ *if and only if* $\varepsilon \in [\tau]_{\wp\mathbf{t}}$.

An important notion related to Kripke structures and the modal μ-calculus is the notion of bisimulation. Indeed this notion is easily extended to semi-algebras of arbitrary signatures.

Definition 6.3.10. A *bisimulation* between semi-algebras $\mathcal{B}$ and $\mathcal{B}'$ over the same signature *Sig* is a binary relation $R \subseteq B \times B'$ such that

- – $\forall b \in B, \exists b' \in B',\ bRb'$,
  - $\forall b' \in B', \exists b \in B,\ bRb'$,
- $\forall f \in Sig, \forall b \in B, b' \in B' : bRb'$,
  - if $b \doteq f^{\mathcal{B}}(b_1, \ldots, b_k)$, then there exist $b'_1, \ldots, b'_k \in B'$, such that $b' \doteq f^{\mathcal{B}'}(b'_1, \ldots, b'_k)$, and $b_i R b'_i$, for $i = 1, \ldots, k$;
  - if $b' \doteq f^{\mathcal{B}'}(b'_1, \ldots, b'_k)$, then there exist $b_1, \ldots, b_k \in B$, such that $b \doteq f^{\mathcal{B}}(b_1, \ldots, b_k)$, and $b_i R b'_i$, for $i = 1, \ldots, k$.

It is easy to check that this definition applied to semi-algebras associated with Kripke structures is exactly the usual definition of a bisimulation relation.

If $(R_i)_{i\in I}$ is a family of bisimulations between $\mathcal{B}$ and $\mathcal{B}'$ then the relation $\bigcup_{i\in I} R_i$ is also a bisimulation between $\mathcal{B}$ and $\mathcal{B}'$. Thus, if there exists a bisimulation between $\mathcal{B}$ and $\mathcal{B}'$, there exists a greatest one.

The following proposition is a straightforward consequence of the above definitions.

Proposition 6.3.11. *If $h : \mathcal{B} \to \mathcal{B}'$ is a homomorphism of semi-algebras then the relation $\{\langle b, h(b)\rangle \mid b \in B\} \subseteq B \times B'$ is a bisimulation if and only if h is surjective and reflective.*

We will call such a homomorphism a *bisimulation homomorphism.*

If $h_1 : \mathcal{B}' \to \mathcal{B}$ and $h_2 : \mathcal{B}'' \to \mathcal{B}$ are two bisimulation homomorphisms, the composition product $h_1 \cdot h_2^{-1}$ of the two binary relations h_1 and h_2^{-1} is a bisimulation between $\mathcal{B}'$ and $\mathcal{B}''$.

Conversely, we can generalize a result of [3].

Proposition 6.3.12. *If R is a bisimulation between $\mathcal{B}'$ and $\mathcal{B}''$, then there exists a semi-algebra $\mathcal{B}$ and two bisimulation homomorphisms $h : \mathcal{B}' \to \mathcal{B}$ and $h_2 : \mathcal{B}'' \to \mathcal{B}$ such that $R \subseteq h_1 \cdot h_2^{-1}$.*

Proof. Obviously the relation $(R.R^{-1})^*$ is an equivalence over the domain B' of $\mathcal{B}'$, denoted by $\equiv_R$. This relation is also a bisimulation between $\mathcal{B}'$ and itself, since for any n, $(R.R^{-1})^n$ is a bisimulation between $\mathcal{B}$ and itself. Let $B = B'/\equiv_R$ be the set of equivalence classes of B'. We give B a structure of semi-algebra over *Sig* by setting $b \doteq f^{\mathcal{B}}(b_1, \dots, b_n)$ if and only if there exist $b' \in b, b'_1 \in b_1, \dots, b'_n \in b_n$ such that $b' \doteq f^{\mathcal{B}'}(b'_1, \dots, b'_n)$. Let $h_1 : B' \to B$ be the mapping defined by: $h_i(b')$ is the equivalence class of b'. Let us show that it is a bisimulation homomorphism. By construction h_1 is surjective, and, by definition of $\mathcal{B}$, $b' \doteq f^{\mathcal{B}'}(b'_1, \dots, b'_n) \Rightarrow h_1(b') \doteq f^{\mathcal{B}}(h_1(b'_1), \dots, h_1(b'_n))$. It remains to prove that h_1 is reflective. Let $h_1(b') \doteq f^{\mathcal{B}}(b_1, \dots, b_n)$. By definition of $\mathcal{B}$ there exist $c', c'_1, \dots, c'_n$ such that $c' \doteq f^{\mathcal{B}'}(c'_1, \dots, c'_n)$, $h_1(c') = h_1(b')$ and $h_1(c'_i) = b'_i$ for $i = 1, \dots, n$. Since $c' \equiv_R b'$ and since $\equiv_R$ is a bisimulation, there exist $b'_i \equiv_R c'_i$ (and thus $h_1(b'_i) = h_1(c'_i)$) such that $b' \doteq f^{\mathcal{B}'}(b'_1 \dots, b'_n)$.

Now let us remark that for any $b'' \in B''$ there exists at least one $b' \in B'$ such that $b'Rb''$, and that if b'_1Rb'' and b'_2Rb'' then $b_1 \equiv_R b_2$. Therefore we can define a surjective mapping $h_2 : B'' \to B$ by $h_2(b'') = h_1(b')$ for any b' such that $b'Rb''$. It immediately follows that $R \subseteq h_1 \cdot h_2^{-1}$. Let us show that h_2 is a reflective homomorphism. If $b'' \doteq f^{\mathcal{B}''}(b''_1, \dots, b''_n)$, then $h_2(b'') = h_1(b')$ for some b' such that $b'Rb''$. Since R is a bisimulation, there exist $b'_iRb''_i$ such that $b' \doteq f^{\mathcal{B}'}(b'_1, \dots, b'_n)$. But then $h_2(b'') = h_1(b') \doteq f^{\mathcal{B}}(h_1(b'_1), \dots, h_1(b'_n)) = f^{\mathcal{B}}(h_2(b''_1), \dots, h_2(b''_n))$. If $b \doteq f^{\mathcal{B}}(b_1, \dots, b_n)$ and if $h_2(b'') = b$, there exists $b'Rb''$ such that $h_1(b') = h_2(b'') = b$. Since h_1 is reflective, there exist $b'_1, \dots, b'_n$ such that $b' \doteq f^{\mathcal{B}'}(b'_1, \dots, b'_n)$ and $h_1(b'_i) = b_i$. Since R is a bisimulation there exist $b''_1, \dots, b''_n$ such that $b'' \doteq f^{\mathcal{B}''}(b''_1, \dots, b''_n)$ with $b'_iRb''_i$, hence $h_2(b''_i) = h_1(b'_i) = b_i$. □

Together with Proposition 6.3.7, it follows, in particular, that if there is a bisimulation between $\mathcal{B}_1$ and $\mathcal{B}_2$, then for any closed fixed-point term t, $[t]_{\wp\mathcal{B}_1} \neq \emptyset$ if and only if $[t]_{\wp\mathcal{B}_2} \neq \emptyset$.

Another consequence of the previous proposition is that if there is a bisimulation between $\mathcal{B}'$ and $\mathcal{B}''$, there exist two bisimulation homomorphisms $h_1 : \mathcal{B}' \to \mathcal{B}$ and $h_2 : \mathcal{B}'' \to \mathcal{B}$ such that $h_1 \cdot h_2^{-1}$ is the greatest bisimulation between $\mathcal{B}'$ and $\mathcal{B}''$.

6.4 Bibliographic notes and sources

The idea of a powerset algebra can be traced back to the work by Jónnson and Tarski [47, 48]. The definition presented in this chapter extends the concept of a powerset algebra (without dualities) considered in [77, 10, 78]. Similar structures were considered in the work of McAllester, Givan, Witty and Kozen [60].

The notion of bisimulation for semi-algebras as well as its characterization by bisimulation homomorphisms are straightforward extensions of similar notions for transition systems [81, 63, 3].

7. The μ-calculus vs. automata

We have seen in Chapter 5 that finite automata running over finite or infinite words can be adequately characterized by a suitably designed μ-calculi. We will now set up a similar characterization on a much more general level – for automata interpreted in arbitrary semi-algebras. To this end, we will view automata as syntactic objects which, like terms, can have multiple interpretations. Technically, we will define the semantics of automata over semi-algebras in terms of parity games played by Adam and Eva as in Chapter 4. We shall see that this setting comprises many familiar examples, in particular the nondeterministic and alternating automata on trees.

We will further organize our automata into a μ-calculus similar to the μ-calculus of fixed-point terms of Section 2.3. As there, we shall see that an interpretation is a homomorphism from the μ-calculus of automata into a functional μ-calculus. Then, we will establish a connection between automata and fixed-point terms, by showing that a natural translation from terms to automata is a homomorphism which, failing surjectivity, captures all the automata up to semantical equivalence. We conclude that automata and fixed-point terms are two notations for one and the same thing.

7.1 Automata over semi-algebras

Basically, we view an automaton as a finite set of *transitions*, i.e. equations of the form $x = f(y_1, \dots, y_k)$, where f is a function symbol in Sig, and the symbols $x, y_i, \dots$ are the automaton's *states*. We qualify each state as *existential* or *universal*, and associate with it a natural number called its *rank*. A proviso is made that each state is a head (i.e., the left-hand side) of *exactly one* transition. This may appear as a drastic restriction, but we shall see that it is not the case, since any Boolean combination of transitions can be easily expressed in such a form thanks to the operation eq.

Classical automata take as input special objects, e.g. words or trees, possibly infinite. A computation typically consists in matching the positions of an input object (e.g., word or tree) with the transitions of an automaton.

This gives rise to a marking of the positions by the automaton's states, usually called a *run*. Such a run is sometimes viewed as a strategy in a game that an automaton plays in favor of the acceptance of the input object. We will adopt the game perspective while generalizing the above situation to arbitrary semi-algebras.

7.1.1 Informal description

Let us fix an automaton, and a semi-algebra $\mathcal{B}$. We define a game of Eva and Adam, thinking that Eva plays for the automaton's benefit.

The play consists of a possibly infinite sequence of rounds in which the players move alternately. Each round starts at a position of the form (b, x), where x is a state of the automaton, and b is an element of B. The rules of the game applicable to the position (b, x) are as follows. Suppose the transition for x is $x = f(y_1, \ldots, y_k)$. Then, if x is universal, Eva has to find some elements $d_1, \ldots, d_k \in B$ such that $b \doteq f^{\mathcal{B}}(d_1, \ldots, d_k)$. Adam answers by selecting $i \in \{1, \ldots, k\}$, and the starting position of the next round is (d_i, y_i). If x is existential, the development is symmetric. Thus, Adam starts by choosing $d_1, \ldots, d_k \in B$ such that $b \doteq f^{\mathcal{B}}(d_1, \ldots, d_k)$, and then Eva picks an i in $\{1, \ldots, k\}$. Again, the next-round position is (d_i, y_i). If any of the players cannot make her or his move, the other player wins the game. Note that it may happen, e.g., if there is no decomposition $b \doteq f^{\mathcal{B}}(d_1, \ldots, d_k)$; on the other hand, if the arity of f is $\rho(f) = 0$, and $b \doteq f^{\mathcal{B}}$ (i.e., a decomposition is possible), then Eva wins if x is universal, and Adam wins if x is existential as, obviously, the other player cannot pick an $i \in \emptyset$. If the players play *ad infinitum*, the win is determined by the parity criterion applied to the sequence of (ranked) states subsequently assumed in the play. Thus, if the highest rank occurring infinitely often is even, Eva wins the game, otherwise Adam is the winner.

We say that the automaton ***accepts*** an element $b \in B$ if Eva has a winning strategy at the position (b, x_0), where x_0 is a distinguished initial state.

In the next section, we will present these concepts more formally, in terms of the parity games defined in Chapter 4. However, aiming at a parallel between automata and fixed-point terms, we will allow the former, as well as the latter, to define operators $(\wp(B))^k \to \wp(B)$ rather than merely subsets of B. To this end, we will equip our automata with an additional feature, namely *variables*. (In some sense, these variables can be seen as free variables of the automaton, while the states are its boun variables.)

7.1.2 Basic definitions

We fix a countably infinite set of variables $Var = \{v_0, v_1, \ldots\}$, and a finite signature Sig, assuming the following

Proviso. The symbol eq of arity 2 belongs to Sig and, in any semi-algebra $\mathcal{B}$, is interpreted by $eq^{\mathcal{B}}(a,b) \doteq c$ if and only if $a = b = c$ (see Sections 6.1.2 and 6.1.3).

Definition 7.1.1. An *automaton* over a signature Sig is presented as a tuple

$$A = \langle Sig, Q, V, x_I, Tr, qual, rank \rangle,$$

where

- Q is a finite set of *states*.
- $V \subseteq Var$, $V \cap Q = \emptyset$, is a finite set, the set of the automaton's *variables*. The elements of $Q \cup V$ are referred to as the automaton's *symbols*. The automaton is said to be *closed* if its set V of variables is empty.
- $x_I \in Q \cup V$ is an *initial symbol* referred to as *initial state* if $x_I \in Q$.
- Tr is a set of *transitions*, which are pairs of the form $\langle x, f(y_1, \ldots, y_{\rho(f)}) \rangle$ also presented as equations $x = f(y_1, \ldots, y_{\rho(f)})$ where $f \in Sig$, $x \in Q$, and $y_1, \ldots, y_{\rho(f)} \in Q \cup V$.
  We assume that for each state $x \in Q$, there is *exactly one* transition of the form $x = t$; we call x the *head* of the transition, and write $t = Tr(x)$.
- $qual : Q \to \{\exists, \forall\}$ is *qualification function*. We call those states $x \in Q$ for which $qual(x) = \exists$ *existential*, and those for which $qual(x) = \forall$, *universal*.
- $rank : Q \to \mathbb{N}$ is a mapping that with each state x associates its *rank*, $rank(x)$.

We will often refer to the components of an automaton A by the superscript 'A', thus $A = \langle Sig, Q^A, V^A, x_I^A, Tr^A, qual^A, rank^A \rangle$.

Definition 7.1.2. The semantics of automata will be given in terms of parity games of Chapter 4. Let A be an automaton as above, and let $\mathcal{B}$ be a semi-algebra. Let $val : V \to \mathcal{P}(B)$ be a mapping that associates a subset $val(z) \subseteq B$ with each variable $z \in V$. With these data, we associate a game $G(A, \mathcal{B}, val) = (Pos_e, Pos_a, Mov, rank)$ as follows.

The set of positions $Pos = Pos_e \cup Pos_a$ is a disjoint union of

- the set of *state positions* $B \times Q^A$,
- the set of *transition positions* that is the set of all elements

  $$(x = f(y_1, \ldots, y_k), w) \in Tr^A \times B^*$$

  such that $w = \langle d_1, \ldots, d_k, b \rangle$ with $b \doteq f^{\mathcal{B}}(d_1, \ldots, d_k)$. We sometimes denote such a position by

  $$((b, x) = f((d_1, y_1), \ldots, (d_k, y_k))).$$

– the set of *variable positions* $B \times V^A$.

Note that the symbols $x, y_1, \ldots, y_k$ in the transition position above need not be different, and that an equality $x = y_i$ (or $y_i = y_j$) does not imply $b = d_i$ (or $d_i = d_j$). For this reason, we cannot view the tuple $\langle d_1, \ldots, d_k, b\rangle$ as a valuation of the variables $y_1, \ldots, y_k, x$.

The positions are distributed to the players as follows.

– A state position (b, x) is a position of Eva, whenever $qual^A(x) = \forall$.
– A transition position $(x = f(y_1, \ldots, y_k), \langle d_1, \ldots, d_k, b\rangle)$ is a position of Eva, whenever $qual^A(x) = \exists$.
– A variable position (b, z) is a position of Eva, whenever $b \notin val(z)$.
– The remaining are positions of Adam.

The relation *Mov* is defined by the following rules.

– There is an edge from a state position (b, x) to a transition position
$$(x = f(y_1, \ldots, y_k), \langle d_1, \ldots, d_k, b\rangle).$$
– There is an edge from a transition position
$$(x = f(y_1, \ldots, y_k), \langle d_1, \ldots, d_k, b\rangle)$$
to the position (d_i, y_i), for each $i = 1, \ldots, k$. (Note that the latter can be a state or a variable position.)
– There are no other edges.

Note that no edge comes out from a variable position (b, z), $z \in V^A$, and, whenever a play reaches such a position, Eva wins if and only if $b \in val(z)$.

Finally, we define the function *rank* over positions, according to the rank of the state part of the position. That is, for a state position (b, x), we let $rank(b, x) = rank^A(x)$, and we let $rank(p) = 0$, for all other positions.

Definition 7.1.3. An element b of $\mathcal{B}$ is *recognized* (or *accepted*) by A with respect to a mapping $val : V \to \mathcal{P}(B)$ if the position (b, x_I) is winning for Eva in the game $G(A, \mathcal{B}, val)$. We denote the set of all such elements by $[A]_{\wp\mathcal{B}}(val)$. Consequently, $[A]_{\wp\mathcal{B}}$ denotes the mapping $\mathcal{P}(B)^V \to \mathcal{P}(B)$ that sends val to $[A]_{\wp\mathcal{B}}(val)$.

Two concepts of equivalence between automata will be in order. We say that two automata A and A' with the same set of variables are *semantically equivalent* if, for any semi-algebra $\mathcal{B}$, the mappings $[A]_{\wp\mathcal{B}}$ and $[A']_{\wp\mathcal{B}}$ are identical.

A closer relation is defined via the concept of *isomorphism*. We call two automata over *Sig*, A and A', *isomorphic* if they have the same set of variables, i.e. $V^A = V^{A'}$, and there exists a bijective mapping $h : Q^A \to Q^{A'}$, which we call *isomorphism*, such that

– for each $x \in Q^A$, $rank^A(x) = rank^{A'}(h(x))$, and $qual^A(x) = qual^{A'}(h(x))$,
– for any $f \in Sig$, there is a transition $x = f(x_1, \ldots, x_k)$ in Tr^A if and only if there is a transition $h(x) = f(h^*(x_1), \ldots, h^*(x_k))$ in $Tr^{A'}$, where

$$\begin{aligned} h^*(y) &= h(y) \text{ if } y \in Q^A \\ &= y \text{ if } y \in V^A \end{aligned}$$

Clearly, the inverse mapping h^{-1} satisfies a symmetric condition and so h^{-1} is an isomorphism from A' to A.

The following is an easy consequence of the definitions.

Proposition 7.1.4. *If two automata are isomorphic, they are also semantically equivalent.*

We also consider a useful construction which eliminates the useless symbols of an automaton, i.e. the symbols not reachable from the initial one.

Let $A = \langle Sig, Q, V, x_I, Tr, qual, rank \rangle$ be an automaton. Consider a directed graph whose set of nodes is $Q \cup V$, and there is an edge from x to y if and only if there is a transition of the form $x = f(\ldots, y, \ldots)$, for some $f \in Sig$. Let $red(Q)$ (respectively, $red(V)$) be the set of the states (resp. variables) y such that there is a path in this graph from x_I to y. Note that $x_I \in red(Q) \cup red(V)$, and, if $x_I \in V$ then the set $red(Q)$ is empty and $red(V) = \{x_I\}$.

Definition 7.1.5. An automaton A is in *reduced form* (or simply: *reduced*) if $red(Q^A) = Q^A$ and $red(V^A) = V^A$.

For any automaton A, we define the *reduced version* $red(A)$ as an automaton whose set of states is $red(Q^A)$, the set of variables is $red(V^A)$, and the set of transitions as well as the functions *qual* and *rank* are obtained from those of A by restriction to $red(Q^A)$ and $red(V^A)$ in the obvious way.

Note that the set of variables of $red(A)$ can be smaller than that of A. However, it is easy to see that the construction preserves semantics in the following sense.

Proposition 7.1.6. *For any semi-algebra $\mathcal{B}$, and any mapping val from V to $\mathcal{P}(B)$, $[A]_{\wp\mathcal{B}}(val) = [red(A)]_{\wp\mathcal{B}}(val \restriction red(V^A))$.*

7.1.3 Hierarchy of indices

We will show in Section 7.3 of this chapter that automata are, in some precise sense, equivalent to fixed-point terms. We now define a parameter of automata which, in this equivalence, will correspond to the alternation depth of terms.

First, let us remark that an automaton whose minimal value of *rank* is greater than 2 is equivalent to the one with the function *rank* modified by $rank(q) := rank(q) - 2$. Therefore, from now on, we assume that *all the automata we consider satisfies the property:* $\min(rank(Q)) \in \{0, 1\}$ (provided Q is not empty!).

Definition 7.1.7. For any automaton A which has a nonempty set of states we define the *Mostowski index* of A as the pair (ι, k), where

$$k = \max\ (rank(Q))$$

and $\iota \in \{0, 1\}$ is given by

$$\iota = \min\ (rank(Q))$$

so that it is the pair $(\min(rank(Q))\ ,\ \max(rank(Q))\ )$.

Essentially, the Mostowski index measures the amplitude of the rank of an automaton.

As we have remarked it in Section 5.2.3 (page 118), the Büchi acceptance criterion is a special case of parity criterion characterized by the fact that the rank function ranges over the set $\{1, 2\}$. Therefore, a Büchi automaton can be presented as a parity automaton of Mostowski index $(1, 2)$.

It is sometimes convenient to compare indices of automata. We will say that an index (ι, k) is *compatible* with an index (ι', k') if either $\iota' \leq \iota$ and $k \leq k'$ or $\iota = 0$, $\iota' = 1$, and $k + 2 \leq k'$. It is easy to see that if (ι, k) is *compatible* with (ι', k') then, for any automaton of index (ι, k) there exists a semantically equivalent automaton of index (ι', k').

This concept gives rise to a formal hierarchy of automata (see Figure 7.1). We shall see later in Chapter 8 that the expressive power of the automata increases indeed with the level in the hierarchy.

7.1.4 Dual automata

We have seen in Proposition 6.1.6 (page 144) that any term t has a dual $\tilde{t}$ such that $[\tilde{t}]_{\wp B}$ is the dual mapping of $[t]_{\wp B} : \mathcal{P}(B)^{ar(t)} \to \mathcal{P}(B)$. We define in this section the notion of a dual automaton which enjoys a similar property.

Definition 7.1.8. Let $A = \langle Sig, Q, V, x_I, Tr, qual, rank \rangle$ be an automaton, and let (ι, k) be its Mostowski index. The dual automaton of A is $\tilde{A} = \langle Sig, Q, V, x_I, Tr, qual', rank' \rangle$ where

- $qual'(q) = \forall$ if and only if $qual(q) = \exists$,
- $rank'(q) = \begin{cases} rank(q) + 1 & \text{if } \iota = 0 \\ rank(q) - 1 & \text{if } \iota = 1 \end{cases}$

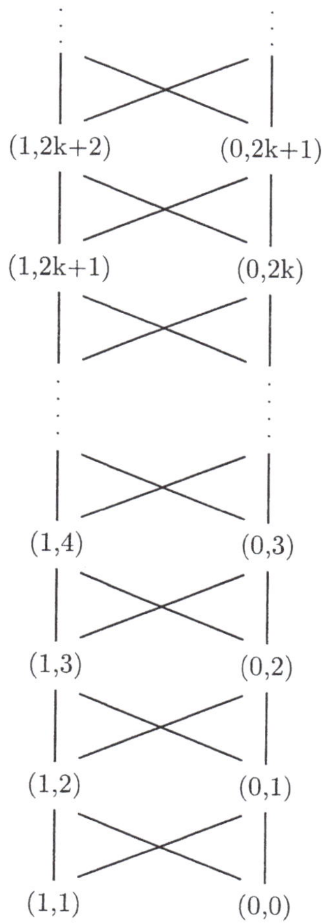

Fig. 7.1. The hierarchy of Mostowski indices

It follows that the Mostowski index of $\tilde{A}$ is $(1, k+1)$ if $\iota = 0$ and $(0, k-1)$ if $\iota = 1$ so that duality exchanges the two classes on a same row of the diagram of Figure 7.1. The following proposition is direct consequence of the definition.

Proposition 7.1.9. *The dual of $\tilde{A}$ is exactly A.*

Proposition 7.1.10. *For any automaton A, the mapping*

$$[\tilde{A}]_{\wp\mathcal{B}} : \mathcal{P}(B)^{ar(A)} \to \mathcal{P}(B)$$

is the dual of $[A]_{\wp\mathcal{B}}$.

Proof. Let val be a mapping from $ar(A)$ to $\mathcal{P}(B)$ and let $\overline{val}$ be defined by $\overline{val}(v) = \mathcal{P}(B) - val(v)$. Let G (resp. G') be the game associated with A and val (resp. $\tilde{A}$ and $\overline{val}$), as in Definition 7.1.2, page 157.

These two games have the same sets of positions and moves, but the position of Adam and Eva are interchanged as well as the parities of the rank of the state positions. Thus a play in G is also a play in G' (and vice-versa) and a play is won by Eva in G if and only if it is won by Adam in G'. Therefore, if Eva has a winning strategy at some position in G, this very strategy is winning for Adam at the same position in G', and, by the determinacy theorem (Theorem 4.3.10, page 92), we get as a consequence of Definition 7.1.3 (page 158), $[A]_{\wp\mathcal{B}}(val) = \overline{[\tilde{A}]_{\wp\mathcal{B}}(\overline{val})}$. □

7.1.5 Relation to classical automata

We will now illustrate our general concept of automata by some familiar examples presented in our setting. To this end it is convenient to slightly modify the definition of an automaton in order to provide a possibility of choice between multiple transitions, in existential or universal mode. We will later see that this modification can be easily incorporated in our setting.

Alternating automata. Rather than to release our requirement that each state is the head of only one transition, we extend the structure of transitions by means of the operations $\vee$ and $\wedge$. In an alternating automaton A, we let transitions be of the form $x = \tau$, where $Tr(x) = \tau$ is any element of the least set of expressions D^A containing the *base terms* $f(y_1, \cdots, y_k) \in D^A$, for $f \in Sig$ and $y_1, \cdots, y_k \in Q^A \cup V^A$, and closed under (binary) $\vee$ and $\wedge$, i.e. if τ_1 and τ_2 are in D^A, so are $(\tau_1 \vee \tau_2)$ and $(\tau_1 \wedge \tau_2)$. Note that these terms are all guarded functional terms as defined in Section 9.2.2, page 208.

Given a semi-algebra $\mathcal{B}$ and a mapping $val : V^A \to \mathcal{P}(B)$, the semantics is again defined by the game $G(A, \mathcal{B}, val) = (Pos_e, Pos_a, Mov, rank)$, but with slightly modified positions and moves. We extend the concept of state positions, by letting a *state position* be any pair $(b, x = \tau')$, where τ' is a subterm of $Tr(x)$ different from a variable; in other words, a subterm of $Tr(x)$ in D^A. (Recall that, in the previous game, we have presented the state positions by (b, x), but we could unambiguously present them also by $(b, x = Tr(x))$.) Next, we let the *transition positions* be of the form $(x = f(y_1, \ldots, y_k), v)$, where $f(y_1, \ldots, y_k)$ is a base subterm of $Tr(x)$. A state position of the form $(b, x = (\tau_1 \circ \tau_2))$ is a position of Eva for $\circ = \vee$, and of Adam for $\circ = \wedge$. A state position of the form $(b, x = f(y_1, \ldots, y_k))$ belongs to Eva or Adam depending on whether x is universal or existential, respectively. (Exactly as for the position (b, x) in the previous game.) Dually, a transition position $(x = f(y_1, \ldots, y_k), v)$ belongs to Eva or Adam depending on whether x is existential or universal, respectively. There is a move from a position $(b, x = (\tau_1 \circ \tau_2))$ to $(b, x = \tau_i)$, $i = 1, 2$. The remaining moves are like in the previous game: there is a move from $(b, x = f(y_1, \ldots, y_k))$ to $(x = f(y_1, \ldots, y_k), \langle d_1, \ldots, d_k, b\rangle)$, for any $d_1, \ldots, d_k, b \in B$ such that $b \doteq f(d_1, \ldots, d_k)$, and from $(x = f(y_1, \ldots, y_k), \langle d_1, \ldots, d_k, b\rangle)$ to (d_i, y_i) if y_i is a variable, and to $(d_i, y_i = Tr(y_i))$ if y_i is a state. The rank of a state position $(b, x = \tau')$ is $rank^A(x)$, and the rank of the remaining positions is 0.

Finally, $[A]_{\wp B}(val)$ is defined as the set of all $b \in B$ such that the state position $(b, x_I^A = Tr(x_I^A))$ is winning for Eva.

It should be clear that if an alternating automaton A happens to be in the form of Section 7.1.2, then the above game coincides with the game defined there, up to identification of positions $(b, x = \tau)$ with (b, x). We will show that, in general, any alternating automaton A can be transformed to an automaton A' in the previous sense, semantically equivalent to A.

The construction consists in consecutively eliminating $\vee$ and $\wedge$. We will show the case of $\vee$. Suppose the transition for x is $x = (\tau_1 \vee \tau_2)$. Let us modify A by removing this transition, and adding instead the three transitions

$$x = eq(x_1, x_2), \quad x_1 = \tau_1, \quad x_2 = \tau_2,$$

where x_1 and x_2 are some fresh symbols, now considered as states of A, with $qual(x_1) = qual(x_2) = qual(x)$, and $rank(x_1) = rank(x_2) = rank(x)$. We additionally reset the qualification of x by $qual(x) := \exists$.

The graph of the game is modified accordingly. Note that in the former game we have had a position $(b, x = (\tau_1 \vee \tau_2))$, for $b \in B$, from which Eva could move to $(b, x = \tau_1)$ or to $(b, x = \tau_2)$. In the new game we have a position $(b, x = eq(x_1, x_2))$, from which Adam can move to $(x = eq(x_1, x_2), \langle b, b, b\rangle)$ and then Eva can move to $(b, x_1 = \tau_1)$ or to $(b, x_2 = \tau_2)$. Clearly, the resulting state position is determined solely by Eva. Now, to see that the modified automaton is equivalent to the original one, it is enough to show that any positional strategy for Eva in the original game, winning at a position $(a, x_I^A = Tr(x_I^A))$, can be transformed into a strategy for Eva in the modified game winning at the position $(a, x_I^A = Tr'(x_I^A))$, and *vice versa*, for $a \in B$. (Here $Tr'(x_I^A) = Tr(x_I^A)$, unless x_I^A coincides with the state x for which the transition has been actually modified as above.) The transformation is fairly easy: for example, if a strategy in the original game, at the position $(b, x = (\tau_1 \vee \tau_2))$, chooses a move to $(b, x = \tau_1)$ then the corresponding strategy in the modified game will choose $(b, x_1 = \tau_1)$, at the position $(x = eq(x_1, x_2), \langle b, b, b \rangle)$. Conversely, given a strategy for Eva in the modified game, we determine her move at the position $(b, x = (\tau_1 \vee \tau_2))$ in the original game, depending on that strategy's choice at a position $(x = eq(x_1, x_2), \langle b, b, b \rangle)$. It is easy to see that the modified strategies behave as expected.

If the transition for x in the original game is $x = (\tau_1 \wedge \tau_2)$, the modification is similar except for that the new qualification of x shall become $\forall$. The justification is similar, and we leave it to the reader.

Now, by a repeated application of these two steps, we eventually obtain an automaton in the required form, semantically equivalent to the original one.

Remark. Alternating automata considered above look more close to classical automata than our generic automata defined in Section 7.1.2. For example, the transitions of a finite automaton over finite words, say, $p \xrightarrow{a} q$, and $p \xrightarrow{b} r$, can be represented as a single alternating transition $p = (aq \vee br)$ (see below). But of course, we could also use here, e.g., the transition $p = (br \vee aq)$, and it seems natural not to distinguish between these two possibilities. More generally, we could let the right-hand sides of our transitions be the elements of a *free distributive lattice* generated by the base terms $f(y_1, \ldots, y_k)$, rather than as formal expressions. Thus we could also identify, e.g., $((\tau_1 \vee \tau_2) \vee \tau_3)$ with $(\tau_1 \vee (\tau_2 \vee \tau_3))$, $((\tau_1 \vee \tau_2) \wedge \tau_3)$ with $((\tau_1 \wedge \tau_3) \vee (\tau_2 \wedge \tau_3))$, etc. Such an approach may appear more natural, although it is less convenient for a formal presentation. However, is easy to see that if the terms τ and τ' are equivalent over the free distributive lattice, then an exchange of τ for τ' in an alternating automaton A defined as above shall not alter the semantics of A. Therefore, in the examples below, we will abuse the notation, writing freely, e.g., $x = \tau_1 \vee \tau_2 \vee \tau_3$, by which we mean that the transition in question can be $x = ((\tau_1 \vee \tau_2) \vee \tau_3)$ or $x = (\tau_1 \vee (\tau_2 \vee \tau_3))$, or perhaps $x = ((\tau_3 \vee \tau_1) \vee \tau_2)$, and we leave to the reader the verification that an actual choice is unimportant.

Nondeterministic automata. A *nondeterministic automaton* over *Sig* is an alternating automaton whose transitions are in the form $x = \tau_1 \vee \ldots \vee \tau_m$, where $m \geq 1$, and each τ_i is a base term of the form $f(y_1, \ldots, y_{\rho(f)})$, for some $y_1, \ldots, y_k \in Q^A \cup V^A$, and some $f \in Sig - \{eq\}$. (The special symbol *eq* is excluded from transitions for technical reasons. Note that it shall reappear if we convert a nondeterministic automaton into the general form of Section 7.1.2, by the procedure described above for alternating automata.) The qualification of all the states is *universal*, while no constraint is imposed on the rank.

For such automata, we will simplify the game $G(A, \mathcal{B}, val)$ by letting the state positions be only in the form $(b, x = \tau_1 \vee \cdots \vee \tau_m)$ or $(b, x = \tau_i)$, where $Tr^A(x) = \tau_1 \vee \cdots \vee \tau_m$. We assume that there is a move of Eva from $(b, x = \tau_1 \vee \cdots \vee \tau_m)$ (directly) to $(b, x = \tau_i)$, for $i = 1, \ldots, m$. It is easy to see that this simplification does not alter the semantics of the automata.

By calling the above automata "nondeterministic", we have wished to contrast them with the general alternating automata, rather than to emphasize nondeterminism. To complete the picture, let us call a nondeterministic automaton *deterministic* if, in each transition $x = \tau_1 \vee \ldots \vee \tau_m$, the function symbols in the τ_i's are pairwise different. (Such a contradictory phrasing is common in automata theory since *nondeterministic* is understood as *not necessarily deterministic*.) We note however that the qualification of such an automaton as deterministic is problematic if the semi-algebra on which it is interpreted is not itself deterministic in the following sense: no element can be decomposed in two different ways. In Section 10.4.1 (page 242) we introduce formally this property, called there *codeterminism*, and show that the semi-algebras of syntactic trees (see Example 6.1.3, page 142) have this property. Therefore an adequate notion of determinism should involve both an automaton and an interpretation. However, since determinism is not an issue in this chapter, we do not go further in these considerations.

In the examples below, the variables of automata do not play any essential role, and we can always assume $V^A = \emptyset$. Therefore we will simply omit this item in the presentation of automata.

Automata on words. Let Σ be a finite alphabet. The set of finite or infinite words over Σ, $\Sigma^\infty = \Sigma^* \cup \Sigma^\omega$ can be organized into a semi-algebra (in fact, an algebra) over the signature $Sig = \Sigma \cup \{e\} \cup \{eq\}$, where each symbol $\sigma \in \Sigma$ is interpreted by the operation $u \mapsto \sigma u$ of Definition 5.1.2 (page 106), e is a constant symbol interpreted by the empty word ε, and eq is interpreted as usual. Then a finite nondeterministic automaton over *finite words*, $\mathcal{A} = \langle S, T, I, F \rangle$ (see Section 5.2.1, page 115) can be presented in our setting as follows. Suppose first that there is a single initial state $I = \{x_I\}$. Then we can keep S as the set of states, and let the transitions be of the form $x = \sigma_1 y_1 \vee \ldots \vee \sigma_m y_m$ or $x = \sigma_1 y_1 \vee \ldots \vee \sigma_m y_m \vee e$, where $x \xrightarrow{\sigma_i} y_i$ is a transition of the original automaton, and the summand e occurs whenever x is an accepting state. Note that in the new automaton, there is a single transition $x = \ldots$, for each state x. We let $rank(x) = 1$, for all states $x \in S$, so that no infinite play can be won by Eva. The qualification of states is unimportant, because of the monadicity of operations. (There is at most one move from a position $(u, x = \sigma y)$, and thus the next state position is determined, whoever moves the first.)

If there is more than one initial state, say $I = \{z_1, \ldots, z_\ell\}$, we add a fresh initial state x_I and a transition $x_I = z_1 \vee \ldots \vee z_\ell$.

An automaton over *infinite words* with parity acceptance condition defined as in Section 5.2.2 (page 117) can be presented similarly, except for that the constant e does not occur on the right-hand sides of transitions. Readily, we can extend the definition and, by allowing e, to let an automaton accept finite as well as infinite words.

Let us consider a simple example. To avoid confusion, we will refer to the presentation of a classical word automaton in our setting as to an *abstract automaton*. Let $\Sigma = \{a, b\}$, and suppose an automaton (in classical presentation) has transitions $p \xrightarrow{a} p$, $p \xrightarrow{b} q$, $q \xrightarrow{a} q$, $p \xrightarrow{b} r$, where the state p is initial, and the state q accepting. The abstract automaton has the transitions $p = ap \vee bq \vee br$, $q = aq \vee e$. The following sequence of positions forms a possible play in the game associated with the corresponding abstract automaton interpreted in the algebra of words: $(aba, p = ap \vee bq), (aba, p = ap)$, $(p, aba) = a(p, ba)$, $(ba, p = bq \vee br), (ba, p = bq)$, $(p, ba) = b(q, a)$, $(a, q = aq \vee e), (a, q = aq)$, $(q, a) = a(q, \varepsilon)$, $(\varepsilon, q = aq \vee e), (\varepsilon, q = e)$. The play is won by Eva since Adam is not able to move from the last position. Intuitively, we can think that Eva has used a strategy suggested by an accepting run on the word aba, $p \xrightarrow{a} p \xrightarrow{b} q \xrightarrow{a} q$. That is, at a position of the form $(w, x = \tau_1 \vee \tau_2)$, the strategy told Eva which transition to choose: $x = \tau_1$, or $x = \tau_2$. Here the choice at the position $(aba, p = ap \vee bq)$ was obviously indicated by the first letter of aba; while at the position $(ba, p = bq \vee br)$ the advice of the strategy was essential, because of the nondeterminism of our automaton.

It is not hard to prove that the accepting runs of a classical word automaton always induce the winning strategies for Eva in the game associated with the abstract automaton, and conversely, the strategies induce runs. Consequently, the game semantic of the abstract automata interpreted in the algebra of words coincides with the classical semantics of word automata. We will show it in detail for automata on trees, which can be viewed as a generalization of words to arbitrary signature.

Automata on trees. The reader familiar with tree automata may have found our presentation of nondeterministic automata quite close to the usual presentation of such automata (assuming the parity acceptance condition), except for that a transition $x = \tau_1 \vee \cdots \vee \tau_k$ is sometimes presented as a set of transitions $\{x = \tau_1, \ldots, x = \tau_k\}$.

Consider a nondeterministic automaton A over a signature Sig. Let $A = \langle Sig, Q^A, x_I^A, Tr^A, qual^A, rank^A \rangle$ be a nondeterministic automaton. With the classical interpretation, a tree automata takes as an input a syntactic tree over $Sig - \{eq\}$, i.e. an element t of $T_{Sig-\{eq\}}$ (see Example 6.1.3, page 142) A computation consists in examining the nodes of the tree, level by level, according to the transition table, starting from the root. Formally, a *run* of A on t is defined as a mapping r : dom $t \to Q^A$, such that $r(\varepsilon) = x_I^A$, and, for each $w \in$ dom t, if $r(w) = x$, $t(w) = f$, and the transition for x is $x = \tau_1 \vee \cdots \vee \tau_k$, then $f(r(w1), \ldots, r(w\rho(f)))$ coincides with τ_i, for some i (recall that $w1, \ldots, wk$ are the successors of w in dom t). A run is considered *accepting* if, for each infinite *path* in dom t, i.e. a sequence $w_0, w_1, \ldots$ such that each $w_{\ell+1}$ is a successor of w_ℓ in dom t, $\limsup_{\ell\to\infty} rank^A(r(w_\ell))$ is even. The tree language recognized by A consists of those trees t for which there exists an accepting run of A on t.

For example, an automaton with transitions $x = a(x,x) \vee b(z,z)$ and $z = a(x,x) \vee b(z,z)$, where $rank(x) = 0$ and $rank(z) = 1$, recognizes those trees over the signature $\{a, b\}$ (with $ar(a) = ar(b) = 2$) which do not have a path with infinitely many b's.

This example can be generalized to an automaton over a signature $\{a_0, a_1, \ldots, a_k\}$ (again with all the symbols binary) whose states are $\{x_0, x_1, \ldots, x_k\}$ with $rank(x_i) = i$, and the transitions are of the form

$$x_i \quad = \quad a_0(x_0, x_0) \vee a_1(x_1, x_1) \vee \ldots \vee a_k(x_k, x_k)$$

It is plain to see that this automaton accepts a tree t if and only if, for each infinite path $w_0, w_1, \ldots$, $\limsup_{\ell\to\infty} \sharp t(w_\ell)$ is even, where $\sharp a_i = i$.

An important class of tree automata is induced by the Büchi acceptance criterion (see Section 5.2.2, page 117). A *Büchi automaton* differs from the automaton above only in that the acceptance condition is given by a set of states $F \subseteq Q$. A run on a tree t is considered accepting if for each infinite path $(w_0, w_1, \ldots)$, $r(w_n) \in F$, for infinitely many n's. It is plain to see that the Büchi automata can be equivalently presented as the parity automata of Mostowski index $(1, 2)$ (the states of F are ranked 2, and the remaining states are ranked 1).

In Example 6.1.3 (page 142), we have introduced two semi-algebras related to syntactic trees; now on we shall reconsider them enriched by the operation eq. Hence, there are two ways of presenting a tree automaton in our general setting: it can be interpreted in the algebra of all trees $\mathcal{T}_{Sig}$, or, for any tree $t \in T_{Sig-\{eq\}}$, in the semi-algebra $\mathbf{t}$. We shall see that both ways are in some precise sense equivalent to the classical semantics of automata on trees.

Consider first the semi-algebra $\mathbf{t}$. Recall that its universe is dom t, and $w \doteq f^{\mathbf{t}}(v_1, \ldots, v_k)$, whenever $t(w) = f$ and $v_1, \ldots, v_k$ are the successors of w. Consider the game associated with A and $\mathbf{t}$. Note that the positions of Eva are of the form $(w, x = \tau_1 \vee \cdots \vee \tau_m)$ or $(w, x = \tau_i)$, and the positions of Adam of the form $(x = \tau_i, v)$, where $\tau_1 \vee \cdots \vee \tau_m = Tr(x)$. At a position of the first kind, Eva selects one τ_i by moving to $(w, x = \tau_i)$. Suppose $\tau_i = f(y_1, \ldots, y_k)$.

Then Eva can make a next move only if $f = t(w)$, and then she has no choice: she moves to $(w, x) = f((w1, y_1), \ldots, (wk, y_k))$. From there, Adam can move to any $(wi, y_i = Tr(y_i))$, $i = 1, \ldots, k$ (provided $k > 0$). Thus, in course of a play, the first components of the positions of Eva form a path in dom t. In fact, the path is selected by the moves of Adam — in the early version of this game, due to Gurevich and Harrington [40], the player corresponding to our Adam was called *Pathfinder*.

Now suppose there is an accepting run $r : \text{dom } t \to Q^A$. We construct an induced positional strategy s_r for Eva, as follows. For each position $p = (w, x = \tau_1 \vee \cdots \vee \tau_m)$ such that $r(w) = x$, we let $s_r : p \mapsto (w, x = \tau_i)$, where τ_i is the term chosen by the automaton in the run, i.e., it is of the form $f(y_1, \ldots, y_k)$, such that $f = t(w)$ and $y_i = r(wi)$, for $i = 1, \ldots, k$. For the remaining positions, we let s_r be defined arbitrarily. (Note that the move of Eva from a position $(w, x = \tau)$ is determined by the unique decomposition of w in $\mathbf{t}$, $w \doteq f^{\mathbf{t}}(w1, \ldots, wk)$, with $f = t(w)$.) Now it is plain to see that in each play starting from the position $(\varepsilon, x_I^A = Tr(x_I^A))$ and consistent with the strategy s_r, the positions of Eva are of the form $(w, x = \ldots)$, where $x = r(w)$. Therefore Eva cannot loose in a finite play, and each infinite play corresponds in a natural way to a path in r and is won by Eva since r is accepting. Thus, the strategy s_r is winning for Eva at the position $(\varepsilon, x_I^A = Tr(x_I^A))$.

Conversely, suppose Eva has a positional strategy s which is winning at the position $(\varepsilon, x_I^A = Tr(x_I^A))$. We construct a run r_s on t by induction on $w \in \text{dom } t$. Let $r_s(\varepsilon) = x_I^A$. Next, suppose $r_s(w) = x$, and s maps the position $(w, x = Tr(x))$ to $(w, x = f(y_1, \ldots, y_k))$. Since s is winning, obviously $t(w) = f$. We let $r(wi) = y_i$, for $i = 1, \ldots, k$. Again, it is easy to see that any infinite path in r_s corresponds to some play consistent with the strategy s, and thus it satisfies the parity acceptance condition.

We conclude that a tree t is accepted by the automaton A according to the classical definition if and only if the position $(\varepsilon, x_I^A = Tr(x_I^A))$ is winning for Eva in the game associated with A and $\mathbf{t}$.

A similar characterization can be shown for the interpretation of A in the algebra of trees $\mathcal{T}_{Sig}$: t is accepted by A if and only if Eva has a winning strategy from the position $(t, x_I^A = Tr(x_I^A))$. To see it, it is enough to realize a connection between the previous game, and the game associated with A and $\mathcal{T}_{Sig}$. Let us denote these games by $G(A, \mathbf{t})$ and $G(A, \mathcal{T}_{Sig})$, respectively. Given a positional strategy s_1 for Eva in $G(A, \mathbf{t})$, we define a strategy s_2 in $G(A, \mathcal{T}_{Sig})$ as follows. For each $w \in \text{dom } t$, whenever s_1 maps $(w, x = Tr(x))$ to $(w, x = f(y_1, \ldots, y_k))$, we let s_2 map the position $(t.w, x = Tr(x))$ to $(t.w, x = f(y_1, \ldots, y_k))$, where $t.w$ is the subtree of t induced by w (see Example 6.3.3, page 148) i.e., dom $t.w = \{v : wv \in \text{dom } t\}$, and $t.w(v) = t(wv)$, for $v \in \text{dom } t.w$). For the remaining positions, s_2 can be defined arbitrarily.

It is easy to show that if s_1 is winning at $(\varepsilon, x_I^A = Tr(x_I^A))$ then s_2 is winning at $(t, x_I^A = Tr(x_I^A))$.

Similarly, given a positional strategy for Eva in $G(A, \mathcal{T}_{Sig})$, say s_2, and given a tree $t \in T_{Sig}$, we consider a strategy s_1 in $G(A, \mathbf{t})$, such that, whenever s_2 maps $(t.w, x = Tr(x))$ to $(t.w, x = f(y_1, \ldots, y_k))$, s_1 maps $(w, x = Tr(x))$ to $(w, x = f(y_1, \ldots, y_k))$. Again, it is easy to see that whenever s_2 is winning at $(t, x_I^A = Tr(x_I^A))$, so is s_1 at $(\varepsilon, x_I^A = Tr(x_I^A))$ in $G(A, \mathbf{t})$. Consequently, the tree language recognized by A in the classical sense coincides precisely with the interpretation of A in the algebra $\mathcal{T}_{Sig}$.

Finally, let us remark that the deterministic automata over trees are essentially weaker that nondeterministic automata.

Example 7.1.11. Consider a nondeterministic automaton with the transitions $x_I = f(x_0, x_1) \vee f(x_1, x_0)$, $x_0 = a$, and $x_1 = b$, where a and b are constant symbols. Interpreted over trees, this automaton accepts precisely the trees $f(a, b)$ and $f(b, a)$, but it is easy to see than any deterministic automaton accepting these two trees should also accept the trees $f(a, a)$ and $f(b, b)$.

The use of a non-unary symbol f was essential here. If we restrict ourselves to signatures with symbols of arity at most 1 then automata over trees can be easily identified with automata over (finite or infinite) words, and McNaughton's theorem can be applied.

The concept of a nondeterministic tree automaton can be naturally extended to that of an *alternating automaton on trees*. We will consider such automata in Chapter 8. However, the alternating automata on trees are not essentially simpler than the alternating automata in any other semi-algebra (in contrast to nondeterministic automata); therefore we do not discuss them here.

Other acceptance conditions. In Section 5.2.2 (page 117), we have considered automata over infinite words with a somewhat weaker Büchi acceptance criterion, but also with acceptance criteria apparently more general than the parity condition, as, e.g., the Rabin criterion, given by a set of pairs of sets of states, and the yet more general Muller criterion, given by any family of sets of states. We have seen however that any nondeterministic automaton over infinite words with Muller (or Rabin) condition can be transformed into an equivalent *deterministic* automaton with a parity condition (by Theorem 5.2.8, page 121, and McNaughton's theorem). As far as determinism is concerned, this analogy does not extend to automata over semi-algebras, and even not to automata over trees, as we have seen in the Example 7.1.11.

On the other hand, the parity acceptance condition turns out to preserve its universal power also for our most general concept of automata, and it is in fact an easy consequence of the aforementioned properties of the automata on words.

To see it, let us consider an automaton A defined as in Section 7.1.2 (page 156), except for that the rank function is replaced by a *recognizable* set of infinite words $C^A \subseteq (Q^A)^\omega$. The game $G(A, \mathcal{B}, val)$ is defined as previously except for that the winning criterion for infinite plays is given by C^A as follows. Consider a play $p_0, p_1, \ldots$ starting from a state position $p_0 = (b, x)$. By definition, the play forms an alternating sequence of state positions and transition positions. We can derive an infinite sequence of states consisting of the second components of the state positions, i.e., of $p_0, p_2, p_4, \ldots$. We let, by definition, the play be won by Eva if this sequence is in C^A; otherwise Adam is the winner. A parity acceptance condition can be easily presented in this way, by taking $C^A = \{u \in Q^\omega : \limsup_{n\to\infty} rank^A(u_n) \text{ is even }\}$. Similarly, a Muller acceptance condition given by a set $\mathcal{F} \subseteq \mathcal{P}(Q^A)$ can be presented by the set of strings $u \in Q^\omega$ such that the set of states that occur infinitely often in u is an element of $\mathcal{F}$; clearly, this set is ω-regular.

Now, by the McNaughton theorem and Theorem 5.2.8 (page 121), there exists a *deterministic* automaton D with the parity acceptance condition such that $L(D) = C^A$. We construct a parity automaton $A \times D$ equivalent to A as a kind of product of A and D. (For simplicity, we consider automata without variables.) We let $Q^{A\times D} = Q^A \times Q^D$, $x_I^{A\times D} = \langle x_I^A, x_I^D\rangle$, and the transitions of $A \times D$ be of the form

$$\langle x, p\rangle \quad = \quad f(\langle y_1, q\rangle, \ldots, \langle y_k, q\rangle)$$

where $x = f(y_1, \ldots, y_k)$ is a transition of A, and $p \xrightarrow{x} q$ is a transition of D. (By this writing, we allow a transition of the form $\langle x, p\rangle = c$, whenever $x = c$ is a transition of A.) We let $qual^{A\times D}(x, p) = qual^A(x)$, and $rank^{A\times D}(x, p) = rank^D(p)$.

Now, let $\mathcal{B}$ be a semi-algebra, and let $b \in B$. Suppose Eva has a winning strategy in the game $G(A, \mathcal{B})$ from a position (b, x_I^A). Then it is not difficult to see that she can also win while playing the game associated with the product automaton, $G(A \times D, \mathcal{B})$, from the position $(b, \langle x_I^A, x_I^D\rangle)$, with essentially the same strategy (the missing states of D are determined by the states of A and the transition table of D). Indeed, any infinite play consistent with the strategy shall derive a sequence of states in C^A, hence accepted by D. So the play will satisfy the parity condition induced by $rank^{A\times D}$.

Conversely, suppose Eva has a strategy s winning at position $(b, \langle x_I^A, x_I^D\rangle)$ in the parity game $G(A \times D, \mathcal{B})$. We can assume that s is positional. Then Eva can use the strategy s while playing from (b, x_I^A) in $G(A, \mathcal{B})$. More specifically, suppose an initial play in $G(A, \mathcal{B})$ consists of a sequence of positions $p_0, p_1, \ldots, p_{2m}$, and $p_{2m} = (a, x)$ is a position of Eva. Let the sequence of states of A derived from the state positions be $x_I^A = y_0, y_1, \ldots, y_m = x$, and suppose $x_I^D \stackrel{y_0 y_1 \ldots y_m}{\longrightarrow} q$, i.e., the automaton D enters the state q after reading the word $y_0 y_1 \ldots y_m$. Then Eva uses the advice of the strategy s at the position $(a, \langle x, q \rangle)$ (neglecting the second components of the states). The argument for the case of an initial play of odd length is similar. Again, it is plain to see that the resulted strategy is winning for Eva. (Note however, that it need not be positional. In general, in a game with the winning condition given by the Muller criterion, the winner may not have a positional strategy, and an analogue to Corollary 4.4.3 (page 97) fails to hold [62].)

Thus, the parity automaton $A \times D$ is indeed semantically equivalent to A, as desired.

7.2 Automata in the μ-calculus perspective

7.2.1 The μ-calculus of automata

We will now organize the automata into a μ-calculus in the sense of Chapter 2. The basic idea goes back to the classical construction of finite automata for regular expressions. Given automata for languages L and M, one constructs automata for the languages $L \cup M$, LM, and L^* by a kind of operations uniformly defined over automata. This can be further extended to a construction of a nondeterministic automaton for an ω-language L^ω, and more generally for $\bigcup_i M_i L_i^\omega$ (where L, M_i, L_i are regular sets of finite words), i.e. for any rational set of infinite words. As we have seen in Chapter 5, the two kinds of iteration: L^* and L^ω correspond to the least and greatest fixed-point operator respectively. Situating automata in the μ-calculus frame will allow us to express this correspondence both generally and precisely.

However, to organize the automata into a μ-calculus is slightly more difficult than was an analogous task for fixed-point terms in Section 2.3.2 (page 47). Difficulty will arise in composition. Intuitively, while composing an automaton A with a substitution ρ into an automaton, say $A[\rho]$, one should convert each variable z of A into the initial symbol of the automaton $\rho(z)$. Thus, if in course of a computation of the automaton $A[\rho]$ the symbol z is assumed, the computation (or the play, if we think in terms of games) will not stop but switch to the automaton $\rho(z)$, and proceed. Now, a confusion may obviously arise if the states of the automaton A and those of the automata $\rho(z)$ are not different. An easy remedy to this is to make all the sets of symbols in question pairwise disjoint. However, if we wish to satisfy the Axiom 7 of the μ-calculus, we need to do it very carefully.

We will define the universe of our μ-calculus as the set of reduced automata presented in some canonical way. The idea behind it comes from a graphical representation which one can naturally associate with any automaton in reduced form. Intuitively, as a first step, we unravel the automaton into a possibly infinite tree labeled by the automaton's symbols. That is, we put the initial symbol x_I at the root, and, whenever the label of a node is x, x is a state, and the unique transition with the head x is $x = f(x_1, \ldots, x_k)$, we create k successors of this node labeled by $x_1, \ldots, x_k$, respectively. At the second step, we prune this tree leaving only the first (lexicographically) occurrence of each state.

More specifically, let $\sqsubseteq$ be a linear ordering of $\mathbb{N}^*$ that orders sequences first by length, and then lexicographically. That is, $u \sqsubseteq v$ if $|u| < |v|$, or $|u| = |v|$, and there exists $w, u_1, v_1 \in \mathbb{N}^*$, and $i, j \in N$, such that $u = wiu_1$, $v = wjv_1$, and $i < j$ (as natural numbers).

Definition 7.2.1. An automaton A is in canonical form if

- A is reduced,
- $Q^A \subseteq \mathbb{N}^*$,
- if the initial symbol is a state, it is ε,
- if $x = f(x_1, \ldots, x_k)$ is a transition in Tr^A, and x_i is a state (not a variable), then

$$x_i \sqsubseteq xi,$$

- and, moreover, if Q^A is not empty then $\min(rank(Q^A)) \in \{0, 1\}$.

We denote the set of all automata over Sig in canonical form by $\mathbf{Aut}_{Sig}$.

The following can be seen as a justification of the term "canonical".

Proposition 7.2.2. *For any automaton A, one can compute an automaton in canonical form isomorphic to $red(A)$.*

Proof. By Proposition 7.1.6, we may assume that the automaton $A = \langle Sig, Q, V, x_I, Tr, qual, rank \rangle$ is already in reduced form. If x_I is a variable then, by virtue of reduced form, we have $Q = Tr = \emptyset$, and thus A is obviously in canonical form. Suppose x_I is a state. Define a tree $t_A : \text{dom } t_A \to Q$, with dom $t_A \subseteq \mathbb{N}^*$, inductively as follows.

- $t_A(\varepsilon) = x_I$.
- If $t_A(w) = x$, and the transition with head x is $x = f(x_1, \ldots, x_k)$ then, for each i such that x_i is a state, w has a successor wi in dom t_A valued $t_A(wi) = x_i$. The node w has no other successors.

Note that by virtue of reduced form, each $x \in Q$ is a value of t_A. Now, for each $x \in Q$, let $h(x)$ be the least (with respect to the ordering $\sqsubseteq$) node $w \in$ dom t_A, such that $t(w) = x$. Clearly, h is one to one, and $h(x_I) = \varepsilon$. Let C be the unique automaton with the set of states $Q^C = \{h(x) : x \in Q\}$, such that h is an isomorphism from the original automaton A to C. It is straightforward to check that this C is in canonical form, and by Proposition 7.1.4, it is semantically equivalent to A.

The condition concerning *rank* is easily obtained by a translation (see Section 7.1.3, page 159). □

We are going to organize the set $\mathbf{Aut}_{Sig}$ into a μ-calculus.

The first two items of Definition 2.1.1 (page 42) are easy to define: For each variable $z \in Var$, we let $\hat{z}$ be a trivial automaton $\langle Sig, \emptyset, \{z\}, z, \emptyset, \emptyset, \emptyset, \rangle$. We also define the *arity* of any automaton A in $\mathbf{Aut}_{Sig}$ by $ar(A) = V^A$. Note that by assumption, all variables in $ar(A)$ are accessible from the initial symbol.

Now let $A = \langle Sig, Q, V, x_I, Tr, qual, rank\rangle$, be an automaton in $\mathbf{Aut}_{Sig}$, and let $\rho : Var \to \mathbf{Aut}_{Sig}$ be a substitution. We define the automaton $comp(A, \rho) = A[\rho]$ by the following construction. At first, for each $z \in V$, we define an auxiliary set $\mathrm{add}(z) \subseteq \mathbb{N}^*$ whose elements we call ***addresses*** of z. If the initial symbol x_I is a variable then we let $\mathrm{add}(x_I) = \{\varepsilon\}$. Note that, since A is reduced, we have in this case $V = \{x_I\}$, and so the definition of the addresses is completed. Suppose the initial symbol x_I is a state. For each $z \in V$, we let $\mathrm{add}(z) \subseteq \mathbb{N}^*$ be the set of all words of the form xi, such that $x \in Q$, the transition of head x is $x = f(y_1, \ldots, y_k)$, and $y_i = z$. Note that, since A is reduced, we have $\mathrm{add}(z) \neq \emptyset$, for each $z \in V$. Also, if $z \neq z'$ then $\mathrm{add}(z) \cap \mathrm{add}(z') = \emptyset$. It is also convenient to distinguish those variables z in V for which the initial symbol of $\rho(z)$ is a state (not a variable); we will call these variables *active* and note $V_\rho = \{z \in V : x_I^{\rho(z)} \in Q^{\rho(z)}\}$. The automaton $comp(A, \rho) = A[\rho]$ is defined by the following items. (Recall that, by the superscript convention, $Q^{\rho(z)}$ denotes the set of states of the automaton $\rho(z)$, etc.)

- $Q^{A[\rho]} = Q \cup \bigcup_{z \in V} \mathrm{add}(z) Q^{\rho(z)}$.
  That is, the states of $A[\rho]$ are either states of A, or sequences of the form xy, where x is an address of z, and y is a state of the automaton $\rho(z)$.
- $V^{A[\rho]} = \bigcup_{z \in V} V^{\rho(z)}$.
- If x_I is a variable then $x_I^{A[\rho]}$ is the initial symbol of the automaton $\rho(x_I)$. Otherwise, $x_I^{A[\rho]} = x_I = \varepsilon$.
- The transitions of $A[\rho]$ are of two kinds.
  - For each transition $x = f(x_1, \ldots, x_k)$ of A, we have a transition $x = f(x'_1, \ldots, x'_k)$ of $A[\rho]$, where

$$\begin{aligned} x'_i &= x_i \text{ if } x_i \in Q \\ &= xi \text{ if } x_i \in V_\rho \\ &= x_I^{\rho(x_i)} \text{ if } x_i \in V - V_\rho \end{aligned}$$

That is, whenever x_i is a variable in V_ρ, we replace it by its address xi, and whenever $x_i \in V - V_\rho$, we replace it by the initial symbol (a variable) of the corresponding automaton $\rho(x_i)$.

- For each transition $y = f(y_1, \ldots, y_k)$ of $\rho(z)$, where $z \in V_\rho$, and for each $x \in \text{add}(z)$, we have a transition $xy = f(y'_1, \ldots, y'_k)$ of $A[\rho]$, where

$$\begin{aligned} y'_i &= xy_i \text{ if } y_i \in Q^{\rho(z)} \\ &= y_i \text{ if } y_i \in V^{\rho(z)} \end{aligned}$$

That is, for each address x of z, we have a separate copy of the set of transitions of $\rho(z)$, obtained by prefixing the states of $\rho(z)$ by x.

- For $x \in Q$, $qual^{A[\rho]}(x) = qual(x)$, and $rank^{A[\rho]}(x) = rank(x)$. For $z \in V_\rho$, $y \in Q^{\rho(z)}$, and $w \in \text{add}(z)$, we let $qual^{A[\rho]}(wy) = qual^{\rho(z)}(y)$, and $rank^{A[\rho]}(wy) = rank^{\rho(z)}(y)$.

It follows from the construction that the automaton $A[\rho]$ is in canonical form. Note that if the initial symbol x_I of A is a variable then $A[\rho] = \rho(x_I)$.

In order to define operators μ and ν over automata, it will be convenient first to fix two automata of arity 0, A_{tt} and A_{ff}, with the property that, for any semi-algebra $\mathcal{B}$, $[A_{tt}]_{\wp\mathcal{B}} = B$, while $[A_{ff}]_{\wp\mathcal{B}} = \emptyset$.

Using the proviso that the symbol $eq \in Sig$ is always interpreted in the standard way (Section 6.1.2), we let

$$\begin{aligned} A_{tt} &= \langle Sig, \{\varepsilon\}, \emptyset, \varepsilon, \{\varepsilon = eq(\varepsilon, \varepsilon)\}, qual : \varepsilon \mapsto \exists, rank : \varepsilon \mapsto 0\rangle \\ A_{ff} &= \langle Sig, \{\varepsilon\}, \emptyset, \varepsilon, \{\varepsilon = eq(\varepsilon, \varepsilon)\}, qual : \varepsilon \mapsto \forall, rank : \varepsilon \mapsto 1\rangle \end{aligned}$$

so that A_{ff} is the dual of A_{tt}. Note that substituting $\forall$ for $\exists$ in the qualification of ε would not change the semantics of these two automata.

Now let $A = \langle Sig, Q, V, x_I, Tr, qual, rank\rangle$ be an automaton in $\mathbf{Aut}_{Sig}$, and let $z \in Var$. We define the automaton $\mu z.A$ as follows.

If $z \notin V$, we let $\mu z.A = A$.

If $z \in V$ and $x_I \in V$ then, by virtue of reduced form, we must have $x_I = z$. In this case we let $\mu z.A = A_{ff}$.

Now suppose $z \in V$ and x_I is a state. Let $\text{add}(z)$ be the set of addresses of z defined as above. Again, by virtue of reduced form, $\text{add}(z) \neq \emptyset$. Let $\text{min-add}(z)$ be the least element of $\text{add}(z)$ with respect to the ordering $\sqsubseteq$.

We define $\mu z.A$ by the following items.

- $Q^{\mu z.A} = Q \cup \{\text{min-add}(z)\}$.
- $V^{\mu z.A} = V - \{z\}$.
- $x_I^{\mu z.A} = x_I = \varepsilon$.
- For each transition $x = \tau$ of A, we have a transition $x = \tau'$ in $Tr^{\mu x.A}$, where τ' is obtained from τ by replacing each occurrence of z (if any) by min-add(z). Additionally, we have in $Tr^{\mu x.A}$ a transition min-add$(z) = \Delta$, where Δ is obtained from $Tr(x_I)$ by replacing each occurrence of z by min-add(z).
- The mapping $qual^{\mu x.A}$ is an extension of $qual$ by $qual^{\mu x.A}(\text{min-add}(z)) = qual(\varepsilon)$.
- Let $k = \max\{rank(x) : x \in Q\}$. The mapping $rank^{\mu x.A}$ is the extension of $rank$ by

$$rank^{\mu z.A}(\text{min-add}(z)) \quad = \quad 2 \cdot \lfloor \frac{k}{2} \rfloor + 1$$

  That is, $rank^{\mu z.A}(\text{min-add}(z))$ is k if k is odd, and $k+1$ otherwise.

It is easy to see that $\mu z.A$ is in canonical form.

The automaton $\nu z.A$ is defined in similar manner. Namely, if $z \notin V$, we let $\nu z.A = A$, and if $z = x_I$, we let $\nu z.A = A_{tt}$. If $z \in V$ and x_I is a state then the automaton $\nu z.A$ is defined analogically to $\mu z.A$ above, with the only difference that the mapping $rank^{\nu x.A}$ now extends $rank$ by the equation

$$rank^{\nu z}(\text{min-add}(z)) \quad = \quad 2 \cdot \lceil \frac{k}{2} \rceil$$

That is, $rank^{\mu z.A}(z)$ is k if k is even, and $k+1$ otherwise.

Proposition 7.2.3. $\widetilde{\mu z.A} = \nu z.\tilde{A}$, $\widetilde{\nu z.A} = \mu z.\tilde{A}$.

Proof. By Proposition 7.1.9 (page 162), it is sufficient to show that $\widetilde{\mu z.A} = \nu z.\tilde{A}$.

If $z \notin ar(A)$ or if $z = x_I$, the result is obvious. Otherwise, we have just to check that the ranks of min-add(z) are equal in $\widetilde{\mu z.A}$ and $\nu z.\tilde{A}$. If A is of Mostowski index $(0, k)$ then

- $\mu z.A$ is of Mostowski index $(0, k')$ with $k' = 2 \cdot \lfloor \frac{k}{2} \rfloor + 1$,
- and the rank of min-add(z) in $\mu z.A$ is k',
- the rank of min-add(z) in $\widetilde{\mu z.A}$ is $k' + 1$,
- $\tilde{A}$ is of Mostoski index $(1, k+1)$,
- $\nu z.\tilde{A}$ is of Mostoski index $(1, k'')$ with $k'' = 2 \cdot \lceil \frac{k+1}{2} \rceil$,
- and the rank of min-add(z) in $\nu z.\tilde{A}$ is k''.

It is easy to check that $k' + 1 = k''$, i.e., $2 \cdot \lfloor \frac{k}{2} \rfloor + 2 = 2 \cdot \lceil \frac{k+1}{2} \rceil$.

The proof is similar if A is of Mostowski index $(1, k)$ and amounts to checking that $2 \cdot \lfloor \frac{k}{2} \rfloor = 2 \cdot \lceil \frac{k-1}{2} \rceil$. □

We are ready to show that our system satisfies the axioms of Definition 2.1.1.

Theorem 7.2.4. *The tuple* $\langle \mathbf{Aut}_{Sig}, \mathrm{id}, ar, \mathrm{comp}, \mu, \nu \rangle$ *defined above is a μ-calculus.*

Proof. The satisfaction of Axioms 1–5 follows directly from definitions. To verify Axiom 6, we need to show that the automata $(A[\rho])[\pi]$ and $A[\rho \star \pi]$ are identical (and not only isomorphic), where the substitution $\rho \star \pi$ is defined by $\rho \star \pi(x) = \rho(x)[\pi]$.

We will prove that these automata have the same sets of states.

Since we consider several automata, we will now distinguish notationally the set of addresses of a variable z in an automaton, say, B, by writing $\mathrm{add}^B(z)$. By definition, $Q^{A[\rho]} = Q \cup \bigcup_{z \in V} \mathrm{add}^A(z) Q^{\rho(z)}$, and $V^{A[\rho]} = \bigcup_{z \in V} V^{\rho(z)}$. Therefore,

$$Q^{(A[\rho])[\pi]} = Q \cup \bigcup_{z \in V} \mathrm{add}^A(z) Q^{\rho(z)} \cup \bigcup_{\tilde{z} \in \bigcup_{z \in V} V^{\rho(z)}} \mathrm{add}^{A[\rho]}(\tilde{z}) Q^{\pi(\tilde{z})}$$

Now, it is easy to see that

$$\mathrm{add}^{A[\rho]}(\tilde{z}) = \bigcup_{z \in V} \mathrm{add}^A(z)\, \mathrm{add}^{\rho(z)}(\tilde{z})$$

and hence, using some basic set algebra, we get

$$Q^{(A[\rho])[\pi]} = Q \cup \bigcup_{z \in V} \mathrm{add}^A(z) Q^{\rho(z)} \cup \bigcup_{z \in V} \bigcup_{\tilde{z} \in V^{\rho(z)}} \mathrm{add}^A(z)\, \mathrm{add}^{\rho(z)}(\tilde{z})\, Q^{\pi(\tilde{z})}$$

On the other hand, we have

$$Q^{A[\rho \star \pi]} = Q \cup \bigcup_{z \in V} \mathrm{add}^A(z) Q^{\rho \star \pi(z)}$$

where $Q^{\rho \star \pi(z)} = Q^{\rho(z)[\pi]} = Q^{\rho(z)} \cup \bigcup_{\tilde{z} \in V^{\rho(z)}} \mathrm{add}^{\rho(z)}(\tilde{z})\, Q^{\pi(\tilde{z})}$. Hence we clearly get $Q^{(A[\rho])[\pi]} = Q^{A[\rho \star \pi]}$. The proof that the remaining items of $(A[\rho])[\pi]$ and $A[\rho \star \pi]$ coincide can be carried out similarly and we omit the details.

Finally, in order to satisfy Axiom 7, we have to show that, for some variable y, $(\theta x.A)[\rho] = \theta y.A[\rho\{\hat{y}/x\}]$. It is easy to check that this equality holds indeed for any variable y, provided $y \notin ar((\theta x.A)[\rho]) = \bigcup_{z \in V - \{x\}} ar(\rho(z))$. □

In the sequel, by abuse of notation, we use the symbol $\mathbf{Aut}_{Sig}$ also for denoting the μ-calculus of Theorem 7.2.4 (page 176).

7.2.2 The interpretation as homomorphism

We will now show that the μ and ν constructions over automata behave indeed as one should expect. That is, the semantics of the automaton $\mu x.A$ is the least fixed point of the interpretation associated with A, etc. In our general perspective of the μ-calculi this amounts to the fact that interpretation of automata is a homomorphism of the μ-calculi. We have here a property analogous to that of fixed-point terms, Proposition 2.3.10 (page 50), reminding the definition of the functional μ-calculus over a complete lattice (Section 2.2, page 44).

Theorem 7.2.5. *Let $\mathcal{B}$ be a semi-algebra over Sig. The mapping that with each automaton A in* $\mathbf{Aut}_{Sig}$ *associates its interpretation $[A]_{\wp\mathcal{B}}$ is a homomorphism from the μ-calculus of automata to the functional μ-calculus over* $\langle \mathcal{P}(B), \subseteq \rangle$.

Proof. We need to verify the four conditions of the definition of homomorphism (page 44). For each condition, we first explain what does it actually mean for the mapping $[.]_{\wp\mathcal{B}}$, and then verify the claim.

Identity. For any variable z in Var, $[\hat{z}]_{\wp\mathcal{B}}$ coincides with the $\hat{z}$ of the functional μ-calculus, i.e. with the identity mapping $\hat{z} : \mathcal{P}(B)^z \to \mathcal{P}(B)$ that sends each mapping $val : \{z\} \to \mathcal{P}(B)$ to $val(z)$.

Recall that $\hat{z}$ amounts to the trivial automaton $\langle Sig, \emptyset, \{z\}, z, \emptyset, \emptyset, \emptyset, \rangle$. Then the game $G(\hat{z}, \mathcal{B}, val)$ has only variable positions and a position (b, z) is winning for Eva if and only if $b \in val(z)$. Thus $[\hat{z}]_{\wp\mathcal{B}}(val) = val(z)$, as required.

Arity. The arity of $[A]_{\wp\mathcal{B}}$ equals to the arity of A. This property is an immediate consequence of the definition.

Composition. Let $A = \langle Sig, Q, V, x_I, \mathrm{Tr}, qual, rank \rangle$, be an automaton in $\mathbf{Aut}_{Sig}$, *and let $\rho : Var \to \mathbf{Aut}_{Sig}$ be a substitution. Let $V' = ar(A[\rho])$. Then the mapping $[A[\rho]]_{\wp\mathcal{B}} : \mathcal{P}(B)^{V'} \to \mathcal{P}(B)$ coincides with $[A]_{\wp\mathcal{B}}[[\rho]_{\wp\mathcal{B}}]$, where $[\rho]_{\wp\mathcal{B}} : Var \to \mathcal{P}(B)$ is defined by $[\rho]_{\wp\mathcal{B}}(z) = [\rho(z)]_{\wp\mathcal{B}}$.*

Note that, by definition of the functional μ-calculus,

$$[A]_{\wp\mathcal{B}}[[\rho]_{\wp\mathcal{B}}] : \mathcal{P}(B)^{V'} \to \mathcal{P}(B)$$

sends a mapping $val' : V' \to \mathcal{P}(B)$ to $[A]_{\wp\mathcal{B}}(val)$, where $val : V \to \mathcal{P}(B)$ is defined by $val(z) = [\rho(z)]_{\wp\mathcal{B}}(val' \upharpoonright V^{\rho(z)})$.

We have to verify

$$[A[\rho]]_{\wp\mathcal{B}}(val') = [A]_{\wp\mathcal{B}}(val)$$

If x_I is a variable then $A[\rho] = \rho(x_I)$, and, by virtue of the reduced form, $A = \hat{x_I}$. Hence the right-hand side amounts to $[\hat{x_I}]_{\wp\mathcal{B}}(val) = val(x_I) = [\rho(x_I)]_{\wp\mathcal{B}}(val')$, and so the claim is satisfied.

Suppose x_I is a state. Let $a \in B$. So, we have two games, and we need to show the following.

Claim. Eva can win the game $G(A[\rho], \mathcal{B}, val')$ from a position $(a, x_I^{A[\rho]})$ if and only if she can win the game $G(A, \mathcal{B}, val)$ from a position (a, x_I).

The following observation will be useful (see Section 7.2.1).

Lemma 7.2.6. *Suppose that $x \in add_A(z)$ is an address of $z \in V_\rho$ and that w is a state of the automaton $\rho(z)$. Then Eva wins the game $G(A[\rho], \mathcal{B}, val')$ from a state position (b, xw) if and only if she wins the game $(\rho(z), \mathcal{B}, val')$ from the position (b, w).*

Proof. It follows from the fact that the automaton $\rho(z)$ is isomorphic to the automaton (not in normal form!) obtained from $A[\rho]$ by restricting the states to $xQ^{\rho(z)}$ and taking x as the initial state. □

We are ready to prove the claim.

($\Rightarrow$) Suppose s' is a globally winning positional strategy for Eva in the game $G(A[\rho], \mathcal{B}, val')$. We define a positional strategy s for Eva in the game $G(A, \mathcal{B}, val)$ as follows. Let $x \in Q$ and let the transition in A with the head x be $x = f(x_1, \ldots, x_k)$. Therefore the corresponding transition in $A[\rho]$ is $x = f(x'_1, \ldots, x'_k)$, where the $x'_1, \ldots, x'_k$ are defined as on the page 173. Suppose (b, x) is a position of Eva in $G(A, \mathcal{B}, val)$, hence also in $G(A[\rho], \mathcal{B}, val')$ (recall $Q \subseteq Q^{A[\rho]}$). So, s' is defined for a position (b, x) and maps it to $(x = f(x'_1, \ldots, x'_k), v)$, for some $v = \langle d_1, \ldots, d_k, b\rangle$. We let s be defined for the position (b, x), and map it to $(x = f(x_1, \ldots, x_k), v)$.

Now suppose that, for an x as above, s' is defined for a transition position $(x = f(x'_1, \ldots, x'_k), \langle d_1, \ldots, d_k, b\rangle)$, and maps it to (x'_i, d_i). We let s be defined for the position $(x = f(x_1, \ldots, x_k), \langle d_1, \ldots, d_k, b\rangle)$ and map it to (x_i, d_i). Note that whenever x_i is a state, the positions (x'_i, d_i) and (x_i, d_i) coincide. We will show that if x_i is a variable, and Eva wins the game $G(A[\rho], \mathcal{B}, val')$ from the position (x'_i, d_i) by strategy s', then $d_i \in val(x_i)$, and hence the position (x_i, d_i) is winning for Eva in the game $G(A, \mathcal{B}, val)$.

There are two cases.

— If x'_i is a state then the claim follows by Lemma 7.2.6. Indeed, $x'_i = xi$ is an address of x_i (see page 173) and $val(x_i)$ is precisely

$$[\rho(x_i)]_{\wp\mathcal{B}}(val' \restriction V^{\rho(x_i)}).$$

— If x'_i is a variable then it must be $x_I^{\rho(x_i)}$, and $\rho(x_i) = \hat{x}'_i$. Hence $val(x_i) = [\hat{x'_i}]_{\wp\mathcal{B}}(val' \restriction V^{\rho(x_i)}) = val'(x'_i)$, and we have $d_i \in val'(x'_i)$, because this position is winning for Eva in $G(A[\rho], \mathcal{B}, val')$.

For all other positions, s can be defined arbitrarily.

Now suppose Eva wins the game $G(A[\rho], \mathcal{B}, val')$ from a position (a, ε) using the strategy s'. Observe that either this position or all its successors are in the domain of s' (depending on the qualification of $x_I^{A[\rho]} = \varepsilon$). Then it follows by the considerations above that every play consistent with s is also won by Eva. Indeed, such a play either coincides with a play in $G(A[\rho], \mathcal{B}, val')$ (in particular, it happens always if the play is infinite), or terminates in a variable position (d, x_i), such that $d \in val(x_i)$.

($\Leftarrow$) Now suppose r is a globally winning positional strategy for Eva in the game $G(A, \mathcal{B}, val)$. We know, moreover, that for each $x_i \in V$, if $b \in val(x_i)$ then Eva wins the game $G(\rho(x_i), \mathcal{B}, val' \restriction V^{\rho(x_i)})$ from the position $(b, x_I^{\rho(x_i)})$. If $\rho(x_i) = \hat{z}$, for some z, we have $b \in val'(z)$, and so (b, z) is a variable position winning for Eva in $G(A[\rho], \mathcal{B}, val')$. If $x_I^{\rho(z)}$ is a state, we have by Lemma 7.2.6 a winning strategy in the game $G(A[\rho], \mathcal{B}, val')$ from a corresponding position. By combining all these strategies we obtain a global winning strategy for Eva in $G(A[\rho], \mathcal{B}, val')$. The argument is similar as in the previous paragraph, and we omit the details.

Fixed points. The condition splits into two cases. We find it convenient to start with the greatest fixed point.

(ν) Let $A = \langle Sig, Q, V, x_I, Tr, qual, rank \rangle$, be an automaton in $\mathbf{Aut}_{Sig}$, and let $z \in Var$. We have to show that for each mapping $val : V - \{z\} \to \mathcal{P}(B)$, $[\nu z.A]_{\wp\mathcal{B}}(val)$ is indeed the greatest fixed point of the mapping that sends $M \in \mathcal{P}(B)$ to $[A]_{\wp\mathcal{B}}(val \oplus \{z \mapsto M\})$. where $val \oplus \{z \mapsto M\} : V \to \mathcal{P}(B)$ is the mapping which coincides with val on all variables $y \in V - \{z\}$, and whose value on z is M.

If $z \notin V$ then, by definition, $\nu z.A = A$, and the above mapping is constant; so the claim is obvious.

If $z = x_I$ then, by definition, $\nu z.A = A_{tt}$, and the mapping above is identity; so the claim is also true.

Suppose $z \in V$ and x_I is a state. Let us abbreviate $[\nu z.A]_{\wp\mathcal{B}}(val)$ by L. By Knaster–Tarski Theorem (see Corollary 1.2.10, page 11), it is enough to show two things.

(i) For all $M \subseteq B$, $M \subseteq [A]_{\wp\mathcal{B}}(val \oplus \{z \mapsto M\})$ implies $M \subseteq L$.

(ii) $L \subseteq [A]_{\wp\mathcal{B}}(val \oplus \{z \mapsto L\})$.

Ad (i). By hypothesis, there exists a globally winning strategy for Eva in the game $G(A, \mathcal{B}, val \oplus \{z \mapsto M\})$, say s, that is winning at each position $(b, x_I^A) = (b, \varepsilon)$, provided $b \in M$. We will show that from each such position (b, ε), Eva can also win the game $G(\nu z.A, \mathcal{B}, val)$ (which will prove $M \subseteq L$).

To this end, we define a strategy s' as follows. Suppose $x \in Q$ and the transition of A with the head x is $x = f(x_1, \ldots, x_k)$. Then the corresponding transition in $\nu z.A$ is $x = f(x'_1, \ldots, x'_k)$, where $x'_i = x_i$ if $x_i \in Q$, and $x'_i =$ min-add(z), whenever $x_i = z$. Now, if s is defined for a position (b, x) and maps it to $(x = f(x_1, \ldots, x_k), v)$, $v = \langle d_1, \ldots, d_k, b\rangle$, we let s' be also defined for (b, x) and map it to $(x = f(x'_1, \ldots, x'_k), v)$. In turn, if s is defined for a position $(x = f(x_1, \ldots, x_k), v)$ (with v as above) and maps it to, say, (x_i, d_i), we let s' be defined for the position $(x = f(x'_1, \ldots, x'_k), v)$ and map it to (x'_i, d_i). Moreover, for any $b \in M$, if $(b, \text{min-add}(z))$ is a position of Eva, we define s' on this position as follows. Note that, by hypothesis, s is defined on the position (b, x_I), say $s : (b, x_I) \mapsto (x_I = f(x_1, \ldots, x_k), v)$. We let $s' : (b, \text{min-add}(z)) \mapsto (\text{min-add}(z) = f(x'_1, \ldots, x'_k), v)$, where the x'_i's are defined as above.

Now suppose that a play starting from a position (b, ε), where $b \in M$, is consistent with s'. We claim that it is won by Eva. There are two possibilities. Either this play visits only finitely many times positions of the form $(d, \text{min-add}(z))$, and so, from some moment on, it coincides with a play in $G(A, \mathcal{B}, val \oplus \{z \mapsto M\})$ consistent with a winning strategy s. Or, it enters such positions infinitely often. But, since the rank of the state min-add(z) in $\nu z.A$ is maximal and even, the play is won by Eva also in this case.

Ad (ii). Now, by hypothesis, we have a globally winning strategy for Eva in the game $G(\nu z.A, \mathcal{B}, val)$, say r, which is winning at each position $(b, x_I^{\nu z.A}) = (b, \varepsilon)$, provided $b \in L$. We wish to show that, from each such position, Eva also wins the game $G(A, \mathcal{B}, val \oplus \{z \mapsto L\})$. The situation is somehow opposite to the previous case: Whenever a play played by Eva according to the strategy s reaches a state position $(d, \text{min-add}(z))$, the corresponding play in $G(A, \mathcal{B}, val \oplus \{z \mapsto L\})$ would reach a variable position (d, z). But clearly we have $d \in L$, since the right-hand side of the transition for min-add(z) coincides with that for ε. So Eva wins in this case. We leave a formal construction of a suitable strategy r' to the reader.

(μ) By duality, we reduce this case to the ν-case by using Proposition 7.2.3 (page 175) and Proposition 7.1.10 (page 162). □

7.3 Equivalences

7.3.1 From fixed-point terms to automata

In view of Theorem 7.2.5 of the previous section and the connection between the functional μ-calculus and games established in Section 4.3 (page 88), automata and fixed-point terms appear very close. We will make this intuition precise by showing that any fixed-point term can be transformed into an automaton which defines the same semantics. For this aim, the concept of homomorphism of μ-calculi will prove beneficial.

We first equip the set $\mathbf{Aut}_{Sig}$ with some operations which actually organize it into an algebra over $Sig_{\sim}$.

Definition 7.3.1. For each $f \in Sig$ with $\rho(f) = k$, let $\mathbf{f}$ be an automaton with the unique state ε, the set of variables $\{z_1, \ldots, z_k\}$, the unique transition $\varepsilon = f(z_1, \ldots, z_k)$, and such that $qual^{\mathbf{f}}(\varepsilon) = \forall$ and $rank^{\mathbf{f}}(\varepsilon) = 0$. Let $\tilde{\mathbf{f}}$ be defined similarly, with the only exception that $qual^{\tilde{\mathbf{f}}}(\varepsilon) = \exists$. (For concreteness, we assume that $z_1, \ldots, z_k$ are the first k variables in *Var*.)

Now, using $\mathbf{f}$ and the composition (see page 173), for any $A_1, \ldots, A_k \in \mathbf{Aut}_{Sig}$, we define an automaton $f^{\mathbf{Aut}_{Sig}}(A_1, \ldots, A_k)$ by $f^{\mathbf{Aut}_{Sig}}(A_1, \ldots, A_k) = \mathbf{f}[id\{A_1/z_1, \ldots, A_k/z_k\}]$.
Similarily, we let $\tilde{f}^{\mathbf{Aut}_{Sig}}(A_1, \ldots, A_k) = \tilde{\mathbf{f}}[id\{A_1/z_1, \ldots, A_k/z_k\}]$.

The following property will be useful.

Lemma 7.3.2. *For any substitution* ρ,
$f^{\mathbf{Aut}_{Sig}}(A_1, \ldots, A_k)[\rho] = f^{\mathbf{Aut}_{Sig}}(A_1[\rho], \ldots, A_k[\rho])$ *and*
$\tilde{f}^{\mathbf{Aut}_{Sig}}(A_1, \ldots, A_k)[\rho] = \tilde{f}^{\mathbf{Aut}_{Sig}}(A_1[\rho], \ldots, A_k[\rho])$

Proof. We consider the case of $\mathbf{f}$, the case of $\tilde{\mathbf{f}}$ is similar. Let us abbreviate $id\{A_1/z_1, \ldots, A_k/z_k\}$ by $id\{\boldsymbol{A}/\boldsymbol{z}\}$. Then, by the axiom 6 of the μ-calculus, we have $f^{\mathbf{Aut}_{Sig}}(A_1, \ldots, A_k)[\rho] = \mathbf{f}[id\{\boldsymbol{A}/\boldsymbol{z}\}][\rho] = \mathbf{f}[id\{\boldsymbol{A}/\boldsymbol{z}\} \star \rho]$, where $id\{\boldsymbol{A}/\boldsymbol{z}\} \star \rho(z_i) = A_i[\rho]$. Hence, the last term amounts to $\mathbf{f}[id\{A_1[\rho]/z_1, \ldots, A_k[\rho]/z_k\}] = f^{\mathbf{Aut}_{Sig}}(A_1[\rho], \ldots, A_k[\rho])$, as required. □

We are ready to define the aforementioned translation from the set of fixed-point terms over $Sig_{\sim}$ to $\mathbf{Aut}_{Sig}$, $auto : fix\mathcal{T}(Sig_{\sim}) \to \mathbf{Aut}_{Sig}$. It is defined by induction on the structure of fixed-point terms, using the construction of the μ-calculus of automata.

- For $z \in Var$, $auto : z \mapsto \hat{z}$,
- $auto : f(t_1, \ldots, t_k) \mapsto f^{\mathbf{Aut}_{Sig}}(auto(t_1), \ldots, auto(t_k))$,
- $auto : \tilde{f}(t_1, \ldots, t_k) \mapsto \tilde{f}^{\mathbf{Aut}_{Sig}}(auto(t_1), \ldots, auto(t_k))$,
- $auto : \theta x.t \mapsto \theta x.auto(t)$.

Proposition 7.3.3. *The mapping auto* : $\mathrm{fix}\mathcal{T}(Sig_{\sim}) \to \mathbf{Aut}_{Sig}$ *defined above is a homomorphism of the* μ*-calculi.*

Proof. The first and the last condition required of a homomorphism in Definition 2.1.4 (page 44) follow directly by the definition of *auto*, and the equation $ar(auto(t)) = ar(t)$ can be easily checked by induction on t. It remains to show that, for any substitution $\rho : Var \to \mathcal{T}(Sig_{\sim})$, $auto(t[\rho]) = auto(t)[auto \circ \rho]$. We will verify this condition by induction on t.

For $t = z$, we have

$$auto(z[\rho]) = auto(\rho(z)) = \hat{z}[auto \circ \rho] = auto(z)[auto \circ \rho].$$

For $t = f(t_1, \ldots, t_k)$, using the definition of substitution of fixed-point terms, the induction hypothesis and Lemma 7.3.2, we have

$$\begin{aligned} auto(f(t_1, \ldots, t_k)[\rho]) &= auto(f(t_1[\rho], \ldots, t_k[\rho])) \\ &= f^{\mathbf{Aut}_{Sig}}(auto(t_1[\rho]), \ldots, auto(t_k[\rho])) \\ &= f^{\mathbf{Aut}_{Sig}}(auto(t_1)[auto \circ \rho], \ldots, auto(t_k)[auto \circ \rho]) \\ &= auto(f(t_1, \ldots, t_k))[auto \circ \rho] \end{aligned}$$

The argument for $t = \tilde{f}(t_1, \ldots, t_k)$ is similar.

Finally, let $t = \theta x.t'$. Recall (see Definition 2.3.4, page 48) that $(\theta x.t')[\rho]$ is equal to $\theta z.t'[\rho\{\hat{z}/x\}]$, where z is the first variable in *Var* that does not belong to $ar'(t, \rho) = \bigcup_{w \in ar(t)-\{z\}} ar(\rho(w))$. Therefore, by the induction hypothesis, we have

$$\begin{aligned} auto((\theta x.t')[\rho]) &= auto(\theta z.t'[\rho\{\hat{z}/x\}]) \\ &= \theta z.auto(t'[\rho\{\hat{z}/x\}]) \\ &= \theta z.auto(t')[auto \circ \rho\{\hat{z}/x\}]. \end{aligned}$$

On the other hand, we have $auto(\theta x.t') = \theta x.auto(t')$. Moreover, while verifying the last axiom of the μ-calculus in the case of $\mathbf{Aut}_{Sig}$, we have remarked that $(\theta x.A)[\rho] = \theta y.A[\rho\{\hat{y}/x\}]$ holds, whenever $y \notin ar((\theta x.A)[\rho])$ (see page 176). Hence, in the present case, we have, by the choice of z,

$$(\theta x.auto(t'))[auto \circ \rho] = \theta z.auto(t')[(auto \circ \rho)\{\hat{z}/x\}]$$

Then the claim eventually follows by the observation that the substitutions $auto \circ \rho\{\hat{z}/x\}$ and $(auto \circ \rho)\{\hat{z}/x\}$ coincide, since $auto(\hat{z}) = auto(z) = \hat{z}$. □

Now, we are ready to show the crucial property of the homomorphism *auto*, namely that it preserves semantics.

Theorem 7.3.4. *For each fixed-point term t over Sig, and each semi-algebra $\mathcal{B}$,*

$$[auto(t)]_{\wp\mathcal{B}} = [t]_{\wp\mathcal{B}}$$

Proof. We have noted in Proposition 2.3.10 (page 50) that the interpretation of fixed-point terms over a set of function symbols F under a μ-interpretation $\mathcal{I}$, $t \mapsto [t]_{\mathcal{I}}$, is the *unique* homomorphism from $fix\mathcal{T}(F)$ to the functional μ-calculus over $D_{\mathcal{I}}$ satisfying the property

$$h(f(x_1, \ldots, x_{ar(f)}))(val) = f^{\mathcal{I}}(val(x_1), \ldots, val(x_{ar(f)})),$$

for each $f \in F$, and any mapping $val : \{x_1, \ldots, x_{ar(f)}\} \to D_{\mathcal{I}}$. So it holds in particular for $F = Sig_{\sim}$ and $\mathcal{I} = \wp\mathcal{B}$. Now, the mapping $t \mapsto [auto(t)]_{\wp\mathcal{B}}$ is a composition of two homomorphisms: $auto : fix\mathcal{T}(Sig_{\sim}) \to \mathbf{Aut}_{Sig}$ (Proposition 7.3.3) and the interpretation of automata (Theorem 7.2.5), and hence it is a homomorphism as well (see the remark after Definition 2.1.4, page 44). It remains to verify the equality $[auto(\varphi(x_1, \ldots, x_k)]_{\wp\mathcal{B}}(val) = \varphi^{\wp\mathcal{B}}(val(x_1), \ldots, val(x_k))$, for φ equal to f or $\tilde{f}$, $f \in Sig$. This property follows easily by the definition of *auto*. □

7.3.2 From automata to fixed-point terms

The reader may have noticed that the automata produced from fixed-point terms by the homomorphism *auto* have some special form, for example the initial state ε does not occur at the right-hand side of transitions. As a matter of fact, the mapping *auto* is not an isomorphism. *A priori* one could expect that the automata, apparently more complex than the fixed point terms, are also more powerful semantically. We shall see however that it is not so: any automaton is semantically equivalent to some fixed-point term.

The construction given below of such a fixed-point term is indeed a natural generalization of the construction given in Section 5.2.4 (page 122) in the case of words.

Let $A = \langle Sig, Q, V, x_I, Tr, qual, rank\rangle$ be an automaton in $\mathbf{Aut}_{Sig}$. We will construct a term t_A, semantically equivalent to A. We will also exhibit a relation between the Mostowski index of A and the level of the term t_A in the alternating fixed-point hierarchy.

If the initial symbol x_I is a variable, we let $t_A = x_I$.

If x_I is a state, we will find our term t_A as a component of a certain vectorial fixed-point term τ_A. Let $n = |Q|$ and $k = max\, rank(Q)$. We fix $n \cdot k$ fresh variables $x_q^{(i)}$ (distinct from V), doubly indexed by $i = 0, 1, \ldots, k$, and $q \in Q$. These variables (some of them) will be "new names" of the states of A. More precisely, we define a mapping $ren : Q \cup V \to Var$, by $ren(x) = x_q^{(rank(q))}$, for $q \in Q$, and $ren(z) = z$, for $z \in V$. Next, for each $q \in Q$, we define a term t_q over $Sig_{\sim}$, by

$$t_q = \begin{cases} f(ren(y_1), \ldots, ren(y_p)) & \text{if } qual(q) = \forall \\ \tilde{f}(ren(y_1), \ldots, ren(y_p)) & \text{if } qual(q) = \exists \end{cases}$$

We fix some ordering of Q (e.g. $\sqsubseteq$ of Definition 7.2.1), and let $\boldsymbol{x}^{(i)}$ be the vector of variables $x_q^{(i)}$, $q \in Q$, listed in this order. Similarly, we let $\boldsymbol{t}$ be the vector of the t_i's.

As the reader may have guessed, we are going to define τ_A by closing the vectorial fixed-point term $\boldsymbol{t}$ by fixed-point operators. However, since we wish to obtain an exact correspondence between the hierarchy of Mostowski indices and the fixed-point hierarchy, we will carefully distinguish two cases.

Let the Mostowski index of A be (ι, k). Then, if $\iota = 0$, we let

$$\tau_A = \theta x^{(k)}. \cdots .\nu x^{(2)}.\mu x^{(1)}.\nu x^{(0)}.t$$

and, if $\iota = 1$, we let

$$\tau_A = \theta x^{(k)}. \cdots .\nu x^{(2)}.\mu x^{(1)}.t$$

where the operator binding $x^{(i)}$ is μ if i is odd and ν if i is even.

Therefore, τ_A is in the normal form in the sense of Section 2.7.4, and moreover,

- if k is even then τ_A is in $\mathcal{P}^n_{k+(1-\iota)}(Sig)$,
- if k is odd then τ_A is in $\mathcal{S}^n_{k+(1-\iota)}(Sig)$.

For example, if the Mostowski index of A is $(1, 3)$ then the fixed-point prefix of τ_A has form $\mu\,\nu\,\mu$, and if the index is $(0, 3)$ then the prefix is $\mu\,\nu\,\mu\,\nu$, see also Figure 7.2, page 188.

Recall that, according to Definition 2.7.2, τ_A is formally a vector of fixed-point terms, and, by remark above and Proposition 2.7.8, its components are in the class $\Sigma_{k+(1-\iota)}(Sig)$ or $\Pi_{k+(1-\iota)}(Sig)$, depending on the parity of k. For $q \in Q$, let $(\tau_A)_q$ be the component of τ_A corresponding to the index q. It follows by construction that the set of free variables of τ_A is $ar(\tau_A) = V$. Since A is reduced (in particular, any variable is reachable from x_I), it can be easily seen that this is also the set of free variables of $(\tau_A)_{x_I}$.

We let $t_A = (\tau_A)_{x_I}$.

Proposition 7.3.5. *For any automaton $A \in \mathbf{Aut}_{Sig}$, and any semi-algebra $\mathcal{B}$,*

$$[t_A]_{\wp\mathcal{B}} = [A]_{\wp\mathcal{B}}$$

Proof. Let $A = \langle Sig, Q, V, x_I, Tr, qual, rank \rangle$. If x_I is a variable, the result is obvious.

Without loss of generality we may assume that A is closed. Indeed, if it is not the case originally, with each variable v of A we associate a constant symbol c_v not in Sig, and we let $Sig' = Sig \cup \{c_v \mid v \in V\}$. We consider the closed automaton A' obtained from A by adding the universal transitions $v = c_v$. With each Sig-semi-algebra $\mathcal{B}$ and any $val : V \to \mathcal{P}(B)$, we associate the Sig'-semi-algebra $\mathcal{B}'$, with the same domain B as $\mathcal{B}$, where $b \doteq c_V^{\mathcal{B}'}$ if and only if $b \in val(v)$. It is easy to see that $[A]_{\wp\mathcal{B}}(val) = [A']_{\wp\mathcal{B}'}$ and that $[t_A]_{\wp\mathcal{B}}(val) = [t_{A'}]_{\wp\mathcal{B}'}$. Therefore, it is enough to prove $[A]_{\wp\mathcal{B}} = [t_A]_{\wp\mathcal{B}}$ for any closed automaton A. We will show this by comparing two vectorial Boolean fixed-point terms: the one associated with $\lceil t_A \rceil_{\wp\mathcal{B}}$, and the one characterizing the game associated with the automaton A and the semi-algebra $\mathcal{B}$.

Let us consider the game associated with the closed automaton A by Definition 7.1.2, page 157. Its set of state positions is $B \times Q$; its set Pos_{Tr} of transition positions is the subset of elements $(q = f(q_1, \ldots, q_p), (b_1, \ldots, b_p, b))$ such that $b \doteq f^{\mathcal{B}}(b_1, \ldots, b_p)$. The vectorial boolean fixed-point term associated with this game (see Definition 4.3.2, page 89) is

$$\boldsymbol{T} = \theta \begin{bmatrix} \boldsymbol{X}^{(k)} \\ \boldsymbol{Y}^{(k)} \end{bmatrix} \cdots . \nu \begin{bmatrix} \boldsymbol{X}^{(2)} \\ \boldsymbol{Y}^{(2)} \end{bmatrix} . \mu \begin{bmatrix} \boldsymbol{X}^{(1)} \\ \boldsymbol{Y}^{(1)} \end{bmatrix} . \nu \begin{bmatrix} \boldsymbol{X}^{(0)} \\ \boldsymbol{Y}^{(0)} \end{bmatrix} . \begin{bmatrix} \boldsymbol{F} \\ \boldsymbol{G} \end{bmatrix}$$

where each $\boldsymbol{F}$ and each $\boldsymbol{X}^{(j)}$ (resp. $\boldsymbol{G}$ and each $\boldsymbol{Y}^{(j)}$) is a vector indexed by $B \times Q$ (resp. Pos_{Tr}). The component of index (b, q) of $\boldsymbol{F}$, where $q = f(q_1, \ldots, q_p)$, is

- $\sum\{\boldsymbol{Y}^{(j)}_{tr} \mid tr = (q = f(q_1, \ldots, q_p), (b_1, \ldots, b_p, b))\}$ if (b, q) is an Eva's position, i.e., if $qual(q) = \forall$,
- $\prod\{\boldsymbol{Y}^{(0)}_{tr} \mid tr = (q = f(q_1, \ldots, q_p), (b_1, \ldots, b_p, b))\}$ if $qual(q) = \exists$.

The component of index $tr = (q = f(q_1, \ldots, q_p), (b_1, \ldots, b_p, b))$ of $\boldsymbol{G}$ is

- $\prod_{i=1}^{p} \boldsymbol{X}^{(rank(q_i))}_{(b_i, q_i)}$ if tr is an Adam's position, i.e., if $qual(q) = \forall$,
- $\sum_{i=1}^{p} \boldsymbol{X}^{(rank(q_i))}_{(b_i, q_i)}$ if $qual(q) = \exists$.

By using Proposition 1.4.5 (page 29), we can substitute the definition of $\boldsymbol{Y}^{(0)}_{tr}$ for $\boldsymbol{Y}^{(0)}_{tr}$ in $\boldsymbol{F}$, so that $\boldsymbol{T}$ is equal to

$$\theta \begin{bmatrix} \boldsymbol{X}^{(k)} \\ \boldsymbol{Y}^{(k)} \end{bmatrix} \cdots . \nu \begin{bmatrix} \boldsymbol{X}^{(2)} \\ \boldsymbol{Y}^{(2)} \end{bmatrix} . \mu \begin{bmatrix} \boldsymbol{X}^{(1)} \\ \boldsymbol{Y}^{(1)} \end{bmatrix} . \nu \begin{bmatrix} \boldsymbol{X}^{(0)} \\ \boldsymbol{Y}^{(0)} \end{bmatrix} . \begin{bmatrix} \boldsymbol{F}' \\ \boldsymbol{G} \end{bmatrix}$$

where the component of index (b, q) of $\boldsymbol{F}'$, where $q = f(q_1, \ldots, q_p)$, is

- $\sum_{b \doteq f^{\mathcal{B}}(b_1, \ldots, b_p)} \prod_{i=1}^{p} \boldsymbol{X}^{(rank(q_i))}_{(b_i, q_i)}$ if $qual(q) = \forall$,
- $\prod_{b \doteq f^{\mathcal{B}}(b_1, \ldots, b_p)} \sum_{i=1}^{p} \boldsymbol{X}^{(rank(q_i))}_{(b_i, q_i)}$ if $qual(q) = \exists$.

Because $\boldsymbol{F}'$ does not depend on the variables in the $\boldsymbol{Y}^{(j)}$, we have, by Proposition 1.4.4 (page 28), that the component of index (b, q) of $\boldsymbol{T}$ is equal to the component of index (b, q) of

$$\boldsymbol{T}' = \theta \boldsymbol{X}^{(k)} . \cdots . \nu \boldsymbol{X}^{(2)} . \mu \boldsymbol{X}^{(1)} . \nu \boldsymbol{X}^{(0)} . \boldsymbol{F}'.$$

Now, by Proposition 3.2.6 (page 74), we have $\lceil \boldsymbol{\tau}_A \rceil_{\wp\mathcal{B}} = h(\chi_{\wp\mathcal{B}}(\boldsymbol{\tau}_A))$, and, by Proposition 3.2.8 (page 75),

$$\chi_{\wp\mathcal{B}}(\boldsymbol{\tau}_A) = \theta u_{\boldsymbol{x}^{(k)}} . \cdots . \nu u_{\boldsymbol{x}^{(2)}} . \mu u_{\boldsymbol{x}^{(1)}} . \nu u_{\boldsymbol{x}^{(0)}} . \chi_{\wp\mathcal{B}}(\boldsymbol{t}).$$

Note that $\chi_{\wp\mathcal{B}}(\boldsymbol{\tau}_A)$ is indexed by $B \times Q$ and that there is a natural bijection between $\boldsymbol{X}^{(j)}$ and $u_{\boldsymbol{x}^{(j)}}$ (namely, $\boldsymbol{X}^{(j)}_{(b,q)}$ is identified with $u_{b,\boldsymbol{x}^{(j)}_q}$). It follows that $[A]_{\wp\mathcal{B}} = [t_A]_{\wp\mathcal{B}}$ if and only if for any $b \in B$, the component of index (b, x_I) of $\boldsymbol{T}'$ is equal to the component of index (b, x_I) of $\theta u_{\boldsymbol{x}^{(k)}}. \cdots .\nu u_{\boldsymbol{x}^{(2)}}.\mu u_{\boldsymbol{x}^{(1)}}.\nu u_{\boldsymbol{x}^{(0)}}.\chi_{\wp\mathcal{B}}(\boldsymbol{t})$. Indeed, we show that $\boldsymbol{F}' = \chi_{\wp\mathcal{B}}(\boldsymbol{t})$, i.e., for any $w : B \times Var \to \mathbb{B}$, we have $\boldsymbol{F}'(w) = \chi_{\wp\mathcal{B}}(\boldsymbol{t})(w)$. By Proposition 3.2.5 (page 74), $h(\chi_{\wp\mathcal{B}}(\boldsymbol{t})(w)) = [\boldsymbol{t}]_{\wp\mathcal{B}}[h^*(w)]$. Therefore, we have to prove that for any b and q, $\boldsymbol{F}'_{(b,q)}(w) = 1$ if and only if $b \in [t_q]_{\wp\mathcal{B}}[h^*(w)]$.

Let $q = f(q_1, \ldots, q_p)$ be the transition of head q. If $qual(q) = \forall$ then

- $\boldsymbol{F}'_{(b,q)}(w) = 1$ if and only if there is $b \doteq f^{\mathcal{B}}(b_1, \ldots, b_p)$ such that for $i = 1, \ldots, p$, $w(\boldsymbol{X}^{(rank(q_i))}_{(b_i,q_i)}) = 1$,
- $b \in [t_q]_{\wp\mathcal{B}}[h^*(w)]$ if and only if there is $b \doteq f^{\mathcal{B}}(b_1, \ldots, b_p)$ such that for $i = 1, \ldots, p$, $b_i \in h^*(w)(\boldsymbol{x}^{(rank(q_i))}_{q_i})$.

But $b_i \in h^*(w)(\boldsymbol{x}^{(rank(q_i))}_{q_i})$ if and only if $w(u_{b,\boldsymbol{x}^{(j)}_q}) = w(\boldsymbol{X}^{(j)}_{(b,q)}) = 1$.

If $qual(q) = \exists$ then

- $\boldsymbol{F}'_{(b,q)}(w) = 0$ if and only if there is $b \doteq f^{\mathcal{B}}(b_1, \ldots, b_p)$ such that for $i = 1, \ldots, p$, $w(\boldsymbol{X}^{(rank(q_i))}_{(b_i,q_i)}) = 0$,
- $b \notin [t_q]_{\wp\mathcal{B}}[h^*(w)]$ if and only if there is $b \doteq f^{\mathcal{B}}(b_1, \ldots, b_p)$ such that for $i = 1, \ldots, p$, $b_i \notin h^*(w)(\boldsymbol{x}^{(rank(q_i))}_{q_i})$,

and we conclude in the same way as above.

□

7.3.3 Conclusion

We summarize the above considerations in the following.

Theorem 7.3.6. *For any automaton A over signature Sig, one can construct a fixed-point term t_A over the signature $Sig_\sim$ with the same semantics, i.e., such that*

$$[A]_{\wp\mathcal{B}} \quad = \quad [t_A]_{\wp\mathcal{B}}$$

for any semi-algebra $\mathcal{B}$. Moreover, if the Mostowski index of A is (ι, k) then t_A can be chosen in $\Pi_{k+(1-\iota)}(Sig_\sim)$, whenever k is even, and in $\Sigma_{k+(1-\iota)}(Sig_\sim)$, whenever k is odd.

Conversely, for any fixed-point term t over signature $Sig_\sim$, an automaton with the same semantics is given by a homomorphism of the μ-calculi auto : $\mathrm{fix}\,\mathcal{T}(Sig_\sim) \to \mathbf{Aut}_{Sig}$. Moreover, if t is in $\Pi_k(Sig_\sim)$, for $k > 0$, then *the Mostowski index of auto(t) is compatible with* $(1, k)$ *or with* $(0, k-1)$*, depending on whether k is even or odd, respectively, and if t is in* $\Sigma_k(Sig_{\sim})$*, for* $k > 0$*, the index is compatible with* $(1, k)$ *or* $(0, k-1)$*, depending on whether k is odd or even, respectively.*

That is, with the assumed semantics, automata and fixed-point terms have the same definability power, and the equivalence refines to the levels of the alternation-depth hierarchy, and the hierarchy of the Mostowski indices, correspondingly, as shown on the Figure 7.2.

Proof. The first part is a consequence Proposition 7.2.2 and Proposition 7.3.5, as well as of the construction of τ_A used there.

The second part is a consequence of Theorem 7.3.4. To see the property regarding indices, recall the definition of the μ-calculus of automata (Section 7.2.1). Observe, in particular, that the composition of automata p-reserves indices, while the operators μ and ν do not alter the first component of the index. Then the claim can be easily checked using the fact that *auto* is a homomorphism.

□

Remark. We can note that although the classes of the hierarchy of Mostowski indices (Figure 7.1, page 161) and those of the alternation-depth hierarchy (Figure 2.1, page 57) coincide, Figure 7.2 is not obtained by superposition of Figures 2.1 and 7.1; instead, one of the hierarchies must be "switched". (The situation is somehow analogous as in the parallel between the Borel hierarchy and arithmetical hierarchy.)

7.4 Bibliographic notes and sources

The concept of a finite automaton took its shape in the fundamental work of Kleene [53]; the equivalence between finite automata and regular expressions (Klenee theorem) established a general paradigm which is confirmed, in particular, by Theorem 7.3.6 above. We refer the reader to the bibliographic notes in the Chapter 5 (page 138) for more references to automata on finite and infinite words. Thatcher and Wright [94] and Doner [28] extended the concept of a finite automaton to (finite) trees, and Rabin [83] used automata on infinite trees in his landmark proof of decidability of the monadic second order arithmetic of n successors. We refer the reader to the exposition of the subject by Thomas [95, 97]. The alternating version of a finite automaton was introduced at the same time as the alternating Turing machine [24]; alternating automata on infinite trees were considered first by Muller and Schupp [71].

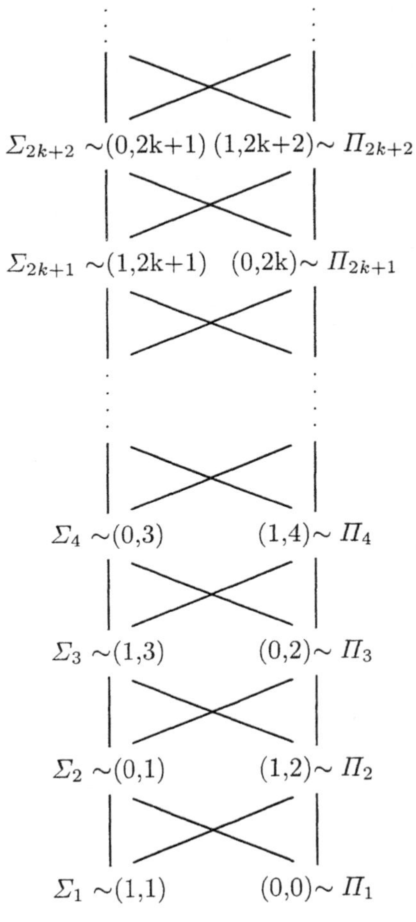

Fig. 7.2. The hierarchy of Mostowski indices and of alternation-depth

A transformation of the formulas of the modal μ-calculus into Rabin tree automata was an essential part of the proof of the elementary decidability of this logic given by Streett and Emerson [91]. A more direct transformation *via* alternating automata was given later by Emerson and Jutla [33]. A converse translation, from automata to the μ-calculus, was shown in [77], by which the *equivalence* in expressive power between the Rabin automata and the μ-calculus of fixed-point terms without intersection interpreted in the powerset algebra of trees was established. (Note that by the Rabin Theorem, adding intersection would not increase the expressive power of the μ-calculus.) In fact, [77] shows an equivalence on a more general level, namely between the μ-calculus over powerset algebras constructed from arbitrary algebras, and for a suitable concept of an automaton over arbitrary algebras. This result was subsequently extended to automata on semi-algebras in [78]. A similar concept of automaton over transition system was introduced by Janin and Walukiewicz [45]. A yet more general concept of automata over complete lattices with monotonic operations was considered by Janin [44].

An equivalence between nondeterministic Büchi automata on trees and a kind of fixed-point expressions (essentially equivalent to fixed-point terms without intersection of the level Π_2) was discovered earlier by Takahashi [92]. The correspondence between fixed-point terms of level Π_2 (with intersection) interpreted in the powerset algebra of trees and nondeterministic Büchi automata was shown by the authors of this book [9]. (Note that this also yields the equivalence between nondeterministic and alternating Büchi automata on trees.) A more direct proof of this result was given later by Kaivola [51].

8. Hierarchy problems

8.1 Introduction

A fundamental question about any μ-calculus is: Are the fixed-point operators redundant? For example, any monotone Boolean function $f : \{0,1\}^n \to \{0,1\}$ can be defined using the operations $\vee$ and $\wedge$, and so any term of the Boolean μ-calculus is equivalent to a term without fixed-point operators. (This example also shows that the redundancy of fixed points does not make a μ-calculus uninteresting.)

It is not hard to give an example of a μ-calculus where actually both extremal fixed-point operators are essential — let us recall the μ-calculus over infinite words discussed in Chapter 5. But then a more subtle question arises, whether the alternations between the least and greatest fixed points add to the expressive power of terms. In Section 2.6 (page 55), we have introduced a formal hierarchy which measures the complexity of objects of a μ-calculus by the number of alternations between μ and ν. While this hierarchy is obviously strict for fixed-point terms considered syntactically (see section 2.6.3, page 56), it need not be so for their interpretations within functional μ-calculi. Note that in the latter case we ask for a minimal number of alternations required in any term t' such that $[t']_{\mathcal{I}} = [t]_{\mathcal{I}}$. Indeed, in the case of the μ-calculus on words, we have seen that the hierarchy actually collapses to the alternation-free level *Comp* (Σ_1, Π_1) in the presence of intersection (see Theorem 5.3.1, page 126) and to the level Π_2 (i.e., $\nu\mu$) without intersection (see Proposition 5.1.17, page 115). *A priori* it is not obvious that there must exist any functional μ-calculus with infinite hierarchy.

A first example of a μ-interpretation with an infinite hierarchy was provided [75] by a powerset algebra of trees $\wp\mathcal{T}_{Sig}$ (c.f. Example 6.1.3, page 142) restricted to the signature $Sig \cup \{\vee\}$ (with $\vee$, standing for $\tilde{eq}$, interpreted as the binary set-union). Note the absence of $\wedge$ and the duals $\tilde{f}$ of $f \in Sig$. The alternation-depth hierarchy over this interpretation corresponds precisely to the hierarchy of Mostowski indices of *nondeterministic* automata. The following terms witness the infinity of the hierarchy:

$$\theta x_k. \ldots \nu x_2.\mu x_1.c_1(x_1,x_1) \vee c_2(x_2,x_2) \vee \ldots \vee c_k(x_k,x_k)$$

The interpretation comprises all trees (over signature $\{c_1, \ldots, c_n\}$) such that, for each path, the maximal subscript i repeating infinitely often is even. (It can be easily seen, for instance, by considering the automaton associated with the above term by the mapping *auto* of Section 7.3.1, page 180.) The above terms can be further encoded within a single finite signature, provided it contains at least two symbols, one of them of arity > 1 (see [78] for details).

This example does not settle, however, the hierarchy problem for the (unrestricted) powerset algebra of trees. As a matter of fact, all the above languages have their complements recognizable by Büchi automata, and therefore are in the Σ_2 class. The problem, better known as the hierarchy problem for the modal μ-calculus, had remained open for a while and was solved in 1996 independently by J. Bradfield and G. Lenzi. One example, given in a subsequent paper by Bradfield [19], consisted of the game terms presented above in Section 4.4.1, page 95 (originally given in a modal form). These formulas, defining the set of winning positions in a parity game, were first introduced by Emerson and Jutla for bi-partite games, and then generalized by I. Walukiewicz for arbitrary parity games (see also Bibliographical notes at the end of this chapter).

8.2 The hierarchy of alternating parity tree automata

8.2.1 Games on the binary tree

A game $G = (Pos_a, Pos_e, Mov, rank)$ is a game on the binary tree if the underlying graph is the full binary tree, i.e., $Pos = Pos_a \cup Pos_e$ is the set $\{1,2\}^*$ of all words over $\{1,2\}$ and $Mov = \{(u, ua) \mid u \in \{1,2\}^*, a \in \{1,2\}\}$.

With this game, assuming that $rank(Pos) \subseteq \{1, \ldots, k\}$, we associate a syntactic tree t_G over the binary signature $S_k = \{c_i \mid 1 \leq i \leq k\} \cup \{d_i \mid 1 \leq i \leq k\}$ defined by: $t_G(u)$ is equal to $c_{rank(u)}$ if $u \in Pos_a$ and to $d_{rank(u)}$ $u \in Pos_e$.

Indeed, this transformation is a bijection: with any tree $t \in T_{S_k}$ we associate the game $G(t) = (Pos_a, Pos_e, Mov, rank)$ on the binary tree defined by $Pos_a = \{u \mid \exists i : t(u) = c_i\}$, $Pos_e = \{u \mid \exists i : t(u) = d_i\}$ and $rank(u) = i$ if and only if $t(u)$ is either c_i or d_i. It is easy to see that $G(t_G) = G$ and that $t_{G(t)} = t$.

We denote by L_k the set of all trees $t \in T_{S_k}$ such that ε is a winning position in the game $G(t)$.

We show that L_k is recognized by a nondeterministic automaton W_k of Mostowski index $(1, k)$ when $k \geq 2$. (see Section 7.1.5, page 164).

The set of states of W_k is $\{q_i \mid 1 \leq i \leq k\} \cup \{q_\top\}$. All these states are universal. The rank of q_i is i, the rank of $q_\top$ is 2 (or any even number less than or equal to k). The transitions of W_k are

- $q_\top = \bigvee_{i=1}^{k} c_i(q_\top, q_\top) \vee \bigvee_{i=1}^{k} d_i(q_\top, q_\top)$,
- $q_j = \bigvee_{i=1}^{k} c_i(q_i, q_i) \vee \bigvee_{i=1}^{k} d_i(q_i, q_\top) \vee \bigvee_{i=1}^{k} d_i(q_\top, q_i)$,
  for any $j \in \{1, \ldots, k\}$.

The initial state is q_1. Since all states but $q_\top$ have the same transition, one could choose any other q_i as initial state without changing the set $[W_k]_{\wp T_{S_k}}$.

If we restrict this automaton to the alphabet $\{c_i \mid 1 \leq i \leq n\}$, the transitions become $q_j = \bigvee_{i=1}^{k} c_i(q_i, q_i)$ and we get the automaton already considered in Section 7.1.5 (page 166).

Proposition 8.2.1. $[W_k]_{\wp T_{S_k}} = L_k$.

Proof. By Definition 7.1.1 (page 157), the game that allows us to define $[W_k]_{\wp T_{S_k}}$ is the following, taking into account some simplifications in the case of nondeterministic automata, (see Section 7.1.5).

- $Pos_e = T_{S_k} \times Q$.
- Pos_a consists of all the following transition positions for all $t_1, t_2 \in T_{S_k}$,
  - $(q_\top = c_i(q_\top, q_\top), \langle t_1, t_2, c_i(t_1, t_2)\rangle)$, $(q_\top = d_i(q_\top, q_\top), \langle t_1, t_2, d_i(t_1, t_2)\rangle)$,
  - $(q_j = c_i(q_i, q_i), \langle t_1, t_2, c_i(t_1, t_2)\rangle)$, $(q_j = d_i(q_i, q_\top), \langle t_1, t_2, d_i(t_1, t_2)\rangle)$, and $(q_j = d_i(q_\top, q_i), \langle t_1, t_2, d_i(t_1, t_2)\rangle)$, for all $j, i \in \{1, \ldots, k\}$
- The Eva's moves are the following:
  - from $(c_i(t_1, t_2), q_j)$ to $(q_j = c_i(q_i, q_i), \langle t_1, t_2, c_i(t_1, t_2)\rangle)$
  - from $(d_i(t_1, t_2), q_j)$ to $(q_j = d_i(q_i, q_\top), \langle t_1, t_2, d_i(t_1, t_2)\rangle)$,
  - from $(d_i(t_1, t_2), q_j)$ to $(q_j = d_i(q_\top, q_i), \langle t_1, t_2, d_i(t_1, t_2)\rangle)$,
  - from $(c_i(t_1, t_2), q_\top)$ to $(q_\top = c_i(q_\top, q_\top), \langle t_1, t_2, c_i(t_1, t_2)\rangle)$
  - from $(d_i(t_1, t_2), q_\top)$ to $(q_\top = d_i(q_\top, q_\top), \langle t_1, t_2, c_i(t_1, t_2)\rangle)$
- The Adam's moves are from $(f(q, q'), \langle t_1, t_2, f(t_1, t_2)\rangle)$ to (t_1, q) and to (t_2, q') for any symbol $f \in S_k$.
- The rank of (t, q) is equal to the rank of q. The rank of an Adam's position is 0.

In position $(c_i(t_1, t_2), q)$, Eva can only move to

$$(q = c_i(q', q'), \langle t_1, t_2, c_i(t_1, t_2)\rangle)$$

where q' is $q_\top$ or q_i, according to whether q is $q_\top$ or not. In position $(d_i(t_1, t_2), q_\top)$, she can only move to $(q_\top = c_i(q_\top, q_\top), \langle t_1, t_2, c_i(t_1, t_2)\rangle)$. In position $(d_i(t_1, t_2), q_j)$, she can only move to $(d_i(q_i, q_\top), \langle t_1, t_2, d_i(t_1, t_2)\rangle)$ or to $(d_i(q_\top, q_i), \langle t_1, t_2, d_i(t_1, t_2)\rangle)$. Furthermore, one may assume that in these last two transition positions, Adam can only move to (t_1, q_i) (resp. (t_2, q_i) because otherwise he looses. Moreover, cancelling Adam's position in a play does not change the win or the loss of this play (since these positions are of rank 0). These remarks allow us to simplify the game as follows.

- $Pos_e = \{(d_i(t_1, t_2), q_j) \mid t_1, t_2 \in T_{S_k}, i, j \in \{1, \dots, k\}\}$.
- $Pos_a = \{(c_i(t_1, t_2), q_j) \mid t_1, t_2 \in T_{S_k}, i, j \in \{1, \dots, k\}\}$.
- There is an Eva's move from $(d_i(t_1, t_2), q)$ to (t_1, q_i) and to (t_2, q_i). There is an Adam's move from $(c_i(t_1, t_2), q)$ to (t_1, q_i) and to (t_2, q_i).
- The rank of (t, q_i) is equal to i.

Let us also associate a rank with each tree $t \in T_{S_k}$, defined by $rank'(t) = i$ if and only if $t = c_i(t_1, t_2)$ or $t = d_i(t_1, t_2)$. Let $(t_1, s_1), (t_2, s_2), \dots (t_n, s_n), \dots$ be an infinite play in this simplified game. It is easy to see that by definition of the moves, $rank(s_{n+1}) = rank'(t_n)$. Therefore Eva wins this play if and only if $\limsup_{n\to\infty} rank'(t_n)$ is even. This remark implies that the states play no role and allows us to yet simplify the previous game to get the game G defined by

- $Pos_e = \{d_i(t_1, t_2) \mid t_1, t_2 \in T_{S_k}, i \in \{1, \dots, k\}\}$.
- $Pos_a = \{c_i(t_1, t_2) \mid t_1, t_2 \in T_{S_k}, i \in \{1, \dots, k\}\}$.
- There is an Eva's move from $d_i(t_1, t_2)$ to t_1 and to t_2. There is an Adam's move from $c_i(t_1, t_2)$ to t_1 and to t_2.
- The rank of t is equal to $rank'(t)$.

Finally, let $G(t)$ be the game on the binary tree associated with the tree t. We claim that t is a winning position in G if and only if ε is a winning position in $G(t)$. The proof is similar to Example 6.3.3 (page 148).

Let us denote by $t.u$, for any $u \in \{1, 2\}^*$, the tree defined by $t.u(v) = t(uv)$. If the position u in $G(t)$ belongs to Eva, then $t(u)$ is equal to some d_i, therefore $t.u$ has the form $d_i(t_1, t_2)$ and $t.u$ is an Eva's position in G. If u is an Adam's position in $G(t)$, then, by a similar argument, $t.u$ is an Adam's position in G. Moreover, the rank of u in $G(t)$ is equal to the rank of $t.u$ in G.

Let s be a memoryless strategy for Eva in G, i.e., for each t' in the form $d_i(t_1, t_2)$, Eva moves to t_j with $j = s(t') \in \{1, 2\}$. We then define the strategy s_t on $G(t)$ by: for any Eva's position u in $G(t)$, $s_t(u) = u \cdot s(t.u)$. If $\varepsilon, u_1, u_2, \dots, \dots$ is a play in $G(t)$ consistent with the strategy s_t then $t, t.u_1, t.u_2, \dots$ is a play in G consistent with the strategy s, and Eva wins the first play if and only if she wins the second one. It follows that if s is winning at t then s_t is winning at ε. Thus, if t is a winning position, then there is a strategy s winning at t, this implies that s_t is winning at ε and that ε is a winning position.

Conversely, if s_t is a memoryless strategy in $G(t)$, we select an arbitrary strategy in G, not necessarily memoryless, that satisfies the following condition.

Let $t, t_1, \dots, t_n$ be the beginning of a play in G where t_n is an Eva position. There exists a word $u = a_1 a_2 \cdots a_n$ such that $t_n = t.u$ and this word is an Eva's position in $G(t)$. Then we set $s(t, t_1, t_2, \dots, t_n) = t_n.s_t(u)$.

Again it is easy to see that if $t_0 = t, t_1, t_2, \ldots$ is a play in G consistent with s, then $\varepsilon, a_1, a_1a_2, \ldots$ is a play in G consistent with s_t, where a_i is the unique element of $\{1, 2\}$ such that $t_i = t_{i-1}.a_i$. For the same reasons as above, it follows that if ε is a winning position in $G(t)$, then t is a winning position in G. □

8.2.2 Alternating automata on trees

Let *Sig* be a signature not containing the symbol *eq* and let Sig_{eq} be the signature $Sig \cup \{eq\}$.

Recall that in the game associated with a closed alternating automaton, in a state position (b, q), where the transition for q is $q = f(q_1, \ldots, q_p)$, the player that decomposes b according to f can be Adam or Eva depending on whether q is universal or existential. However, in the case of trees, the decomposition is unique anyway, and the only thing that matters is the choice which follows afterwards. It is technically convenient to assume that, for f different from *eq*, the decomposition is always performed by Eva (as in a nondeterministic automaton). This motivates the following definition.

Definition 8.2.2. We say that an automaton over the signature Sig_{eq} is *simply alternating* if for any *existential* state q, the transition of head q is $q = eq(q_1, q_2)$.

For example, a nondeterministic automaton is always simply alternating.

Proposition 8.2.3. *For any closed automaton A of Mostowski index $(1, k)$ with $k \geq 2$ there is a simply alternating closed automaton B of same Mostowski index such that $[A]_{\wp\mathcal{T}_{Sig}} = [B]_{\wp\mathcal{T}_{Sig}}$.*

Proof. We construct B by the following transformation. First we add to the states of A a new universal state $q_\top$ of rank 2 and the transition $q_\top = eq(q_\top, q_\top)$.

Next we replace each transition $q = f(q_1, \ldots, q_m)$, where q is existential and $f \in Sig$, by the transition
$q = f(q_1, q_\top, \ldots, q_\top) \vee f(q_\top, q_2, \ldots, q_\top) \vee \cdots \vee f(q_\top, q_\top, \ldots, q_\top, q_m)$
$\vee \bigvee_{g \neq f} g(q_\top, q_\top, \ldots, q_\top)$. Moreover, q is now qualified universal.

By Proposition 7.3.5 (page 184), we have to prove that $[t_A]_{\wp\mathcal{T}_{Sig}} = [t_B]_{\wp\mathcal{T}_{Sig}}$. By construction of t_A and t_B, $[t_A]_{\wp\mathcal{T}_{Sig}}$ is the component of index x_I of the vector $\boldsymbol{L}_A = \theta\boldsymbol{x}^{(k)}.\ldots.\nu\boldsymbol{x}^{(2)}.\mu\boldsymbol{x}^{(1)}.[\boldsymbol{t}_A]_{\wp\mathcal{T}_{Sig}}$ indexed by Q, the set of states of A, and $[t_B]_{\wp\mathcal{T}_{Sig}}$ is the component of index x_I of the vector $\boldsymbol{L}_B = \theta\boldsymbol{y}^{(k)}.\ldots.\nu\boldsymbol{y}^{(2)}.\mu\boldsymbol{y}^{(1)}.[\boldsymbol{t}_A]_{\wp\mathcal{T}_{Sig}}$ indexed by $Q \cup \{q_\top\}$.

Obviously, the component of index $q_\top$ of $\boldsymbol{L}_B$ is T_{Sig}. For the states q of which we have modified the transition, the component of index q of

$[t_A]_{\wp T_{Sig}}$ is $\tilde{f}^{\wp T_{Sig}}(q_1, q_2, \ldots, q_m)$, and the same component of $[t_B]_{\wp T_{Sig}}$ is $f^{\wp T_{Sig}}(q_1, q_\top, \ldots, q_\top) \cup f^{\wp T_{Sig}}(q_\top, q_2, \ldots, q_\top) \cup \cdots \cup f^{\wp T_{Sig}}(q_\top, q_\top, \ldots, q_m) \cup \bigcup_{g \neq f} g^{\wp T_{Sig}}(q_\top, q_\top, \ldots, q_\top)$.

Since T_{Sig} can be substituted for $q_\top$, the result is a consequence of the following equality, obviously true for any $T_1, \ldots, T_m \subseteq T_{Sig}$.

$$\begin{aligned}\tilde{f}^{\wp T_{Sig}}(T_1, T_2, \ldots, T_m) = \quad & f^{\wp T_{Sig}}(T_1, T_{Sig}, \ldots, T_{Sig}) \\ \cup \quad & f^{\wp T_{Sig}}(T_{Sig}, T_2, \ldots, T_{Sig}) \\ \cup \quad & \cdots \\ \cup \quad & f^{\wp T_{Sig}}(T_{Sig}, T_{Sig}, \ldots, T_m) \\ \cup \quad & \bigcup_{g \neq f} g^{\wp T_{Sig}}(T_{Sig}, T_{Sig}, \ldots, T_{Sig}).\end{aligned}$$

This automaton B is an alternating automaton, and we transform it into a " standard" one as explained in Section 7.1.5, (page 162). By construction this automaton is simply alternating.

□

If A is a simply alternating closed automaton, the game $G(A, \mathcal{T}_{Sig})$ that allows us to define the semantics of A in the semi-algebra $\mathcal{T}_{Sig}$ can be simplified by using the fact that from an arbitrary state position (t, q) there is at most one possible move to a transition position: If the transition for q is $q = eq(q_1, q_2)$, the move is to $(q = eq(q_1, q_2), \langle t, t, t \rangle)$; If $q = f(q_1, \ldots, q_m)$, and if $t = f(t_1, \ldots, t_m)$, the move is to $(q = f(q_1, \ldots, q_m), \langle t_1, \ldots, t_m, t \rangle)$.

This remark allows us to simplify this game by cancelling transition positions, as we did it for the automaton W_k, so that with A we associate the following game $G(A)$.

- The set of all positions is the set $T_{Sig} \times Q$ extended by two positions *Loss* and *Win*. The positions *Loss* and *Win* are to Eva and are respectively of rank 1 and 2.
  A position (t, q) is to Eva if and only if q is existential. The rank of this position is the rank of q.
- The only move from *Loss* is to *Loss*, the only move from *Win* is to *Win*. Thus, *Win* is a winning position and *Loss* is not a winning position for Eva.
  The moves from (t, q) are defined as follows.
  - If $q = eq(q_1, q_2)$, there is a move to (t, q_1) and to (t, q_2).
  - If $q = f(q_1, \ldots, q_m)$ and $t = f(t_1, \ldots, t_m)$ (note that in this case, (t, q) is an Adam's position in $G(A)$), there are moves to the m positions (t_i, q_i).
    If $q = f(q_1, \ldots, q_m)$ and t has not the form $f(t_1, \ldots, t_m)$, there is a unique move to *Loss*. Thus, (t, q) is not a winning position for Eva. This

is because in the original game, (t,q) is an Eva's position and Eva from which she cannot move to any transition position.
- If $q = a$, with a a constant symbol, there is a unique move to *Win* if $t = a$, to *Loss* otherwise.

Note that from each Eva's position (t,q) there are 1 or 2 possible moves (according to whether in the transition $q = eq(q_1,q_2)$, $q_1 = q_2$ or not), and that from each Adam's position, the number of moves, always greater than 0, is bounded by the maximal arity of symbols in *Sig*. Thus, any choice of a move by Adam can be done by a sequence of binary choices.

The above remark is the motivation for the following definition.

Definition 8.2.4. Let A be a simply alternating closed automaton of Mostowski index $(1,k)$ with $k \geq 2$.

First we consider the two trees $t_\perp$ and $t_\top$ of T_{S_k} which are the unique two trees that satisfy the equalities $t_\perp = d_1(t_\perp,t_\perp)$ and $t_\top = d_2(t_\top,t_\top)$.

Next, we define the mapping $\gamma_A : T_{Sig} \times Q \to T_{S_k}$ by the following conditions. It is easy to see that there is one and only one mapping satisfying them.

Let $t \in T_{Sig}$ and $q \in Q$. Let $i \in \{1,\dots,k\}$ be the rank of q.

- If $q = eq(q_1,q_2)$ then $\gamma_A(t,q)$ is equal to $d_i(\gamma_A(t,q_1),\gamma_A(t,q_2))$ if $qual(q) = \exists$ and to $c_i(\gamma_A(t,q_1),\gamma_A(t,q_2))$ if $qual(q) = \forall$.
- If $q = f(q_1,\dots,q_m)$ then
  - if $t = f(t_1,\dots,t_m)$ then $\gamma_A(t,q)$ is equal to $c_i(\gamma_A(t_1,q_1),c_i(\gamma_A(t_2,q_2),\cdots,c_i(\gamma_A(t_{m-1},q_{m-1}),\gamma_A(t_m,q_m))\cdots))$ if $m > 1$ and to $c_i(\gamma_A(t_1,q_1),\gamma_A(t_1,q_1))$ if $m = 1$,
  - otherwise $\gamma_A(t,q) = t_\perp$.
- If $q = a$ then $\gamma_A(t,q)$ is equal to $t_\top$ if $t = a$, and to $t_\perp$ otherwise.

By adapting the proof of Proposition 8.2.1, we can prove the following result.

Proposition 8.2.5. *Let A be a simply alternating closed automaton of Mostowski index $(1,k)$ with $k \geq 2$, and let q_I be its initial state.*

Then, $\forall t \in T_{Sig}$, $t \in \lfloor A \rfloor_{\wp T_{Sig}} \Leftrightarrow \gamma_A(t,q_I) \in L_k$.

Finally, let $\Gamma_A : T_{Sig} \to T_{S_k}$ be defined by

$$\Gamma_A(t) = d_1(\gamma_A(t,q_I),\gamma_A(t,q_I)).$$

Since, obviously, $\Gamma_A(t) \in L_k$ if and only if $\gamma_A(t,q_I) \in L_k$, the previous proposition implies the following one.

Proposition 8.2.6. *Let A be a simply alternating closed automaton of Mostowski index $(1,k)$ with $k \geq 2$.*

Then, $\forall t \in T_{Sig}$, $t \in \lfloor A \rfloor_{\wp T_{Sig}} \Leftrightarrow \Gamma_A(t) \in L_k$.

8.2.3 A diagonal argument

Let us recall the definition of the usual ultrametric distance Δ_{Sig} on T_{Sig} which makes T_{Sig} a complete, and even compact, metric space [8].

Definition 8.2.7. If $t = t'$ then $\Delta_{Sig}(t,t') = 0$. If the symbols at the root of t and t' are distinct, then $\Delta_{Sig}(t,t') = 1$. Otherwise, there exists a nonconstant symbol f, such that $t = f(t_1,\dots,t_m)$, $t' = f(t'_1,\dots,t'_m)$ and $t_i \neq t'_i$ for some i; in this case we set $\Delta_{Sig}(t,t') = \frac{1}{2}\max_{i=1}^{m} \Delta_{Sig}(t_i,t'_i)$, which is not equal to 0.

A mapping $h : T_{Sig} \to T_{Sig'}$ is said to be *nonexpansive* if $\forall t,t' \in T_{Sig}$, $\Delta_{Sig'}(h(t),h(t')) \leq \Delta_{Sig}(t,t')$. It is *contracting* if there is a constant $c < 1$ such that $\forall t,t' \in T_{Sig}$, $\Delta_{Sig'}(h(t),h(t')) \leq c\Delta_{Sig}(t,t')$.

Lemma 8.2.8. *Let A be a simply alternating closed automaton of Mostowski index $(1,k)$ with $k \geq 2$. Then $\Gamma_A : T_{Sig} \to T_{S_k}$ is contracting.*

Proof. By construction, for each $q \in Q$ the mapping that sends t on $\gamma_A(t,q))$ is nonexpansive (this statement can be formally proved by induction on $-\log_2(\Delta_{Sig}(t,t'))$, which is always a natural number whenever $t \neq t'$). Hence $\Delta_{S_k}(\Gamma_A(t),\Gamma_A(t')) = \frac{1}{2}\Delta_{S_k}(\gamma_A(t,q_I),\gamma_A(t',q_I)) \leq \frac{1}{2}\Delta_{Sig}(t,t')$. □

Since L_k is the set of trees accepted by the automaton $\mathcal{W}_k$ of Mostowski index $(1,k)$ we know, by Proposition 7.1.10 (page 162), that its complement $\overline{L_k} = T_{S_k} - L_k$ is accepted by the dual automaton $\widetilde{W_k}$ of Mostowski index $(0,k-1)$. The following theorem shows that $\widetilde{W_k}$ cannot be semantically equivalent to an automaton of Mostowski index $(1,k)$.

Theorem 8.2.9. *Let $k \geq 2$. For any closed automaton A over S_k of Mostowski index $(1,k)$, $[A]_{\wp T_{S_k}} \neq \overline{L_k}$.*

Proof. Assume that there is an automaton A of Mostowski index $(1,k)$ such that $[A]_{\wp T_{S_k}} = \overline{L_k}$. By Proposition 8.2.3, we may assume that A is simply alternating, and by Proposition 8.2.6, we have $\forall t \in T_{S_k}$, $t \in \overline{L_k} \Leftrightarrow \Gamma_A(t) \in L_k$. But Γ_A is a contracting mapping from T_{S_k} to T_{S_k}. By the well-known Banach fixed-point theorem, there exists an element $t_A \in T_{S_k}$ such that $t_A = \Gamma(t_A)$. For this t_A, we get $t_A \in \overline{L_k} \Leftrightarrow t_A \in L_k$, an obvious contradiction. □

8.2.4 The hierarchy of Mostowski indices for tree languages

We can formulate in a different way the result stated above, by using a hierarchy of families of tree languages based on the hierarchy of Mostowski indices for automata.

Definition 8.2.10. Let *Sig* be some signature. We say that $L \subseteq T_{Sig}$ is in the family $M_{i,k}(Sig)$ with $0 \leq i \leq 1$ and $k \geq i$ if there is a closed automaton A of Mostowski index (i, k) such that $L = [A]_{\wp T_{Sig}}$. Of course, inclusion between these families are depicted by the same diagram as in Figure 7.1, page 161.

By Proposition 7.1.10 (page 162) we get the following result.

Proposition 8.2.11. *Let* $L \subseteq T_{Sig}$ *and let* $k \geq 1$. *Then* $L \in M_{1,k}(Sig) \Leftrightarrow T_{Sig} - L \in M_{0,k-1}(Sig)$.

Therefore the previous theorem states that for $k \geq 2$, $\overline{L_k} \in M_{0,k-1}(S_k) - M_{1,k}(S_k)$ which is equivalent by the previous proposition to $L_k \in M_{1,k}(S_k) - M_{0,k-1}(S_k)$.

It follows that for each $k \geq 2$, there is a signature S such that the whole Mostowski hierarchy is not included in $M_{1,k}(S)$ or in $M_{0,k-1}(S)$. If it were the case then we would have either $M_{0,k-1}(S) \subseteq M_{1,k}(S)$ or $M_{1,k}(S) \subseteq M_{0,k-1}(S)$, and either $M_{0,k-1}(S_k) - M_{1,k}(S_k)$ or $M_{1,k}(S_k) - M_{0,k-1}(S_k)$ should be empty. This is impossible if $S = S_k$.

However, this is not enough to prove that there is a signature *Sig* for which the Mostowski hierarchy does not collapse (for instance, the whole hierarchy for S_k might be include in $M_{1,2k}(S_k)$).

In the next section we show that such a signature does exist.

8.2.5 Universal languages

Definition 8.2.12. Let *Sig* be a signature, and let $k \geq 2$.
A language $L \subseteq T_{Sig}$ is said to be k-universal, if there is a nonexpansive mapping $h : T_{S_k} \to T_{Sig}$ such that $\forall t \in T_{S_k}$, $t \in L_k \Leftrightarrow h(t) \in L$.

Since the identity mapping from T_{S_k} to T_{S_k} is nonexpansive, L_k is k-universal.

Theorem 8.2.13. *Let Sig be any signature, and let* $k \geq 2$. *If* $L \subseteq T_{Sig}$ *is* k*-universal, then* $L \notin M_{0,k-1}(Sig)$.

Proof. Assume that $L \in M_{0,k-1}(Sig)$. Then $\overline{L} = T_{Sig} - L \in M_{1,k}(Sig)$ and there is a simply alternating closed automaton A of Mostowski index $(1, k)$ such that $\overline{L} = [A]_{\wp T_{Sig}}$. By Proposition 8.2.6 (page 197), we have $t \in \overline{L} \Leftrightarrow \Gamma_A(t) \in L_k$. Since L is universal, there is a nonexpansive mapping $h : T_{S_k} \to T_{Sig}$ such that $t \in L_k \Leftrightarrow h(t) \in L$. It follows that $t \in \overline{L} \Leftrightarrow h(\Gamma_A(t)) \in L$. But the composition of Γ_A and h is contracting. Therefore there exists a tree t_A such that $t_A = h(\Gamma_A(t_A))$, and we get $t \in \overline{L} \Leftrightarrow t_A \in L$, a contradiction. □

Example 8.2.14. For $k \geq 2$, let S'_k be the signature consisting of two binary symbols c and d, and k unary symbols $a_1, \dots, a_k$. The language B_k is the set of trees in $T_{S'_k}$ accepted by the nondeterministic automaton $\mathcal{B}_k$ of Mostowski index $(1, k)$ defined as follows.

- Its set of states is $\{q_1, \dots, q_k, q_\top\}$. The rank of q_i is i. The rank of $q_\top$ is 2. The initial state is q_1.
- There is a transition $q_\top = c(q_\top, q_\top) \vee d(q_\top, q_\top) \vee \bigvee_{j=1}^{k} a_j(q_\top)$, and for each i there is a transition $q_i = c(q_1, q_1) \vee d(q_1, q_\top) \vee d(q_\top, q_1) \vee \bigvee_{j=1}^{k} a_j(q_j)$.

Now let h_k be the tree homomorphism from T_{S_k} into $T_{S'_k}$ defined by $h_k(c_i(x, y)) = c(a_i(x), a_i(y))$ and $h_k(d_i(x, y)) = d(a_i(x), a_i(y))$. Obviously, h_k is nonexpansive, and it is easy to check that $t \in L_k \Leftrightarrow h_k(t) \in B_k$.

Therefore B_k is k-universal. $\square$

Now let Sig be any signature that contains at least one binary symbol b and a constant symbol a. For any $k \geq 2$ we consider the tree homomorphism $h_k : T_{S_k} \to T_{Sig}$ whose value on S_k is defined by:
$h_k(c_1(x, y)) = b(a, b(b(a, b(x, y)), a))$, $\quad h_k(d_1(x, y)) = b(a, b(b(b(x, y), a)), a)$,
$h_k(c_{i+1}(x, y) = b(a, h(c_i(x, y)))$, $\quad h_k(d_{i+1}(x, y) = b(a, h(d_i(x, y)))$
so that $h_k(c_i(x, y)) = \underbrace{b(a, b(a, \dots b(a,}_{i} b(b(a, b(x, y)), a)) \cdots))$

and $h_k(d_i(x, y)) = \underbrace{b(a, b(a, \dots b(a,}_{i} b(b(b(x, y), a), a)) \cdots))$.

It is obvious, by construction, that h_k is nonexpansive. Moreover, it is injective. Hence, $t \in L_k \Leftrightarrow h_k(t) \in h_k(L_k)$. It follows that $h_k(L_k)$ is k-universal.

Proposition 8.2.15. *For each $k \geq 2$, $h_k(L_k) \in M_{1,k}(Sig) - M_{0,k-1}(Sig)$.*

Proof. Since $h_k(L_k)$ is k-universal, by the previous theorem, it is not in $M_{0,k-1}(Sig)$. Let us show that it is in $M_{1,k}(Sig)$.

Let us define the following nondeterministic automaton of Mostowski index $(1, k)$. Its states are

- $s_0, s_1, \dots, s_k, s'_1, s'_2, \dots, s'_k, s_a$, all of rank 1,
- $s_\top$ [illegible] rank 2,
- q_1 $q_2, \dots, q_k, q'_1, q'_2, \dots, q'_k$ where the rank of q_i and q'_i is i.

The initial state is s_0 and the transitions are

- $s_a = a$, $s_\top = \bigvee_{f \in Sig} f(s_\top, \dots, s_\top)$,
- for all i from 1 to $k - 1$, $s_i = b(s_a, s_{i+1}) \vee b(s'_i, s_a)$,
- $s_0 = b(s_a, s_1)$, $s_k = b(s'_k, s_a)$,
- for all i from 1 to k, $s'_i = b(s_a, q_i) \vee b(q'_i, s_a)$
- for all i from 1 to k, $q_i = b(s_0, s_0)$ and $q'_i = b(s_0, s_\top) \vee b(s_\top, s_0)$.

This automaton counts the number of b's having an a as first argument with the states s_i. When the state s_i sees the a as the second argument, the automaton stops counting and memorizes i in the state s'_i. Now it checks whether a is the first or the second argument of b which determines whether it parses an encoding of c_i or of d_i. Then, with q_i or q'_i it goes on (in state s_0) on both successors or on one of them. □

Indeed similar constructions can be performed whenever *Sig* contains at least two symbols, and one of them has arity 2 or more. In such a case the Mostowski hierarchy is strict.

Theorem 8.2.16. *Let Sig be a signature which contains at least two symbols f and g, with $ar(f) \geq 2$. Then for each $k \geq 2$, there is a set $K_k \subseteq T_{Sig}$ which is in $M_{1,k}(Sig) - M_{0,k-1}(Sig)$. Moreover, one can choose K_k so that it is accepted by a nondeterministic tree automaton of index $(1, k)$.*

By Theorem 7.3.6 (page 186) we get the following corollary.

Corollary 8.2.17. *Let Sig be a signature as in the previous theorem. Then the interpretations in the powerset algebra $\wp\mathcal{T}_{Sig}$ of fixed-point terms in the classes $\Sigma_k(Sig_\sim)$ and $\Pi_k(Sig_\sim)$ form a strict hierarchy.*

In particular, there exist functional μ-calculi for which the alternation-depth hierarchy is strict.

8.3 Weak alternating automata

We can easily adapt to automata the notion of a weak parity condition introduced in Section 4.5 (page 99) for games.

An alternating automaton $A = \langle Sig, Q, V, x_I, Tr, qual, rank\rangle$ is said to be *weak* if it has the following additional property:

For any transition $q = f(y_1, \ldots, y_p)$, and for any $i \in \{1, \ldots, p\}$, if $y_i \in Q$ then $rank(q) \leq rank(y_i)$. It is obvious that if A is weak, then its dual $\widetilde{A}$ is weak too.

Restricted to weak simply alternating automata, the Mostowski indices induce a hierarchy on these automata. We prove below, by the same method as before, that this hierarchy is strict, that provides a new proof of a Mostowski's result [67].

Theorem 8.3.1 (Mostowski). *The Mostowski index hierarchy of weak simply alternating automata is strict.*

Proof. Since A is weak, for any $t \in T_{Sig}$, the binary tree $\Gamma_A(t) \in T_{S_k}$ has the following property: For any path $w = u_0u_1 \cdots u_n \cdots \in \{1,2\}^\omega$, the sequence $i_0, i_1, \ldots, i_n, \ldots \in \{1, \ldots, k\}^\omega$ such that i_n is the index of the letter $\Gamma_A(t)(u_0u_1 \cdots u_n)$ is nondecreasing by construction of Γ_A.

Let K_k be the set of all trees in T_{S_k} satisfying this property.

Therefore, if L'_k is any language such that $L'_k \cap K_k = L_k \cap K_k$ then Proposition 8.2.6 (page 197) can be rewritten as $\forall t \in T_{Sig}$, $t \in [A]_{\wp T_{Sig}} \Leftrightarrow \Gamma_A(t) \in L'_k$. We will construct a weak automaton W'_k of Mostowski index $(1,k)$ which recognizes a language with this property. The automaton will be a variant of W_k which takes into account the property of the trees in K_k.

Its states are $\{q_i \mid 1 \le i \le k\} \cup \{q_\top\}$. All these states are universal. The initial state is q_1. The rank of q_i is i, the rank of $q_\top$ is k if k is even and $k-1$ otherwise.

The transitions of W'_k are

- $q_\top = \bigvee_{i=1}^{k} c_i(q_\top, q_\top) \vee \bigvee_{i=1}^{k} d_i(q_\top, q_\top)$,
- for $1 \le j \le k$, $q_j = \bigvee_{i=j}^{k} c_i(q_i, q_i) \vee \bigvee_{i=j}^{k} d_i(q_\top, q_i) \vee \bigvee_{i=j}^{k} d_i(q_i, q_\top)$.

Let L'_k be the set of trees accepted by this automaton. When W_k has to recognize a tree t in K_k, it never sees a letter c_i or d_i in state q_j with $j > i$. Therefore the game associated with t and W_k is the same as the game associated with t and W'_k, and W_k accepts t if and only if W'_k accepts t. Hence, $L_k \cap K_k = L'_k \cap K_k$.

Unfortunately, W'_k is not weak when k is odd; the transition $q_k = c_k(q_k, q_k) \vee d_k(q_k, q_\top) \vee d_k(q_\top, q_k)$ of W'_k contains $q_\top$ of rank $k-1$. But in state q_k, the automaton cannot accept any tree when the rank of q_k is odd, for along at least one path in the tree it will remain for ever in this state. Therefore we can replace this transition by $q_k = c_k(q_k, q_k) \vee d_k(q_k, q_k)$.

This automaton W''_k is now weak. Let L''_k be the language recognized by this automaton when k is odd and be L'_k when k is even.

Then L''_K is recognized by a weak automaton of Mostowski index $(1,k)$. If the complement of L''_k were also recognized by a weak automaton A of Mostowski index $(1,k)$, we would have $t \notin L''_k \Leftrightarrow \Gamma_A(t) \in L''_k$, that is a contradiction when t is a fixed point of Γ_A. □

We have seen in Proposition 8.2.3 (page 195) that every automaton A is equivalent to a simply alternating automaton B of the same Mostowski index. But the construction we gave does not preserve weakness in general: If there is in A a transition $q = f(q_1, \ldots, q_m)$ with q existential, the construction of B replaces this transition by $q = f(q_1, q_\top, \ldots, q_\top) \vee f(q_\top, q_2, \ldots, q_\top) \vee \cdots \vee f(q_\top, q_\top, \ldots, q_\top, q_m) \vee \bigvee_{g \ne f} g(q_\top, q_\top, \ldots, q_\top)$. If A is weak, B is weak if we can assign to $q_\top$ an even rank greater than or equal to the rank of q. In particular, if k is even, we can assign the rank k to $q_\top$ and then B is weak. But if k is odd and if there is transition $q = f(q_1, \ldots, q_m)$ with q existential of rank k, B is not weak unless we assign to $q_\top$ the rank $k+1$ and then B is of Mostowski index $k+1$.

Since we may assume that each weak automaton of Mostowski index $(1,k)$, with k even, is simply alternating, we get as a consequence of the previous theorem:

Theorem 8.3.2. *The Mostowski index hierarchy of weak automata is strict.*

8.4 Bibliographic notes and sources

The fact that the alternation hierarchy collapses to the level Π_2 (i.e., $\nu\mu$) for the powerset algebra of infinite words without intersection, but at the same time Π_2 is different than Σ_2, was discovered by D. Park in collaboration with J. Tiuryn (see [81]). The complexity of reduction of fixed-point terms to a Π_2 form was studied recently in [90]. Collapsing of the analogous hierarchy with intersection to the level $Comp\,(\Sigma_1 \cup \Pi_1)$ was shown in [10]. On the other hand, Wagner [100] showed that (in our present terminology) the hierarchy induced by the Mostowski indices of *deterministic* automata over infinite words is infinite.

A different μ-calculus of finite or infinite words was examined in [76], where the concatenation of ∞-languages (but not intersection) was allowed. That μ-calculus captures the power of context-free grammars (over finite or infinite words), and the alternation hierarchy again collapses at the level $Comp\,(\Sigma_1 \cup \Pi_1)$.

For trees, Rabin [85] showed that (in our present terminology) Büchi automata have less expressive power than Rabin automata; the counter–example was the set of binary trees over alphabet $\{a, b\}$ such that, along each path, b occurs only finitely often. This easily implies that the hierarchy in the powerset algebra of trees without intersection does not collapse at the level Π_2. The strictness of that hierarchy was shown [75] in 1986 (see also [78]). For the powerset algebra of trees with intersection, as well as for a closely related modal μ-calculus, the hierarchy problem has remained open for a decade (it was only known that the hierarchy does not collapse to the level Π_2 or Σ_2 [9]). In 1996, Bradfield [17] showed the strictness of the hierarchy for the modal μ-calculus, by transferring some hierarchy from arithmetic previously studied by Lubarsky [58] (a simpler and self–contained proof of this result was given later by the same author [18]). An independent proof of this result was given in the same year by Lenzi [57] who gave examples of formulas over n–ary trees requiring increasing number of alternations. Both proofs have not automatically yielded the existence of a single powerset algebra of trees (with intersection) with an infinite hierarchy. This fact was shown later independently in [19] and in [5]; it is this last proof that we have presented in the book.

The μ-calculus formulas describing the winning positions in parity games were introduced for bi–partite games by Emerson and Jutla [33] and generalized to arbitrary games by Walukiewicz [102]. The fact that these formulas also witness the strictness of the alternation–depth hierarchy was observed by Bradfield [18].

9. Distributivity and normal form results

We have seen in Chapter 5 that, in the μ-calculus over infinite words, the intersection operation $\wedge$ is redundant: Any closed fixed-point term is semantically equivalent to a fixed-point term without $\wedge$. A similar property can be shown for the powerset algebra of trees, $\wp\mathcal{T}_{Sig}$. It was first discovered by Muller and Schupp [72] and phrased in terms of automata: Any alternating tree automaton can be *simulated* by a nondeterministic automaton. In this chapter, we generalize these results, by providing a condition on μ-interpretations which is sufficient for elimination of intersection. (*Per analogiam*, we continue to call this property *simulation*.) Intuitively, this technical condition states that $\wedge$ commutes with other operators like in the μ-calculus of words, where we have, e.g., $aL \cap aL' = a(L \cap L')$.

9.1 The propositional μ-calculus

Let F be a set of symbols, consisting of two binary symbols $\vee$ and $\wedge$ which will be used as infix operators, and two 0-ary symbols $\perp$ and $\top$. The terms built up from this set of symbols are called *propositional.*

Let $\mathcal{D}$ be the set of all interpretations $\mathcal{I}$ such that the domain $D_{\mathcal{I}}$ is a distributive complete lattice and $\vee$ and $\wedge$ are interpreted as the least upper bound and the greatest lower bound in this domain. We say that two terms t and t' of the same arity are equivalent modulo $\mathcal{D}$, denoted by $t =_{\mathcal{D}} t'$, if for any $\mathcal{I} \in \mathcal{D}$, $[t]_{\mathcal{I}} = [t']_{\mathcal{I}}$. Similarly we say that two vectors of the same length are equivalent if their components are pairwise equivalent.

The following lemma is a straightforward generalization of the Shannon's lemma (Lemma 3.1.1, page 71).

Lemma 9.1.1. *For any functional term* $t \in \text{funct}\mathcal{T}(F)$ *and any variable* $x \in Var$,

$$t =_{\mathcal{D}} t[id\{\perp/x\}] \vee (x \wedge t[id\{\top/x\}]).$$

Proof. The proof is by induction on t, using the distributivity property.

- If $t = \top, \bot$, or if $t = y \neq x$, the result is obvious, since then $t[id\{\bot/x\}] = t[id\{\top/x\}] = t$ and $t \vee (x \wedge t) =_{\mathcal{D}} t$.
- If $t = x$, we have $t[id\{\bot/x\}] \vee (x \wedge t[id\{\top/x\}]) = \bot \vee (x \wedge \top) =_{\mathcal{D}} x$.
- Let $a = t[id\{\bot/x\}], b = t[id\{\top/x\}]$, $a' = t'[id\{\bot/x\}], b' = t'[id\{\top/x\}]$, and let us assume $t =_{\mathcal{D}} a \vee (x \wedge b)$, $t' =_{\mathcal{D}} a' \vee (x \wedge b')$. Then $t \vee t' =_{\mathcal{D}} a \vee (x \wedge b) \vee a' \vee (x \wedge b') =_{\mathcal{D}} (a \vee a') \vee (x \wedge (b \vee b'))$ and

$$\begin{aligned} t \wedge t' &=_{\mathcal{D}} (a \vee (x \wedge b)) \wedge (a' \vee (x \wedge b')) \\ &=_{\mathcal{D}} (a \wedge a') \vee (x \wedge a \wedge b') \vee (a' \wedge x \wedge b) \vee (x \wedge b \wedge b') \\ &=_{\mathcal{D}} (a \wedge a') \vee x \wedge ((a \wedge b') \vee (a' \wedge b) \vee (b \wedge b')). \end{aligned}$$

But, since $a \leq b$ and $a' \leq b'$ we have $(a \wedge b') \vee (a' \wedge b) \vee (b \wedge b') =_{\mathcal{D}} b \wedge b'$.

□

As an immediate consequence of this lemma we get a simple characterization of extremal fixed points.

Proposition 9.1.2. *For any functional term $t \in$ funct$\mathcal{T}(F)$ and any variable $x \in Var$,*

$$\mu x.t =_{\mathcal{D}} t[id\{\bot/x\}], \qquad \nu x.t =_{\mathcal{D}} t[id\{\top/x\}].$$

Proof. Let $\mathcal{I}$ be an interpretation in $\mathcal{D}$.
Let $a = [t[id\{\bot/x\}]]_{\mathcal{I}}$, $b = [t[id\{\top/x\}]]_{\mathcal{I}}$. By the previous lemma, $[t]_{\mathcal{I}} = a \vee (x \wedge b)$. In particular $[t]_{\mathcal{I}}[id\{a/x\}] = a \vee (a \wedge b) = a$ and $[t]_{\mathcal{I}}[id\{b/x\}] = a \vee (b \wedge b) = a \vee b = b$. It follows that $\mu x.[t]_{\mathcal{I}} \leq a \leq b \leq \nu x.[t]_{\mathcal{I}}$.

On the other hand, a and b are respective approximations of $\mu x.[t]_{\mathcal{I}}$ and $\nu x.[t]_{\mathcal{I}}$, thus, $a \leq \mu x.[t]_{\mathcal{I}}$ and $\nu x.[t]_{\mathcal{I}} \leq b$. □

Corollary 9.1.3. *If $\boldsymbol{x}$ is a vector of variables of length n and if $\mathbf{t}$ is a vector of functional terms, of length n, then there is a vector $\mathbf{t}'$ of functional terms, of length n, such that $\theta \boldsymbol{x}.\mathbf{t} =_{\mathcal{D}} \mathbf{t}'$.*

Proof. The proof is by induction on n, using the previous proposition and Definition 2.7.2 (page 59). □

It follows that every fixed point term is equivalent, up to $=_{\mathcal{D}}$ to a functional one and thus, the $\Sigma_n - \Pi_n$ hierarchy collapses.

Note that the above results holds only for distributive lattices. Here is an example of a non distributive lattice where the least fixed point of a propositional functional term is not a functional term.

Example 9.1.4. Let $\overline{\mathbb{N}}$ be the complete lattice $\mathbb{N} \cup \{\infty\}$. Let E be the set $\{\bot, \top, b\} \cup \{a_i, b_i, c_i \mid i \in \overline{\mathbb{N}}\}$.

We order this set by the least ordering such that

- $\forall i, j \in \overline{\mathbb{N}}, i \leq j \Rightarrow$
  - $\bot \leq a_i \leq a_j \leq \top$,
  - $\bot \leq b_i \leq b_j \leq\leq b \leq \top$,
  - $\bot \leq c_i \leq c_j \leq \top$,
- $\forall i \in \overline{\mathbb{N}}$, $b_{2i} \leq a_i$ and $b_{2i+1} \leq c_i$.

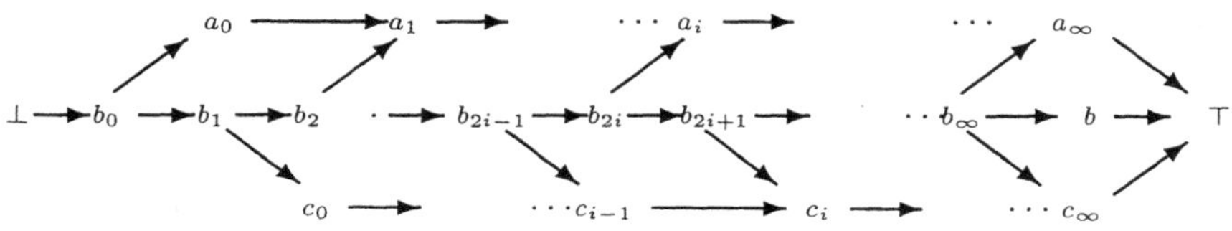

It is easy to see that it is a complete lattice. It is not distributive, since, for instance, $b_2 \wedge (b_1 \vee a_0) = b_2 \wedge a_1 = b_2$ and $(b_2 \wedge b_1) \vee (b_2 \wedge a_0) = b_1 \vee b_0 = b_1$.

Now let $f(x) = a_0 \vee (b \wedge (c_0 \vee (b \wedge x)))$. We claim that $\mu x.f(x) = a_\infty$. For $i \in \mathbb{N}$, let x_i be defined by $x_0 = \bot$ and $x_{i+1} = f(x_i)$. It is easy to prove by induction that $x_{i+1} = a_{i+1}$:

- $x_1 = a_0 \vee (b \wedge (c_0 \vee (b \wedge \bot))) = a_0 \vee (b \wedge c_0)$. But $b \wedge c_0 = b_1$ and $a_0 \vee b_1 = a_1$.
- $x_{i+1} = a_0 \vee (b \wedge (c_0 \vee (b \wedge a_i)))$. But $b \wedge a_i = b_{2i}$, $c_0 \vee b_{2i} = c_i$, $b \wedge c_i = b_{2i+1}$, $a_0 \vee b_{2i+1} = a_{i+1}$.

It follows that $a_\infty = \bigvee_{i \geq 0} a_i \leq \mu x.f(x)$. Indeed a_∞ is a fixed point of f: $b \wedge a_\infty = b_\infty$, $c_0 \vee b_\infty = c_\infty$, $b \wedge c_\infty = b_\infty$, $a_0 \vee b_\infty = a_\infty$.

Now a_∞ is not the value of a functional term built up from a_0, c_0, b. This is because the set $E' = E - \{a_\infty, b_\infty, c_\infty\}$ is closed under $\vee$ and $\wedge$. □

Let $\mathcal{B}$ be the interpretation of the propositional μ-calculus in $\mathbb{B}$. Since $\mathbb{B}$ is distributive, $\mathcal{B} \in \mathcal{D}$. Indeed, we have the following characterization of $=_\mathcal{D}$.

Proposition 9.1.5. *Let t and t' be two terms of the propositional μ-calculus. Then $t =_\mathcal{D} t'$ if and only if $[t]_\mathcal{B} = [t']_\mathcal{B}$.*

Proof. The proof is by induction on $ar(t) = ar(t')$. If this set is empty, there is nothing to prove. Let us assume that $[t]_\mathcal{B} = [t']_\mathcal{B}$. Let $x \in ar(t)$ and let us write $t =_\mathcal{D} a \vee (x \wedge b)$, $t' =_\mathcal{D} a' \vee (x \wedge b')$, where a, b, a', and b' are terms of arity $ar(t) - \{x\}$ (see Lemma 9.1.1, page 205). Since $[t]_\mathcal{B} = [t']_\mathcal{B}$, we have $[t]_\mathcal{B}[id\{0/x\}] = [t']_\mathcal{B}[id\{0/x\}]$ and $[t]_\mathcal{B}[id\{1/x\}] = [t']_\mathcal{B}[id\{1/x\}]$. But $[t]_\mathcal{B}[id\{0/x\}] = [a]_\mathcal{B}$, $[t']_\mathcal{B}[id\{0/x\}] = [a']_\mathcal{B}$, and $[t]_\mathcal{B}[id\{1/x\}] = [b]_\mathcal{B}$, $[t']_\mathcal{B}[id\{1/x\}] = [b']_\mathcal{B}$. By the induction hypothesis, $a =_\mathcal{D} a'$ and $b =_\mathcal{D} b'$, hence, $t =_\mathcal{D} t'$. □

9.2 Guarded terms

If a term contains binary symbols that are to be interpreted as $\vee$ and $\wedge$ in a distributive lattice, we can take advantage of Proposition 9.1.2 (page 206) to simplify it.

Indeed, this simplifications amount to considering *guarded terms*, i.e., those terms where any bound occurrence of a variable is in the scope of a functional symbols different from $\vee$ and $\wedge$.

9.2.1 Introductory example

Let $\boldsymbol{t}$ be the closed vectorial μ-terms $\mu\langle x, y\rangle.\langle x \vee (y \wedge c), f(x, y)\rangle$.

Let $\mathcal{I}$ be any interpretation whose domain $D_{\mathcal{I}}$ is distributive. Let $\langle a, b\rangle = [\![t]\!]_{\mathcal{I}} = \mu\langle x, y\rangle.\langle x \vee (y \wedge c^{\mathcal{I}}), f^{\mathcal{I}}(x, y)\rangle$. By the Gauss elimination principle (Proposition 1.4.7, page 30), we have $a = g(b)$ and $b = \mu y.f^{\mathcal{I}}(g(y), y)$ where $g(y) = \mu x.(x \vee (y \wedge c^{\mathcal{I}}))$. Obviously, $g(y) = y \wedge c^{\mathcal{I}}$, hence, $b = \mu y.f^{\mathcal{I}}(y \wedge c^{\mathcal{I}}, y)$, and $a = b \wedge c^{\mathcal{I}} = f^{\mathcal{I}}(b \wedge c^{\mathcal{I}}, b) \wedge c^{\mathcal{I}}$.

Now, let us consider $\boldsymbol{t}' = \mu\langle x, y\rangle.\langle f(y \wedge c, y) \wedge c, f(y \wedge c, y)\rangle$, and let $\langle a', b'\rangle = [\![t']\!]_{\mathcal{I}} = \mu\langle x, y\rangle.\langle f^{\mathcal{I}}(y \wedge c^{\mathcal{I}}, y) \wedge c^{\mathcal{I}}, f^{\mathcal{I}}(y \wedge c^{\mathcal{I}}, y)\rangle$.

By the Bekič principle (Lemma 1.4.2, page 27), we have $b' = \mu y.f^{\mathcal{I}}(y \wedge c^{\mathcal{I}}, y) = b$, and $a' = \mu x.(f^{\mathcal{I}}(b \wedge c^{\mathcal{I}}, b) \wedge c^{\mathcal{I}}) = a$.

It follows that $[\![\boldsymbol{t}]\!]_{\mathcal{I}} = [\![\boldsymbol{t}']\!]_{\mathcal{I}}$. Indeed as far as $\vee$ and $\wedge$ are respectively interpreted as the least upper bound and the greatest lower bound in a lattice, it is possible to consider only terms like $\boldsymbol{t}'$ which we call *elementary guarded* and which are defined below.

9.2.2 Guarded terms

Let F be a set of symbols, and let $F_{\{\vee,\wedge\}}$ be the set F with two additional binary symbols $\vee$ and $\wedge$ which will be used as infix operators, and two additional 0-ary symbols $\bot$ and $\top$.

We define the two subsets BC and EG of the set $funct\mathcal{T}(F_{\{\vee,\wedge\}})$ of functional terms over $F_{\{\vee,\wedge\}}$ (see Definition 2.3.2, page 47).

The set BC of *positive Boolean combinations of variables* is defined inductively by

- $\bot$ and $\top$ are in BC
- if x is in Var, then x is in BC,
- if t and t' are in BC, so are $t \vee t'$ and $t \wedge t'$.

The set EG of *elementary guarded terms* is defined by: for any symbol f in F, if $t_1, \dots t_{\rho(f)}$ are in BC, then $f(t_1, \dots t_{\rho(f)})$ is in EG.

Example 9.2.1. $x \vee (y \wedge z) \in BC$,
$f(x, y) \in EG$, $f(y \wedge z, y) \in EG$,
$x \wedge f(x, y) \notin EG$. □

Example 9.2.2. A base term t (see Definition 2.3.1, page 47) over $F_{\{\vee,\wedge\}}$ is in BC if it is $\bot$, $\top$, a variable, or has the form $x \vee y$ or $x \wedge y$, and it is in EG if it has the form $f(x_1, \ldots, x_n)$. □

Finally, for any set X of variables, the set $G(X)$ of *guarded functional terms with respect to* X is the set of functional terms where each occurrence of a variable not in X is in the scope of one and only one functional symbol of F and each occurrence of a variable in X is in the scope of at most one functional symbol of F. This set is defined by

- $X \subseteq G(X)$, $EG \subseteq G(X)$,
- if t and t' are in $G(X)$, so are $t \vee t'$ and $t \wedge t'$.

A vector $\boldsymbol{t}$ of terms is said to be in BC, or in EG, or in $G(X)$ if each of its components belongs to this set.

Definition 9.2.3. We say that a fixed-point term $\boldsymbol{\tau} = \theta_k \boldsymbol{x}^{(k)}. \cdots \theta_1 \boldsymbol{x}^{(1)}.\boldsymbol{t}$ is guarded if $\boldsymbol{t} \in G(ar(\boldsymbol{\tau}))$.

By Theorem 2.7.19 (page 68), each vectorial fixed-point term $\boldsymbol{\tau}$ is a variant of a vector consisting of some components of a vectorial fixed-point term in the form $\theta_k \boldsymbol{x}^{(k)}. \cdots \theta_1 \boldsymbol{x}^{(1)}.\boldsymbol{t}$ where each component of $\boldsymbol{t}$ is a base term. We are going to prove that this vectorial fixed-point term is equivalent with respect to the equivalence $=_{\mathcal{D}}$ defined in Section 9.1, (page 205), to a term $\theta_k \boldsymbol{x}^{(k)}. \cdots .\theta_1 \boldsymbol{x}^{(1)}.\boldsymbol{t}'$ where $\boldsymbol{t}'$ is in $G(ar(\boldsymbol{\tau}))$.

Theorem 9.2.4. *Every vectorial fixed-point term is equivalent with respect to* $=_{\mathcal{D}}$ *to a guarded vectorial fixed-point term.*

Proof. First, let us remark that, up to a permutation of indices, one can assume that each vectorial fixed-point term $\boldsymbol{\tau} = \theta_k \boldsymbol{x}^{(k)}. \cdots \theta_1 \boldsymbol{x}^{(1)}.\boldsymbol{t}$ where each component of $\boldsymbol{t}$ is a base term, can be written

$$\theta_k \langle \boldsymbol{y}^{(k)}, \boldsymbol{z}^{(k)} \rangle. \cdots .\theta_1 \langle \boldsymbol{y}^{(1)}, \boldsymbol{z}^{(1)} \rangle.\langle \boldsymbol{s}, \boldsymbol{s}' \rangle$$

with $\boldsymbol{s} \in BC$ and $\boldsymbol{s}' \in EG$.

For $i \in \{0, 1, \ldots, k\}$, let us define $\boldsymbol{\tau}_i$, $\boldsymbol{s}_i$, $\boldsymbol{s}'_i$, and $\boldsymbol{s}''_i$ inductively by

- $\boldsymbol{\tau}_0 = \langle \boldsymbol{s}, \boldsymbol{s}' \rangle$, $\boldsymbol{\tau}_i = \theta_i \langle \boldsymbol{y}^{(i)}, \boldsymbol{z}^{(i)} \rangle.\boldsymbol{\tau}_{i-1}$, so that $\boldsymbol{\tau} = \boldsymbol{\tau}_k$,
- $\boldsymbol{s}_0 = \boldsymbol{s}$, $\boldsymbol{s}_i$ is the term in BC which is equivalent with respect to $=_{\mathcal{D}}$ to $\theta_i \boldsymbol{y}^{(i)}.(\boldsymbol{s}_{i-1}[id\{\boldsymbol{z}^{(i)}/\boldsymbol{z}^{(i-1)}\}])$ (see Corollary 9.1.3, page 206),
- $\boldsymbol{s}'_0 = \boldsymbol{s}'$, $\boldsymbol{s}'_i = \boldsymbol{s}'_{i-1}[id\{\boldsymbol{s}_i/\boldsymbol{y}^{(i)}\}]$,
- $\boldsymbol{s}''_0 = \boldsymbol{s}'_0, \boldsymbol{s}''_i = \theta_i \boldsymbol{z}^{(i)}. \cdots .\theta_1 \boldsymbol{z}^{(1)}.\boldsymbol{s}'_i$.

It is easy to see that $s_i \in BC$, $s'_i \in EG$, $ar(s_i) \subseteq ar(s_{i-1}) - \{y^{(i)}\}$ and thus $ar(s'_i) \subseteq ar(s'_{i-1}) - \{y^{(i)}\}$. We have also $ar(s_i) \subseteq ar(s_{i-1}) - \{z^{(i-1)}\}$.

We are going to prove by induction that $\tau_i =_{\mathcal{D}} \langle s_i[id\{s''_i/z^{(i)}\}], s''_i\rangle$.

This is obviously true for $i = 0$.

Let us assume that $\tau_i =_{\mathcal{D}} \langle t_i[id\{s''_i/z^{(i)}\}], s''_i\rangle$. Then τ_i is also equivalent to $\theta_{i+1}\langle v, v'\rangle.\langle s_i[id\{v'/z^{(i)}\}], s''_i\rangle$. It follows by Proposition 1.3.2 (page 19), that

$$\tau_{i+1} =_{\mathcal{D}} \theta_{i+1}\langle y^{(i+1)}, z^{(i+1)}\rangle.\langle s_i[id\{z^{(i+1)}/z^{(i)}\}], s''_i\rangle.$$

By the Gauss principle (Proposition 1.4.7, page 30)

$$\tau_{i+1} =_{\mathcal{D}} \langle r[id\{r'/z^{(i+1)}\}], r'\rangle$$

where

$$\begin{aligned} r &= \theta_{i+1} y^{(i+1)}.(s_i[id\{z^{(i+1)}/z^{(i)}\}]) =_{\mathcal{D}} s_{i+1}, \\ r' &= \theta_{i+1} z^{(i+1)}.(s''_i[id\{r/y^{(i+1)}\}]). \end{aligned}$$

Thus, it remains to prove that

$$s''_{i+1} = r' = \theta_{i+1} z^{(i+1)}.(s''_i[id\{s_{i+1}/y^{(i+1)}\}]).$$

By definition,

$$s''_i[id\{s_{i+1}/y^{(i+1)}\}] = (\theta_i z^{(i)}.\cdots.\theta_1 z^{(1)}.s'_i)[id\{s_{i+1}/y^{(i+1)}\}].$$

Since s_{i+1} does not depend on $z^{(1)}, \ldots, z^{(i)}$, this is equivalent to

$$\theta_i z^{(i)}.\cdots.\theta_1 z^{(1)}.(s'_i[id\{s_{i+1}/y^{(i+1)}\}]) = \theta_i z^{(i)}.\cdots.\theta_1 z^{(1)}.s'_{i+1}.$$

It follows that $r' =_{\mathcal{D}} s''_{i+1}$.

Thus, we have $\tau =_{\mathcal{D}} \langle s_k[id\{s''_k/z^{(k)}\}], s''_k\rangle$.

Since s'_k is in EG it is in $G(ar(\tau))$. Since s_k is in BC and $ar(s_k) \subseteq ar(\tau) \cup \{z^{(k)}\}$, $s_k[id\{s'_k/z^{(k)}\}]$ is also in $G(ar(\tau))$. We claim that

$$\theta_k\langle y^{(k)}, z^{(k)}\rangle.\cdots.\theta_1\langle y^{(1)}, z^{(1)}\rangle.\langle s_k[id\{s'_k/z^{(k)}\}], s'_k\rangle$$

is equivalent with respect to $=_{\mathcal{D}}$ to $\langle s_k[id\{s''_k/z^{(k)}\}], s''_k\rangle$.

Since $\langle s_k[id\{s'_k/z^{(k)}\}], s'_k\rangle$ does not depend on $y^{(1)}, \ldots, y^{(k)}$, we get, by Proposition 1.4.4 (page 28), that

$$\theta_k\langle y^{(k)}, z^{(k)}\rangle.\cdots.\theta_1\langle y^{(1)}, z^{(1)}\rangle.\langle s_k[id\{s'_k/z^{(k)}\}], s'_k\rangle$$

is equivalent with respect to $=_{\mathcal{D}}$ to

$$\langle s_k[id\{s'_k/z^{(k)}\}][id\{s''_k/z^{(1)}, \ldots, s''_k/z^{(k)}\}], s''_k\rangle.$$

Since s_k does not depend on $z^{(1)}, \ldots, z^{(k-1)}$, we get that

$$s_k[id\{s'_k/z^{(k)}\}][id\{s''_k/z^{(1)}, \ldots, s''_k/z^{(k)}\}]$$

is equivalent with respect to $=_{\mathcal{D}}$ to $s_k[id\{s'_k[id\{s''_k/z^{(1)}, \ldots, s''_k/z^{(k)}\}]/z^{(k)}\}]$. But $s''_k = \theta_k z^{(k)}. \cdots \theta_1 z^{(1)}.s'_k$, hence $s''_k =_{\mathcal{D}} s'_k[id\{s''_k/z^{(1)}, \ldots, s''_k/z^{(k)}\}]$. □

9.3 Intersection-free terms

In the previous section, we have defined the set $G(X)$ of guarded terms. We now define the set of *intersection-free* guarded terms with respect to a set X of variables which are those elementary guarded terms where $\wedge$ does not appear, unless one of its arguments is a variable in X.

9.3.1 Definition

Formally, we define the three sets $IFBC(X)$, $IFEG(X)$ and $IFG(X)$ by

- any variable z is in $IFBC(X)$,
- if t and t' are in $IFBC(X)$ then so is $t \vee t'$,
- if $t \in IFBC(X)$ and $x \in X$ then $x \wedge t$ and $t \wedge x$ are in $IFBC(X)$,
- for any symbol f in F, if $t_1, \ldots t_{\rho(f)}$ are in $IFBC(X)$, then $f(t_1, \ldots t_{\rho(f)})$ is in $IFEG(X)$,
- $IFEG(X) \subseteq IFG(X)$,
- if t and t' are in $IFG(X)$, so is $t \vee t'$.
- if $t \in IFG(X)$ and $x \in X$ then $x \wedge t$ and $t \wedge x$ are in $IFG(X)$.

Let us remark that a term in $IFG(\emptyset)$ does not contain intersection.

We already know (see Theorem 9.2.4 page 209) that every vectorial fixed-point term $\boldsymbol{\tau}$ is equivalent to a vectorial fixed-point term $\theta_k \boldsymbol{x}^{(k)}. \cdots .\theta_1 \boldsymbol{x}^{(1)}.\boldsymbol{t}$ with $\boldsymbol{t} \in G(ar(\boldsymbol{\tau}))$. We are going to prove that for some particular families of interpretations (defined in the next section), $\boldsymbol{\tau}$ is equivalent to some components of a term $\boldsymbol{\tau}' = \theta'_m \boldsymbol{y}^{(m)}. \cdots .\theta'_1 \boldsymbol{y}^{(1)}.\boldsymbol{t}'$ with $\boldsymbol{t}' \in IFG(ar(\boldsymbol{\tau}'))$, where $ar(\boldsymbol{\tau}) = ar(\boldsymbol{\tau}')$. Note that $\boldsymbol{\tau}'$ has not necessarily the same length as $\boldsymbol{\tau}$ and not necessarily the same sequence of fixed-point operators.

9.3.2 A syntactic notion of intersection

Let F be a set of functional symbols, and for $f \in F$ let us denote by $\rho(f)$ the arity of f.

Assume that we have a rule Γ that with each pair (f, g) of functional symbols associates a term $t_{f,g} = t_{f,g}(y_1, \ldots, y_m) \in IFG(\emptyset)$ and, with each

$i \in \{1, \ldots, m\}$ a pair (I_i, J_i) where $I_i \subseteq \{1, \ldots, \rho(f)\}$, $J_i \subseteq \{1, \ldots, \rho(g)\}$, and I_i and J_i are not both empty.

An interpretation $\mathcal{I} \in \mathcal{C}$ is said to be compatible with Γ if

(c1) its domain $D_{\mathcal{I}}$, is a distributive lattice,

(c2) moreover, $D_{\mathcal{I}}$ is *completely* $\vee$*-distributive*, i.e., for any subsets D_1 and D_2 of $D_{\mathcal{I}}$,

$$(\bigvee D_1) \wedge (\bigvee D_2) = \bigvee \{d_1 \wedge d_2 \mid d_i \in D_i\},$$

(c3) for any $f, g \in F$, $d_1, \ldots, d_{\rho(f)} \in D_{\mathcal{I}}$, $d'_1, \ldots, d'_{\rho(g)} \in D_{\mathcal{I}}$,

$$f^{\mathcal{I}}(d_1, \ldots, d_{\rho(f)}) \wedge g^{\mathcal{I}}(d'_1, \ldots, d'_{\rho(g)}) = [\![t_{f,g}]\!]_{\mathcal{I}}(d''_1, \ldots, d''_m)$$

where $d''_i = \bigwedge_{j \in I_i} d_j \wedge \bigwedge_{j \in J_i} d'_j$.

Let $\mathcal{C}_\Gamma$ be the set of all interpretations compatible with Γ and let $=_{\mathcal{C}_\Gamma}$ be the equivalence associated with this set of interpretations.

Example 9.3.1. Let *Sig* be an arbitrary signature, and let $Sig_{\{0\}}$ be the signature obtained by adding to *Sig* a symbol not in *Sig*, of arity 0, denoted by 0.

We consider the rule Γ that associates with $f, g \in Sig_{\{0\}}$

- the term 0 if $g \neq f$,
- otherwise, if $g = f$, the term $f(y_1, \ldots, y_m)$, where m is the arity of f, and for each $i \in \{1, \ldots, m\}$, the sets $I_i = J_i = \{i\}$.

Let T_{Sig} be the set of all syntactic trees, defined in Example 6.1.3 (page 142), and let us define the following interpretation $\mathcal{I}$ of the terms over $Sig_{\{0\}}$.

- The domain of $\mathcal{I}$ is the powerset $\mathcal{P}(T_{Sig})$ ordered by inclusion.
- The interpretation $0^{\mathcal{I}}$ of the symbol 0 is the empty set, and for any $f \in Sig$,

$$f^{\mathcal{I}}(T_1, \ldots, T_m) = \{f^{\mathcal{T}_{Sig}}(t_1, \ldots, t_m) \mid t_1 \in T_1, \ldots, t_m \in T_m\},$$

  where $f^{\mathcal{T}_{Sig}}$ is also defined in Example 6.1.3.

In other words $\mathcal{I}$ restricted to *Sig* is exactly the powerset interpretation $\wp\mathcal{T}_{Sig}$ associated with the semi-algebra $\mathcal{T}_{Sig}$ (see Section 6.1.2, page 143)

This interpretation is compatible with Γ. Obviously $\mathcal{P}(T_{Sig})$ is a completely $\vee$-distributive lattice. If $g \neq f$, and for any $T_1, \ldots, T_m, T'_1, \ldots, T'_{m'} \subseteq T_{Sig}$, with $m = ar(f)$ and $m' = ar(g)$, we have

$$f^{\mathcal{I}}(T_1, \ldots, T_m) \cap g^{\mathcal{I}}(T'_1, \ldots, T'_{m'}) = \emptyset = 0^{\mathcal{I}}$$

because if t is a tree in this intersection, its root must be labelled both by f and g. On the other hand, it is obvious that

$$f^{\mathcal{I}}(T_1, \ldots, T_m) \cap f^{\mathcal{I}}(T'_1, \ldots, T'_m) = f^{\mathcal{I}}(T_1 \cap T'_1, \ldots, T_m \cap T'_m).$$

□

Example 9.3.2. Now, let F be consisting of a single binary symbol f, and let Γ be the rule which associates with (f, f) the term $t = f(y_1, y_2) \vee f(y_3, y_4)$ and the set of pairs (I_1, J_1) (I_2, J_2), (I_3, J_3), and (I_4, J_4) with $I_1 = I_3 = J_1 = J_4 = \{1\}$ and $I_2 = I_4 = J_2 = J_3 = \{2\}$.

Let $\mathcal{I}$ be an interpretation of domain D compatible with Γ. Then, for any $a, a', b, b' \in D$, we have

$$f^{\mathcal{I}}(a, a') \wedge f^{\mathcal{I}}(b, b') = f^{\mathcal{I}}(a \wedge b, a' \wedge b') \vee f^{\mathcal{I}}(a \wedge b', a' \wedge b).$$

It follows that for any $a, a', b, b', c, c' \in D$ we have

$$\begin{aligned} & f^{\mathcal{I}}(a, a') \wedge f^{\mathcal{I}}(b, b') \wedge f^{\mathcal{I}}(c, c') \\ = \; & (f^{\mathcal{I}}(a \wedge b, a' \wedge b') \vee f^{\mathcal{I}}(a \wedge b', a' \wedge b)) \wedge f^{\mathcal{I}}(c, c') \\ = \; & (f^{\mathcal{I}}(a \wedge b, a' \wedge b') \wedge f^{\mathcal{I}}(c, c')) \vee (f^{\mathcal{I}}(a \wedge b', a' \wedge b) \wedge f^{\mathcal{I}}(c, c')) \\ = \; & f^{\mathcal{I}}(a \wedge b \wedge c, a' \wedge b' \wedge c') \vee f^{\mathcal{I}}(a \wedge b \wedge c', a' \wedge b' \wedge c) \vee \\ & f^{\mathcal{I}}(a \wedge b' \wedge c, a' \wedge b \wedge c') \vee f^{\mathcal{I}}(a \wedge b' \wedge c', a' \wedge b \wedge c'). \end{aligned}$$

Therefore, if we denote by t the term $f(y_1, y_2) \vee f(y_3, y_4) \vee f(y_5, y_6) \vee f(y_7, y_8)$, we have $f(x, x') \wedge f(y, y') \wedge f(z, z') =_{C_\Gamma} t[id\{\rho(y_i)/y_i\}_{i=1,\dots,8}]$ where $\rho(y_1) = x \wedge y \wedge z$, $\rho(y_2) = x' \wedge y' \wedge z'$, $\rho(y_3) = x \wedge y \wedge z'$, $\rho(y_4) = x' \wedge y' \wedge z$, $\rho(y_5) = x \wedge y' \wedge z$, $\rho(y_6) = x' \wedge y \wedge z$, $\rho(y_7) = x \wedge y' \wedge z'$, and $\rho(y_8) = x' \wedge y \wedge z'$. □

The following proposition generalizes the construction given in the previous example.

Proposition 9.3.3. *Let $t_1, \dots, t_i, \dots, t_n$ be n terms in $IFG(X)$, and let $Y_i = ar(t_i) - X$. There exist a term $t' \in IFG(X)$, and n mappings $\varphi_i : Y' \to \mathcal{P}(Y_i)$ $(i = 1, \dots, n)$, where $Y' = ar(t') - X$, such that for any substitutions ρ_i $(i = 1, \dots, n)$*

$$\bigwedge_{i=1}^{n} t_i[id\{\rho_i(y)/y\}_{y \in Y_i}] =_{C_\Gamma} t'[id\{\rho'(y')/y'\}_{y' \in Y'}],$$

where $\rho'(y') = \bigwedge_{i=1}^{n} \bigwedge_{y \in \varphi_i(y')} \rho_i(y)$.

Before starting the proof let us remark that the set Y' can be chosen arbitrarily, up to a bijection. Let $\beta : Y' \to Y''$ be a bijection, let $\varphi'_i : Y'' \to \mathcal{P}(Y_i)$ be defined by $\varphi'_i(y'') = \varphi_i(\beta^{-1}(y''))$, and let $t'' = t'[id\{\beta(y')/y'\}_{y' \in Y'}]$. Let $\rho''(y'') = \bigwedge_{i=1}^{n} \bigwedge_{y \in \varphi'_i(y'')} \rho_i(y)$. It follows that $\rho''(\beta(y')) = \bigwedge_{i=1}^{n} \bigwedge_{y \in \varphi_i(y')} \rho_i(y) = \rho(y')$. Hence, $t''[id\{\rho''(y'')/y''\}_{y'' \in Y''}] = t''[id\{\rho'(y')/\beta(y')\}_{y' \in Y'}] = t'[id\{\beta(y')/y'\}_{y' \in Y'}][id\{\rho'(y')/\beta(y')\}_{y' \in Y'}] = t'[id\{\rho'(y')/y'\}_{y' \in Y'}]$.

Proof. The proof is by induction on n.

For $n = 1$, there is nothing to prove: take $t' = t_1$ and $\varphi_1(y) = \{y\}$.

For $n = 2$ the proof is by induction on the size of $t_1 \wedge t_2$.

If one of the t_i (say t_2) is not in $IFEG(X)$, then $t_2 = t_2' \wedge x$ for $x \in X$ or $t_2 = t_2' \vee t_2''$. In the first case, we get t' and $\varphi_i (i = 1, 2)$ which satisfy the proposition for $t_1 \wedge t_2'$. Thus $t' \wedge x$ and $\varphi_i (i = 1, 2)$ satisfy it for $t_1 \wedge t_2$. In the second case, $t_1 \wedge t_2 =_{C_\Gamma} (t_1 \wedge t_2') \vee (t_1 \wedge t_2'')$. By the induction hypothesis, there exists t' and $\varphi_i' (i = 1, 2)$ which satisfy the proposition for $t_1 \wedge t_2'$ and t'' and $\varphi_i'' (i = 1, 2)$ which satisfy the proposition for $t_1 \wedge t_2''$. By our preliminary remark we may assume that the two sets $Y' = ar(t') - X$ and $Y'' = ar(t'') - X$ are disjoint. Then we set $t = t' \vee t''$ and we define φ_i as the disjoint union of φ_i' and φ_i''.

If t_1 and t_2 are both in $IFEG(X)$, then there exist two disjoint sets of variables Z_1, Z_2, also disjoint from X, and a substitution $\sigma : Var \to Var$ satisfying $\forall y \in X, \sigma(y) = y$ and $\forall z \in Z_i, \sigma(z) \in Y_i$, such that $t_1 = f_1(t_1', \ldots, t_n')[\sigma]$ and $t_2 = f_2(t_1'', \ldots, t_{n'}'')[\sigma]$, where the t_i' and the t_j'' are in $IFBC(X)$, and $ar(t_i') - X \subseteq Z_1$, $ar(t_j'') - X \subseteq Z_2$. Let $t_0' = f_1(t_1', \ldots, t_n')$ and $t_0'' = f_2(t_1'', \ldots, t_{n'}'')$. By definition of C_Γ, there exists $t(y_1, \ldots, y_m)$ such that $t_0' \wedge t_0'' =_{C_\Gamma} t(s_1, \ldots, s_m)$ with $s_\ell = \bigwedge_{i \in I_\ell} t_i' \wedge \bigwedge_{j \in J_\ell} t_j''$. Now, each s_ℓ is in BC and thus can be written as a disjunction of conjunctions of variables in X, in Z_1, and in Z_2. Therefore $t(s_1, \ldots, s_m)$ can be written in the form $t'(s_1', \ldots, s_{m'}')$ where $t'(z_1, \ldots, z_{m'})$ is in $IFEG(X)$, and each s_ℓ' is a conjunction $\bigwedge_{z_1 \in \psi_1(\ell)} z_1 \wedge \bigwedge_{z_2 \in \psi_2(\ell)} z_2$ for two mappings $\psi_i : \{1, \ldots, m'\} \to \mathcal{P}(Z_i)$. If we define the mappings $\varphi_i : \{1, \ldots, m'\} \to \mathcal{P}(ar(t_i) - X)$ by

$$\varphi_i(y_\ell) = \{\sigma(y) \mid y \in \psi_i(\ell)\},$$

we get

$$\begin{gathered} t_1[id\{\rho_1(y)/y\}_{y \in Y_1}] \wedge t_2[id\{\rho_2(y)/y\}_{y \in Y_2}] = \\ t_0'[id\{\rho_1(\sigma(z))/z\}_{z \in Z_1}] \wedge t_0''[id\{\rho_2(\sigma(z))/z\}_{z \in Z_2}] =_{C_\Gamma} \\ (t_0' \wedge t_0'')[id\{\rho'(z)/z\}_{z \in Z_1 \cup Z_2}] \end{gathered}$$

where $\rho'(z) = \rho_i(\sigma(z))$ when $z \in Z_i$.
But $(t_0' \wedge t_0'')[id\{\rho'(z)/z\}_{z \in Z_1 \cup Z_2}] =_{C_\Gamma} t'(s_1'', \ldots, s_{m'}'')$ where

$$s_\ell'' = s_\ell'[id\{\rho'(z)/z\}_{z \in Z_1 \cup Z_2}] = \bigwedge_{z_1 \in \psi_1(\ell)} \rho_1(\sigma(z_1)) \wedge \bigwedge_{z_2 \in \psi_2(\ell)} \rho_2(\sigma(z_2)).$$

Since $\bigwedge_{z \in \psi_i(\ell)} \rho_i(\sigma(z)) = \bigwedge_{y \in \varphi_i(\ell)} \rho_i(y)$, the result follows.

Let t and $\varphi_i : Y \to \mathcal{P}(Y_i)$ which satisfy the property for $t_1, \ldots, t_n$ and let t', $\varphi : Y' \to \mathcal{P}(Y)$, $\varphi_{n+1} : Y' \to \mathcal{P}(Y_{n+1})$ which satisfy the property for t, t_{n+1}. Let, $\varphi_{n+1}' = \varphi_{n+1}$, and, for $i \leq n$, $\varphi_i' : Y' \to \mathcal{P}(Y_i)$ be defined by $\varphi_i'(y') = \bigcup_{y \in \varphi(y')} \varphi_i(y)$. We have $\bigwedge_{i=1}^{n+1} (t_i[id\{\rho_i(y)/y\}_{y \in Y_i}]) =_{C_\Gamma}$

$t[id\{t_y/y\}_{y\in Y}] \wedge t_{n+1}[id\{\rho_{n+1}(y)/y\}_{y\in Y_{n+1}}]$ with $t_y = \bigwedge_{i=1}^{n} \bigwedge_{y_i \in \varphi_i(y)} \rho_i(y_i)$. This is also equal to $t'[id\{t'_{y'}/y'\}_{y'\in Y'}]$ with

$$t'_{y'} = (\bigwedge_{y\in\varphi(y')} t_y) \wedge (\bigwedge_{y_{n+1}\in\varphi_{n+1}(y')} \rho_{n+1}(y_{n+1})).$$

By definition of t_y,

$$\bigwedge_{y\in\varphi(y')} t_y = \bigwedge_{y\in\varphi(y')} \bigwedge_{i=1}^{n} \bigwedge_{y_i\in\varphi_i(y)} \rho_i(y_i) = \bigwedge_{i=1}^{n} \bigwedge_{y\in\varphi(y')} \bigwedge_{y_i\in\varphi_i(y)} \rho_i(y_i)$$

and $\bigwedge_{y\in\varphi(y')} \bigwedge_{y_i\in\varphi_i(y)} \rho_i(y_i) = \bigwedge_{y_i\in\varphi'_i(y)} \rho_i(y_i)$. □

Theorem 9.3.4. *For any term t in $G(X)$, there exist $t' \in IFG(X)$ and $\varphi : ar(t') - X \to \mathcal{P}(ar(t) - X)$ such that $t =_{C_\Gamma} t'[id\{t_y/y\}_{y\in ar(t')-X}]$ where $t_y = \bigwedge_{z\in\varphi(y)} z$.*

Proof. The proof is by induction on t When t is in EG the result has been proved in the proof of the previous proposition. If $t = t_1 \wedge t_2$ the result is a consequence of the previous proposition when $\rho_1 = \rho_2 = id$. If $t = t_1 \vee t_2$ we can choose t'_1 and t'_2 such that $ar(t'_1) - X$ and $ar(t'_2) - X$ are disjoint, and then we take $t' = t'_1 \vee t'_2$ and φ as the disjoint union of φ_1 and φ_2. □

Let $\boldsymbol{\tau} = \theta_k \boldsymbol{x}^{(k)}. \cdots .\theta_1 \boldsymbol{x}^{(1)}.\boldsymbol{t}$ be a guarded vectorial fixed-point term of length n (i.e., each component $\boldsymbol{t}_i$ is in $G(Z)$ where $Z = ar(\boldsymbol{\tau})$). By the previous theorem, we may assume that there exist terms t'_i in $IFG(Z)$ and mappings $\psi_i : Y_i = ar(t'_i) - Z \to \mathcal{P}(ar(\boldsymbol{t}_i) - Z)$ such that $\boldsymbol{t}_i =_{C_\Gamma} t'_i[id\{(\bigwedge_{z\in\psi_i(y)} x)/y\}_{y\in Y_i}]$.

By Proposition 9.3.3, we can deduce the following result.

Corollary 9.3.5. *Let $\boldsymbol{\tau} = \theta_k \boldsymbol{x}^{(k)}. \cdots .\theta_1 \boldsymbol{x}^{(1)}.\boldsymbol{t}$ be a guarded vectorial fixed-point term of length n, let $Z = ar(\boldsymbol{\tau})$ and $X = \bigcup_{i=1}^{k}\{\boldsymbol{x}^{(i)}\}$.*

For any nonempty subset I of $\{1, \ldots , n\}$ there exists a term t_I in $IFG(Z)$ and for each $i \in I$ a mapping $\varphi_{I,i} : ar(t_I) - Z \to \mathcal{P}(X)$ such that for any substitutions ρ_i $(i \in I)$, $\bigwedge_{i\in I} \boldsymbol{t}_i[id\{\rho_i(x)/x\}_{x\in X}] =_{C_\Gamma} t_I[id\{t_y/y\}_{y\in ar(t_I)-Z}]$ where $t_y = \bigwedge_{i\in I} \bigwedge_{x\in\varphi_{I,i}(y)} \rho_i(x)$.

Proof. By the previous remark on the form of $\boldsymbol{t}_i$, there exist $t'_i \in IFG(Z)$ and mappings ψ_i such that

$$\bigwedge_{i\in I} \boldsymbol{t}_i[id\{\rho_i(x)/x\}_{x\in X}] = \bigwedge_{i\in I} t'_i[id\{\bigwedge_{x\in\psi_i(y)} \rho_i(x)/y\}_{y\in Y_i}].$$

By Proposition 9.3.3, there exists t_I with $ar(t_I) - Z = Y$ and for any $i \in I$ a mapping $\varphi_i : Y \to \mathcal{P}(Y_i)$ such that $\bigwedge_{i\in I} t'_i[id\{\bigwedge_{x\in\psi_i(y)} \rho_i(x)/y\}_{y\in Y_i}] =_{\mathcal{C}_\Gamma} t_I[id\{t_y/y\}_{y\in Y}]$ with $t_y = \bigwedge_{i\in I} \bigwedge_{z\in\varphi_i(y)} \bigwedge_{x\in\psi_i(z)} \rho_i(x)$. Thus we can take $\varphi_{I,i}(y) = \bigcup_{z\in\varphi(y)} \psi_i(z)$. □

9.4 The powerset construction

In presence of only one level of fixed-point operators, the elimination of intersection is performed by a powerset construction, quite similar to the classical construction which determinizes any nondeterministic word automata.

Let $\boldsymbol{\tau} = \theta \boldsymbol{x}.\boldsymbol{t}$ be a vectorial fixed-point term of length n, where $\boldsymbol{x} = \langle x_1, \dots, x_n \rangle$. We denote by X the set $\{\boldsymbol{x}\} = \{x_i \mid 1 \le i \le n\}$, and by Z the set $ar(\boldsymbol{t})$.

With every nonempty subset $I \subseteq \{1, \dots, n\}$, we associate a variable $v_I \notin Z$. Let V be this set of variables, and let ρ be an arbitrary substitution such that $\rho(v_I) = \wedge_{i\in I} x_i$, and $\rho(z) = z$ for $z \in Z$.

For each t_I defined in the previous corollary, we define t'_I obtained by substituting for $y \in ar(t_I) - Z$ in t_I the unique variable v_J such that $\rho(v_J) = \wedge_{i\in I} \wedge_{x\in\varphi_{I,i}(y)} x$. It follows that $\bigwedge_{i\in I} t_i = t'_I[\rho]$.

Now, let $\boldsymbol{v}$ be the vector of variables of length $2^n - 1$, indexed by the nonempty subsets of $\{1, \dots, n\}$ whose component of index I is v_I, and let $\boldsymbol{t}'$ be the vector of terms whose component of index I is t'_I, which is in $IFG(Z)$. Let $\boldsymbol{\tau}' = \theta \boldsymbol{v}.\boldsymbol{t}'$.

Proposition 9.4.1. *For any interpretation $\mathcal{I} \in \mathcal{C}_\Gamma$ and any $i \in \{1, \dots, n\}$, the component of index i of $[\![\boldsymbol{\tau}]\!]_{\mathcal{I}}$ is equal to the component of index $\{i\}$ of $[\![\boldsymbol{\tau}']\!]_{\mathcal{I}}$*

Proof. Let E be the domain of $\mathcal{I}$. We have to prove that for any $\sigma : Var \to E$, the component of index i of $[\![\boldsymbol{\tau}]\!]_{\mathcal{I}}[\sigma]$ is equal to the component of index $\{i\}$ of $[\![\boldsymbol{\tau}']\!]_{\mathcal{I}}[\sigma]$, i.e., the component of index i of $\theta\boldsymbol{x}.([\![\boldsymbol{t}]\!]_{\mathcal{I}}[id\{\sigma(z)/z\}_{z\in Z}])$ is equal to the component of index $\{i\}$ of $\theta\boldsymbol{v}.([\![\boldsymbol{t}']\!]_{\mathcal{I}}[id\{\sigma(z)/z\}_{z\in Z}])$.

Let $\boldsymbol{f} = [\![\boldsymbol{t}]\!]_{\mathcal{I}}[id\{\sigma(z)/z\}_{z\in Z}]$ which we can see as a mapping from $E^X \to E^X$ and let $\boldsymbol{F} = [\![\boldsymbol{t}']\!]_{\mathcal{I}}[id\{\sigma(z)/z\}_{z\in Z}]$ which we can see as a mapping from $E^V \to E^V$. Finally, let $\boldsymbol{C} : E^X \to E^V$ be defined by $\boldsymbol{C}(\boldsymbol{a})(v_I) = \bigwedge_{i\in I} \boldsymbol{a}(x_i) = ([\![\rho]\!]_{\mathcal{I}} \star (id\{\boldsymbol{a}/\boldsymbol{x}\}))(v_I)$.

We show that for any $\boldsymbol{a} \in E^X$, $\boldsymbol{C}(\boldsymbol{f}(\boldsymbol{a})) = \boldsymbol{F}(\boldsymbol{C}(\boldsymbol{a}))$. Indeed, $\boldsymbol{C}(\boldsymbol{f}(\boldsymbol{a}))(v_I) = \bigwedge_{i\in I} \boldsymbol{f}(\boldsymbol{a})(x_i)$.
By definition of ρ and $\boldsymbol{f}$, this is equal to $\bigwedge_{i\in I}([\![t_i]\!]_{\mathcal{I}}[id\{\sigma(z)/z\}_{z\in Z}\{\boldsymbol{a}/\boldsymbol{x}\}]) = (\bigwedge_{i\in I} [\![t_i]\!]_{\mathcal{I}})[id\{\sigma(z)/z\}_{z\in Z}\{\boldsymbol{a}/\boldsymbol{x}\}]$ which is also equal to

$$([\![t'_I[\rho]]\!]_{\mathcal{I}}[id\{\sigma(z)/z\}_{z\in Z}\{\boldsymbol{a}/\boldsymbol{x}\}]).$$

By definition of $\boldsymbol{C}$ this is equal to

$$[t'_I]_{\mathcal{I}}[id\{\sigma(z)/z\}_{z\in Z}\{\boldsymbol{C}(\boldsymbol{a})/\boldsymbol{v}\}]) = \boldsymbol{F}(\boldsymbol{C}(\boldsymbol{a}))(v_I).$$

It is easy to see that $\boldsymbol{C}$ has the properties required in Lemma 1.2.15 (page 13). Indeed it is to guarantee the β sup-continuity of $\boldsymbol{C}$ that we assume that $D_{\mathcal{I}}$ is completely $\vee$-distributive. This lemma allows us to deduce that $\boldsymbol{C}(\theta\boldsymbol{x}.\boldsymbol{f}(\boldsymbol{x})) = \theta\boldsymbol{v}.\boldsymbol{F}(\boldsymbol{y})$.

Obviously, the component of index $\{i\}$ of $\boldsymbol{C}(\theta\boldsymbol{x}.\boldsymbol{f}(\boldsymbol{x})$ is equal to the component of index i of $\theta\boldsymbol{x}.\boldsymbol{f}(\boldsymbol{x})$. □

9.5 The $\nu\mu$ case

.

Another case where the elimination of intersection does not change the position of a term in the alternation-depth hierarchy is the case of terms in Π_2.

The construction used to prove the next theorem is a slight generalization of the construction used in [9] to prove the same result in the case of Büchi alternating tree automata. Let us recall shortly the underlying idea.

Let us consider the game associated with such an automaton on a tree t. Let $P(s)$ be the set of plays consistent with an Eva's strategy s, which can be arranged in the form of a finitely branching tree. If s is a winning strategy, each play p in $P(s)$ contains infinitely many positions of rank 2. It follows that for every $n \geq 1$, there exists a number m_n such that the prefix of length n of each p contains at least n positions of rank 2.

In each position of a play, let us keep track, in a counter, of the number of positions of rank 2 visited so far. When for some m all positions reached by paths of length m have their counter strictly positive, we know that $m = m_1$ and we can reset all counters to 0, and start again the process to reach $m = m_2$, and so on .

Instead of considering counters, we can simply split the set of positions reached after m moves into two parts: those which have their counter equal to 0, say J_m, and the others, say I_m. If $J_m = \emptyset$, we know that we have seen at least one more position of rank 2 on each path, and at step $m+1$, J_{m+1} is the set of positions of rank 1 while I_{m+1} is the set of positions of rank 2. If $J_m \neq \emptyset$ then J_{m+1} consists only of the positions of rank 1 that come from a position in J_m, and I_{m+1} consists of all other reachable positions.

The generalization of this idea to fixed-point terms is described in the first two paragraphs of the proof below.

Theorem 9.5.1. *Let $\boldsymbol{\tau} = \nu \boldsymbol{x}.\mu \boldsymbol{y}.\boldsymbol{t}$ be a vectorial fixed-point term of length n, with $\boldsymbol{t} \in G(ar(\boldsymbol{\tau}))$. Then there exists a vectorial fixed-point term $\boldsymbol{\tau}' = \nu \boldsymbol{X}.\mu \boldsymbol{Y}.\boldsymbol{t}'$ of length $3^n - 1$, with $\boldsymbol{t}' \in IFG(ar(\boldsymbol{\tau}))$, and for any $i \in \{1, \ldots, n\}$ there is $i' \in \{1, \ldots, 3^n - 1\}$ such that for any interpretation $\mathcal{I} \in \mathcal{C}_\Gamma$ the component of index i of $[\![\boldsymbol{\tau}]\!]_\mathcal{I}$ is equal to the component of index i' of $[\![\boldsymbol{\tau}']\!]_\mathcal{I}$.*

Proof. Let us assume that $\boldsymbol{x} = \langle x_1, \ldots, x_n \rangle$ and $\boldsymbol{y} = \langle y_1, \ldots, y_n \rangle$. With any nonempty subset I of $\{1, \ldots, n\}$ (there are $2^n - 1$ such subsets) we associate two variables v_I and w_I and with every ordered pair (I, J) of disjoint subsets of $\{1, \ldots, n\}$ such that J is not empty (there are $3^n - 2^n$ such pairs), we associate two variables $v_{I,J}$ and $w_{I,J}$.

By Proposition 9.3.3, for each subset I there exists a term t_I and for each $i \in I$ a mapping $\varphi_{I,i} : Y_I = ar(t_I) - Z \to \mathcal{P}(\{\boldsymbol{x}\} \cup \{\boldsymbol{y}\})$ such that $\bigwedge_{i \in I} \boldsymbol{t}_i =_{\mathcal{C}_\Gamma} t_I[id\{t_y/y\}_{y \in Y_I}]$ where $t_y = \bigwedge_{i \in I} \bigwedge_{x \in \varphi_{I,i}(y)} x$. By the same proposition, there exist $t_{I,J}$ and $\varphi_{I,J,i} : Y_{I,J} = ar(t_{I,J}) - Z \to \mathcal{P}(\{\boldsymbol{x}\} \cup \{\boldsymbol{y}\})$ such that $\bigwedge_{i \in I} \boldsymbol{t}_i[id\{\boldsymbol{x}/\boldsymbol{y}\}] \wedge \bigwedge_{j \in J} \boldsymbol{t}_j =_{\mathcal{C}_\Gamma} t_{I,J}[id\{t_y/y\}_{y \in Y_{I,J}}]$ where t_y is defined as above.

Now in some conjunction t_y we can have both x_i and y_i for some i. In this case, we simply delete x_i and we get t'_y. The following result is a straightforward consequence of the definition of t'_y.

Lemma 9.5.2. *For any interpretation $\mathcal{I}$ of domain D, for any $\boldsymbol{a}, \boldsymbol{b} \in D^n$, if $\boldsymbol{b} \leq \boldsymbol{a}$ then $[\![t_y]\!]_\mathcal{I} id\{\boldsymbol{a}/\boldsymbol{x}, \boldsymbol{b}/\boldsymbol{y}\}] = [\![t'_y]\!]_\mathcal{I} id\{\boldsymbol{a}/\boldsymbol{x}, \boldsymbol{b}/\boldsymbol{y}\}]$.*

Next, we define t'_I and $t'_{I,J}$ by substituting for y either v_I or $w_{I,J}$ according to whether t'_y is equal to $\bigwedge_{i \in I} x_i$ or $\bigwedge_{i \in I} x_i \wedge \bigwedge_{j \in J} y_j$.

Now we consider $\boldsymbol{\tau}' = \nu \boldsymbol{v}.\mu \boldsymbol{w}.\boldsymbol{t}'$ where the vectors are of length $3^n - 1$ and indexed by I and (I, J). We claim that for any $\mathcal{I} \in \mathcal{C}_\Gamma$, the component of index i of $[\![\boldsymbol{\tau}]\!]_\mathcal{I}$ is equal to the component of index $\{i\}$ (not to be confused with the index $(\emptyset, \{i\})$) of $[\![\boldsymbol{\tau}']\!]_\mathcal{I}$.

By using a technique similar to the one used in the previous section, and using the Gauss elimination principle, (Proposition 1.4.7, page 30) since $\boldsymbol{\tau}'$ depends only on the variables v_I and $w_{I,J}$, the result is a consequence of the following proposition (9.5.4). □

Before stating and proving this following proposition, we give an example of the above construction.

Example 9.5.3. Let us consider the μ-calculus on infinite words (see Chapter 5, page 105) over the alphabet$A = \{a, b\}$. The standard interpretation of this μ-calculus is compatible with the rule

- $cx \cap cy = c(x \cap y)$ for any letter c,
- $cx \cap c'y = \emptyset$ if $c \neq c'$.

Let $\boldsymbol{\tau} = \nu \begin{bmatrix} x_1 \\ x_2 \\ x_3 \end{bmatrix} .\mu \begin{bmatrix} y_1 \\ y_2 \\ y_3 \end{bmatrix} . \begin{bmatrix} ax_1 \cup by_1 \\ ay_2 \cup bx_2 \\ a(x_1 \cap x_2) \end{bmatrix}$.
The standard interpretation of $\boldsymbol{\tau}$ is the triple $\langle L_1, L_2, L_3 \rangle$ where $L_1 = (b^*a)^\omega$ is the standard interpretation of $\nu x_1.\mu y_1.(ax_1 \cup by_1)$, $L_2 = (a^*b)^\omega$ is the standard interpretation of $\nu x_2.\mu y_2.(ay_2 \cup ax_1)$, and $L_3 = (a^+b^+)^\omega$ is equal to $a(L_1 \cap L_2)$.

Since the component of index 3 of $\boldsymbol{t}$ is $a(x_1 \cap x_2)$, the component of index $\{3\}$ of $\boldsymbol{t}'$ is $av_{\{1,2\}}$.

Since $(ax_1 \cup by_1) \cap (ay_2 \cup bx_2)$ is equivalent to

$$a(x_1 \cap y_2) \cup b(y_1 \cap x_2),$$

the component of index $\{1, 2\}$ of $\boldsymbol{t}'$ is $aw_{\{1\},\{2\}} \cup bw_{\{2\},\{1\}}$. To define the component of index $(\{1\}, \{2\})$ of $\boldsymbol{t}'$, we have to consider

$$(ax_1 \cup by_1)[id\{x_1/y_1\}] \cap (ay_2 \cup bx_2) = (ax_1 \cup bx_1) \cap (ay_2 \cup bx_2)$$

which is equivalent to $a(x_1 \cap y_2) \cup b(x_1 \cap x_2)$. Hence $\boldsymbol{t}'_{\{1\},\{2\}} = aw_{\{1\},\{2\}} \cup bv_{\{1,2\}}$. Similarly, the component of index $(\{2\}, \{1\})$ of $\boldsymbol{t}'$ is $av_{\{1,2\}} \cup bw_{\{2\},\{1\}}$. The components of index $(\{1\}, \{2\})$ and $(\{2\}, \{1\})$ of $\mu\boldsymbol{w}.\boldsymbol{t}'$ are the two components of $\mu\langle w_{\{1\},\{2\}}, w_{\{2\},\{1\}}\rangle.\langle aw_{\{1\},\{2\}} \cup bv_{\{1,2\}}, av_{\{1,2\}} \cup bw_{\{2\},\{1\}}\rangle$. It follows that, in the standard interpretation, these components are $a^*bv_{\{1,2\}}$ and $b^*av_{\{1,2\}}$. If we report these values in the component of $\boldsymbol{t}'$ of index $\{1, 2\}$, this component becomes $a^+bv_{\{1,2\}} \cup b^+av_{\{1,2\}}$, and thus, the component of index $\{1, 2\}$ of $\nu\boldsymbol{v}.\mu\boldsymbol{w}.\boldsymbol{t}'$ is equivalent to $\nu v_{\{1,2\}}.a^+bv_{\{1,2\}} \cup b^+av_{\{1,2\}}$ and also to $(a^+b \cup b^+a)^\omega$. Finally, the component of index $\{3\}$ of $\nu\boldsymbol{v}.\mu\boldsymbol{w}.\boldsymbol{t}'$ is $a(a^+b \cup b^+a)^\omega$, which is also equal to L_3. □

Proposition 9.5.4. *Let E be a completely $\bigvee$-distributive lattice. Let $\boldsymbol{f} : E^n \times E^n \to E^n$, $\boldsymbol{C} : E^n \to E^{2^n-1}$, $\boldsymbol{D} : E^n \times E^n \to E^{3^n-2^n}$, $\boldsymbol{F}_1 : E^{2^n-1} \times E^{3^n-2^n} \to E^{3^n-2^n}$, and $\boldsymbol{F}_2 : E^{2^n-1} \times E^{3^n-2^n} \to E^{2^n-1}$ be such that*

- *each component of $\boldsymbol{C}$ and $\boldsymbol{D}$ is a nonempty conjunction of some of their arguments,*
- *if $\boldsymbol{y} \le \boldsymbol{x}$ then*
  - $\boldsymbol{F}_1(\boldsymbol{C}(\boldsymbol{x}), \boldsymbol{D}(\boldsymbol{x}, \boldsymbol{y})) = \boldsymbol{C}(\boldsymbol{f}(\boldsymbol{x}, \boldsymbol{y}))$,
  - $\boldsymbol{F}_2(\boldsymbol{C}(\boldsymbol{x}), \boldsymbol{D}(\boldsymbol{x}, \boldsymbol{y})) = \boldsymbol{D}(\boldsymbol{f}(\boldsymbol{x}, \boldsymbol{x}), \boldsymbol{f}(\boldsymbol{x}, \boldsymbol{y}))$.

Then $\boldsymbol{C}(\nu\boldsymbol{x}.\mu\boldsymbol{y}.\boldsymbol{f}(\boldsymbol{x}, \boldsymbol{y})) = \nu\boldsymbol{X}.\boldsymbol{F}_1(\boldsymbol{X}, \mu\boldsymbol{Y}.\boldsymbol{F}_2(\boldsymbol{X}, \boldsymbol{Y}))$.

Proof. Let $\boldsymbol{g}(\boldsymbol{x}) = \mu\boldsymbol{y}.\boldsymbol{f}(\boldsymbol{x}, \boldsymbol{y})$ and $\boldsymbol{h} = \nu\boldsymbol{x}.\boldsymbol{g}(\boldsymbol{x})$. Let $\boldsymbol{G}(\boldsymbol{X}) = \mu\boldsymbol{Y}.\boldsymbol{F}_2(\boldsymbol{X}, \boldsymbol{Y})$ and $\boldsymbol{H} = \nu\boldsymbol{X}.\boldsymbol{F}_1(\boldsymbol{X}, \boldsymbol{G}(\boldsymbol{X}))$. We have to prove $\boldsymbol{H} = \boldsymbol{C}(\boldsymbol{h})$. The proof is in four steps.

Step 1. Let $\boldsymbol{d} \in D^n$. If $\boldsymbol{f}(\boldsymbol{d},\boldsymbol{d}) \leq \boldsymbol{d}$ then $\boldsymbol{G}(\boldsymbol{C}(\boldsymbol{d})) \leq \boldsymbol{D}(\boldsymbol{d},\boldsymbol{g}(\boldsymbol{d}))$.

If $\boldsymbol{f}(\boldsymbol{d},\boldsymbol{d}) \leq \boldsymbol{d}$ then $\boldsymbol{g}(\boldsymbol{d}) \leq \boldsymbol{d}$ and we have

$$\begin{aligned} \boldsymbol{F}_2(\boldsymbol{C}(\boldsymbol{d}),\boldsymbol{D}(\boldsymbol{d},\boldsymbol{g}(\boldsymbol{d}))) &= \boldsymbol{D}(\boldsymbol{f}(\boldsymbol{d},\boldsymbol{d}),\boldsymbol{f}(\boldsymbol{d},\boldsymbol{g}(\boldsymbol{d})) \\ &= \boldsymbol{D}(\boldsymbol{f}(\boldsymbol{d},\boldsymbol{d}),\boldsymbol{g}(\boldsymbol{d})) \\ &\leq \boldsymbol{D}(\boldsymbol{d},\boldsymbol{g}(\boldsymbol{d})). \end{aligned}$$

Hence the result.

Step 2. If $\boldsymbol{g}(\boldsymbol{d}) \leq \boldsymbol{d} \leq \boldsymbol{f}(\boldsymbol{d},\boldsymbol{d})$ then $\boldsymbol{D}(\boldsymbol{d},\boldsymbol{g}(\boldsymbol{d})) \leq \boldsymbol{G}(\boldsymbol{C}(\boldsymbol{d}))$.

Let us define, as usual, $\boldsymbol{g}_0 = \bot$, $\boldsymbol{g}_{\alpha+1} = \boldsymbol{f}(\boldsymbol{d},\boldsymbol{g}_\alpha)$, $\boldsymbol{g}_\beta = \bigvee_{\alpha<\beta} \boldsymbol{g}_\alpha$. We prove by induction that $\boldsymbol{D}(\boldsymbol{d},\boldsymbol{g}_\alpha) \leq \boldsymbol{G}(\boldsymbol{C}(\boldsymbol{d}))$.

- By definition of $\boldsymbol{D}$, there is always a variable in $\{y\}$ in each conjunction which is a component of $\boldsymbol{D}$, thus $\boldsymbol{D}(\boldsymbol{d},\bot) = \bot$.
- Assume $\boldsymbol{D}(\boldsymbol{d},\boldsymbol{g}_\alpha) \leq \boldsymbol{G}(\boldsymbol{C}(\boldsymbol{d}))$. Since $\boldsymbol{g}_\alpha \leq \boldsymbol{g}(\boldsymbol{d}) \leq \boldsymbol{d}$, we have

$$\begin{aligned} \boldsymbol{D}(\boldsymbol{d},\boldsymbol{g}_{\alpha+1}) &= \boldsymbol{D}(\boldsymbol{d},\boldsymbol{f}(\boldsymbol{d},\boldsymbol{g}_\alpha)) \\ &\leq \boldsymbol{D}(\boldsymbol{f}(\boldsymbol{d},\boldsymbol{d}),\boldsymbol{f}(\boldsymbol{d},\boldsymbol{g}_\alpha)) \\ &= \boldsymbol{F}_2(\boldsymbol{C}(\boldsymbol{d}),\boldsymbol{D}(\boldsymbol{d},\boldsymbol{g}_\alpha)) \\ &\leq \boldsymbol{F}_2(\boldsymbol{C}(\boldsymbol{d}),\boldsymbol{G}(\boldsymbol{C}(\boldsymbol{d}))) \\ &= \boldsymbol{G}(\boldsymbol{C}(\boldsymbol{d})). \end{aligned}$$

- By definition of $\boldsymbol{D}$, and because D is completely distributive, $\boldsymbol{D}(\boldsymbol{d},\boldsymbol{g}_\beta) = \boldsymbol{D}(\boldsymbol{d},\bigvee_{\alpha<\beta} \boldsymbol{g}_\alpha) = \bigvee_{\alpha<\beta} \boldsymbol{D}(\boldsymbol{d},\boldsymbol{g}_\alpha) \leq \boldsymbol{G}(\boldsymbol{C}(\boldsymbol{d}))$.

Step 3. $\boldsymbol{C}(\boldsymbol{h}) \leq \boldsymbol{H}$.

By definition of $\boldsymbol{h}$ we have $\boldsymbol{h} = \boldsymbol{g}(\boldsymbol{h}) = \boldsymbol{f}(\boldsymbol{h},\boldsymbol{h})$. By the two previous results, we have $\boldsymbol{D}(\boldsymbol{h},\boldsymbol{h}) = \boldsymbol{G}(\boldsymbol{C}(\boldsymbol{h}))$. Hence,

$$\begin{aligned} \boldsymbol{F}_1(\boldsymbol{C}(\boldsymbol{h}),\boldsymbol{G}(\boldsymbol{C}(\boldsymbol{h}))) &= \boldsymbol{F}_1(\boldsymbol{C}(\boldsymbol{h}),\boldsymbol{D}(\boldsymbol{h},\boldsymbol{h})) \\ &= \boldsymbol{C}(\boldsymbol{f}(\boldsymbol{h},\boldsymbol{h})) \\ &= \boldsymbol{C}(\boldsymbol{h}). \end{aligned}$$

It follows that $\boldsymbol{C}(\boldsymbol{h}) \leq \boldsymbol{H}$.

Step 4. $\boldsymbol{H} \leq \boldsymbol{C}(\boldsymbol{h})$.

Let us approximate $\boldsymbol{h}$ by $\boldsymbol{h}_0 = \top, \boldsymbol{h}_{\alpha+1} = \boldsymbol{g}(\boldsymbol{h}_\alpha), \boldsymbol{h}_\beta = \bigwedge_{\alpha<\beta} \boldsymbol{h}_\alpha$. Let us remark that $\boldsymbol{f}(\boldsymbol{h}_\alpha,\boldsymbol{h}_{\alpha+1}) = \boldsymbol{f}(\boldsymbol{h}_\alpha,\boldsymbol{g}(\boldsymbol{h}_\alpha)) = \boldsymbol{g}(\boldsymbol{h}_\alpha) = \boldsymbol{h}_{\alpha+1}$.

Firstly, we prove by induction that $\boldsymbol{f}(\boldsymbol{h}_\alpha,\boldsymbol{h}_\alpha) \leq \boldsymbol{h}_\alpha$.

- $\boldsymbol{f}(\top,\top) \leq \top$,
- $\boldsymbol{f}(\boldsymbol{h}_{\alpha+1},\boldsymbol{h}_{\alpha+1}) \leq \boldsymbol{f}(\boldsymbol{h}_\alpha,\boldsymbol{h}_{\alpha+1}) = \boldsymbol{h}_{\alpha+1}$,
- $\boldsymbol{f}(\boldsymbol{h}_\beta,\boldsymbol{h}_\beta) \leq \bigwedge_{\alpha<\beta} \boldsymbol{f}(\boldsymbol{h}_\alpha,\boldsymbol{h}_\alpha) \leq \bigwedge_{\alpha<\beta} \boldsymbol{h}_\alpha = \boldsymbol{h}_\beta$.

By the result proved in Step 1, we get $\boldsymbol{G}(\boldsymbol{C}(\boldsymbol{h}_\alpha)) \leq \boldsymbol{D}(\boldsymbol{h}_\alpha, \boldsymbol{g}(\boldsymbol{h}_\alpha)) = \boldsymbol{D}(\boldsymbol{h}_\alpha, \boldsymbol{h}_{\alpha+1})$.

Next, we prove by induction that $\boldsymbol{H} \leq \boldsymbol{C}(\boldsymbol{h}_\alpha)$.

- By definition of $\boldsymbol{C}$, $\boldsymbol{C}(\top) = \top$.
- Now, let us assume $\boldsymbol{H} \leq \boldsymbol{C}(\boldsymbol{h}_\alpha)$. we get

$$\begin{aligned} \boldsymbol{H} &= \boldsymbol{F}_1(\boldsymbol{H}, \boldsymbol{G}(\boldsymbol{H})) \\ &\leq \boldsymbol{F}_1(\boldsymbol{C}(\boldsymbol{h}_\alpha), \boldsymbol{G}(\boldsymbol{C}(\boldsymbol{h}_\alpha))) \\ &\leq \boldsymbol{F}_1(\boldsymbol{C}(\boldsymbol{h}_\alpha), \boldsymbol{D}(\boldsymbol{h}_\alpha, \boldsymbol{h}_{\alpha+1}) \\ &= \boldsymbol{C}(\boldsymbol{f}(\boldsymbol{h}_\alpha, \boldsymbol{h}_{\alpha+1})) \\ &= \boldsymbol{C}(\boldsymbol{h}_{\alpha+1}). \end{aligned}$$

- By definition of $\boldsymbol{C}$, $\boldsymbol{C}(\bigwedge_{\alpha<\beta} \boldsymbol{h}_\alpha) = \bigwedge_{\alpha<\beta} \boldsymbol{C}(\boldsymbol{h}_\alpha)$.

□

9.6 The simulation theorem

In this section we generalize the results of the previous sections and we show that any component of a guarded vectorial fixed-point term

$$\tau = \theta_k \boldsymbol{x}^{(k)}. \cdots .\theta_1 \boldsymbol{x}^{(1)}.\boldsymbol{t}$$

is equivalent to some component of a term $\tau' = \theta'_m \boldsymbol{y}^{(m)}. \cdots .\theta'_1 \boldsymbol{y}^{(1)}.\boldsymbol{t}'$ with $\boldsymbol{t}' \in \mathit{IFG}(ar(\tau'))$, where $ar(\tau) = ar(\tau')$.

The proof of this result is based on the determinization McNaughton theorem which we present in a form convenient for our needs in the next section.

9.6.1 McNaughton's Theorem revisited

Let I be a finite set and let $\mathcal{R} = \mathcal{P}(I \times I)$.

Let $\rho = R_0 R_1 \cdots R_k \cdots \in \mathcal{R}^\omega$. An *infinite path* of ρ is an infinite sequence $i_0, i_1, \ldots, i_k, \ldots$ such that for any $j \geq 0$, $(i_j, i_{j+1}) \in R_j$.

Now, let $r : I \to \mathbb{N}$. An infinite path of ρ is *accepted with respect to* r if $\limsup_{j\to\infty} r(i_j)$ is even.

A word $\rho \in \mathcal{R}^\omega$ is *existentially accepted with respect to* r if there exists an infinite path of ρ which is accepted. It is *universally accepted with respect to* r if all its infinite paths are accepted.

Let us denote by $L^\exists(r)$ (resp. $L^\forall(r)$) the words of $\mathcal{R}^\omega$ which are existentially accepted (resp. universally accepted).

Proposition 9.6.1. *Let* $r' : I \to \mathbb{N}$ *be defined by* $r'(i) = r(i) + 1$. *Then* $L^\exists(r) = \mathcal{R}^\omega - L^\forall(r')$

Proof. An infinite path of ρ is accepted with respect to r if and only if it is not accepted with respect to r'. Therefore, $\rho \in \mathcal{R}^\omega - L^\forall(r')$ if and only if there is an infinite path of ρ which is not accepted with respect to r' if and only if $\rho \in L^\exists(r)$. □

The following theorem is just McNaughton theorem combined with Theorem 5.2.8 (page 121).

Theorem 9.6.2. *There exists a deterministic complete parity word automaton $A = \langle Q_A, T_A, q_A, r_A\rangle$ over $\mathcal{R}$ which recognizes $L^\exists(r)$, and a deterministic complete parity word automaton $B = \langle Q_B, T_B, q_B, r_B\rangle$ over $\mathcal{R}$ which recognizes $L^\forall(r)$.*

Proof. Obviously $L^\exists(r)$ is recognized by the nondeterministic word parity automaton $\langle S, T, S\, r'\rangle$ where $S = \{s_i \mid i \in I\}$, $T = \{(s_i, R, s_j) \mid (i,j) \in R\}$, and $r'(s_i) = r(i)$. □

Since A and B are deterministic and complete, we can replace T_A and T_B by two mappings $\delta_A : Q_A \times \mathcal{R} \to Q_A$ and $\delta_B : Q_B \times \mathcal{R} \to Q_B$.

The following remark will be useful later on. Let i be some element of I. The words ρ accepted by $\langle Q_A, T_A, \delta_A(q_A, \{(i,i)\}), r_A\rangle$ are exactly those such that there exists an accepted infinite path starting in i. Similarly, the words accepted by $\langle Q_B, T_B, \delta_B(q_B, \{(i,i)\}), r_B\rangle$ are those such that every infinite path starting in i is accepted.

9.6.2 The simulation theorem

Let $\tau = \theta_k \boldsymbol{x}^{(k)}. \cdots .\theta_1 \boldsymbol{x}^{(1)}.t$ be a guarded vectorial fixed-point term of length n. Let $X = \{\boldsymbol{x}^{(1)}\} \cup \cdots \cup \{\boldsymbol{x}^{(k)}\}$ and let $rk : X \to \{1, \dots, k\}$ (resp. $ind : X \to \{1, \dots, n\}$) be the mapping that with each variable x in X associates its rank (resp. its index). Let $Z = ar(\tau)$.

Let $\mathcal{R} = \mathcal{P}(X \times X)$. Let $A = \langle Q, \delta, q_*, r\rangle$ be a complete deterministic parity automaton which recognizes the language $L^\forall(rk)$. We assume that $r(Q)$ is included in $\{1, \dots, 2m\}$.

With each state $q \in Q$, with each R in $\mathcal{R}$, and with each $j \in \{1, \dots, 2m\}$, we associate a variable $v^{(j)}_{R,q}$, and we denote by $\boldsymbol{v}^{(j)}$ the vector of variables $v^{(j)}_{R,q}$, indexed by the pairs (R, q).

For each $R \in \mathcal{R}$, we define its codomain

$$Cod(R) = \{x' \mid \exists x : (x, x') \in R\}$$

and $Im(R)$ as the set of indices of the codomain of R, i.e.,

$$Im(R) = ind(Cod(R)) = \{ind(x) \mid x \in Cod(R)\}.$$

We are going to define a vector $\boldsymbol{t}'$ in $IFG(Z)$ indexed by the pairs (R, q) such that any component of $\boldsymbol{\tau}$ is equivalent to some component of $\boldsymbol{\tau}' = \nu\boldsymbol{v}^{(2m)}.\mu\boldsymbol{v}^{(2m-1)}.\cdots.\nu\boldsymbol{v}^{(2)}.\mu\boldsymbol{v}^{(1)}.\boldsymbol{t}'$.

Definition 9.6.3. For any nonempty subset I of $\{1, \ldots, n\}$ we consider the term $t_I \in IFG(Z)$ and, for any $i \in I$, $\varphi_i : Y_I = ar(t_i) - Z \to \mathcal{P}(X)$ which satisfy Proposition 9.3.3 (page 213). With any $y \in Y_I$ we associate the relation $R'_{I,y} \subseteq \{1, \ldots, n\} \times X$ defined by $(i, x) \in R'_{I,y}$ if and only if $i \in I$ and $x \in \varphi_i(y)$.

It follows that

$$\bigwedge_{i \in I} (t_i[id\{\rho_i(x)/x\}_{x \in X}]) =_{C_\Gamma} t_I[id\{t_y/y\}_{y \in Y_I}]$$

with $t_y = \bigwedge_{(i,x) \in R'_{I,y}} \rho_i(x)$.

For each $R'_{I,y}$, we denote by $R_{I,y}$ the relation

$$\{(x, x') \mid (ind(x), x') \in R'_{I,y}\}$$

which is in $\mathcal{R}$, and by $v(I, y, q)$, for $q \in Q$ the variable $v^{(j)}_{R,q'}$ where $R = R_{I,y}$, $q' = \delta(q, R)$, and $j = r(q')$. Let $t_{I,q} = t_I[id\{v(I, y, q)/y\}_{y \in Y_I}]$

By construction, $t_{I,q}$ has the following property.

Proposition 9.6.4. *For any $i \in \{1, \ldots, n\}$ let ρ_i be any substitution. Then for any pair (I, q),*

$$\bigwedge_{i \in I} t_i[id\{\rho_i(x)/x\}_{x \in X}] =_{C_\Gamma} t_{I,q}[id\{\bigwedge_{(i,x) \in R'_{I,y}} \rho_i(x)/v(I, y, q)\}_{y \in Y_i}].$$

Finally let $\boldsymbol{t}'$ be the vector that has $t_{Im(R),q}$ as component of index (R, q).

Theorem 9.6.5. *For any interpretation $\mathcal{I} \in \mathcal{C}_\Gamma$ the component of index i of $[\boldsymbol{\tau}]_{\mathcal{I}}$ is equal to the component of index (R_x, q_x) of $[\boldsymbol{\tau}']_{\mathcal{I}}$ where $ind(x) = i$, R_x is the relation $\{(x, x)\}$, and $q_x = \delta(q_*, R_x)$.*

Proof. The proof contains two main steps. The first step is to show that the equality is true for any interpretation if it is true for any interpretation whose domain is a powerset. The second step is to show the equality for such an interpretation.

The first step of the proof. The basis for the first step is Theorem 2.5.2 (page 54). However, we need to modify the construction of $\mathcal{J}$ from $\mathcal{I}$ so that the interpretation of $\wedge$ in $\mathcal{J}$ is the intersection of subsets. Then we need to prove that such a $\mathcal{J}$ is in $\mathcal{C}_\Gamma$ whenever $\mathcal{I}$ is.

Let $\mathcal{I}$ be an interpretation of domain E and let $\mathcal{J}$ be defined as in Section 2.5 (page 53), excepted that $\wedge^{\mathcal{J}}(E_1, E_2) = E_1 \cap E_2$ (instead of $D(S(E_1) \wedge S(E_2))$).

In order to show that at least the first point of Proposition 2.5.1 still holds, we need to prove that $S(E_1 \cap E_2) = S(E_1) \wedge S(E_2)$. This is not true in general. However, it is true when

- E is completely $\vee$-distributive (it is the case when $\mathcal{I}$ is in $\mathcal{C}_\Gamma$),
- the sets $E_i \subseteq E_+$ are downwards closed: if $e \in E_i$ and $\perp \neq e' \leq e$ then $e' \in E_i$.

Lemma 9.6.6. *If the two above hypothesis hold, then* $S(E_1 \cap E_2) = S(E_1) \wedge S(E_2)$.

Proof. Obviously, by monotonicity of S, $S(E_1 \cap E_2) \leq S(E_1) \wedge S(E_2)$. By complete distributivity $S(E_1) \wedge S(E_2) = S(E')$ where $E' = \{e_1 \wedge e_2 \mid e_i \in E_i\}$. Since E_1 and E_2 are downwards closed, E' is included in $E_1 \cap E_2$, hence $S(E') \leq S(E_1 \cap E_2)$. □

Let $\mathcal{P}_{DC}(E_+)$ be the set of downwards closed subsets of E_+, ordered by inclusion. It is clear that it is a complete lattice: If E_i is downwards closed for any i in a set I of indices, then $\bigcup_{i \in I} E_i$ and $\bigcap_{i \in I} E_i$ are downwards closed too.

Moreover, for any $f \in F_{\{\vee, \wedge\}}$, the restriction of $f^{\mathcal{J}}$ to $\mathcal{P}_{DC}(E_+)$ has its values in $\mathcal{P}_{DC}(E_+)$, thus the domain of this new interpretation $\mathcal{J}$ is $\mathcal{P}_{DC}(E_+)$, instead of $\mathcal{P}(E_+)$.

For this interpretation we get

Proposition 9.6.7. *For any term* $t \in G(X)$, *and for any substitution* $v : Var \to \mathcal{P}_{DC}(E_+)$ *such that* $\forall x \in X, v(x) = D(S(v(x)))$ *we have* $[\![t]\!]_{\mathcal{J}}[v] = D([\![t]\!]_{\mathcal{I}}[S \circ v])$.

Proof. The proof is quite similar to the first part of the proof of Proposition 2.5.1 (page 54). First, it is easy to see that if $t \in BC$ then $S([\![t]\!]_{\mathcal{J}}[v]) = [\![t]\!]_{\mathcal{I}}[S \circ v]$. Like in Proposition 2.5.1, if $t \in EG(X)$ then $[\![t]\!]_{\mathcal{J}}[v] = D([\![t]\!]_{\mathcal{I}}[S \circ v])$. For the same reason, if $t = t_1 \vee t_2$ with $[\![t_i]\!]_{\mathcal{J}}[v] = D([\![t_i]\!]_{\mathcal{I}}[S \circ v])$ then $[\![t]\!]_{\mathcal{J}}[v] = D([\![t]\!]_{\mathcal{I}}[S \circ v])$. Finally, if $t = t_1 \wedge t_2$ with $[\![t_i]\!]_{\mathcal{J}}[v] = D([\![t_i]\!]_{\mathcal{I}}[S \circ v])$ (note that if $t_i = x \in X$, t_i has this property) then $[\![t]\!]_{\mathcal{J}}[v] = [\![t_1]\!]_{\mathcal{J}}[v] \cap [\![t_2]\!]_{\mathcal{J}}[v]$. By the induction hypothesis, this is equal to $D([\![t_1]\!]_{\mathcal{I}}[S \circ v]) \cap D([\![t_2]\!]_{\mathcal{I}}[S \circ v])$. Since $D(e \wedge e') = D(e) \cap D(e')$, we get $D([\![t_1]\!]_{\mathcal{I}}[S \circ v]) \cap D([\![t_2]\!]_{\mathcal{I}}[S \circ v]) = D([\![t_1]\!]_{\mathcal{I}}[S \circ v] \wedge [\![t_2]\!]_{\mathcal{I}}[S \circ v])$ which is equal to $D([\![t]\!]_{\mathcal{I}}[S \circ v])$. □

From which we deduce, denoting by $D^n : E^n \to \mathcal{P}_{DC}(E_+)^n$ and $S^n : \mathcal{P}_{DC}(E_+)^n \to E^n$ the products of D and S n times:

Theorem 9.6.8. *Let* $\boldsymbol{\tau} = \theta_k \boldsymbol{x}^{(k)}.\cdots.\theta_1 \boldsymbol{x}^{(1)}.\boldsymbol{t}$ *with* $\boldsymbol{t} \in G(ar(\boldsymbol{\tau}))$.

For any $v : Var \to \mathcal{P}_{DC}(E_+)$ *such that* $\forall x \in X, v(x) = D(S(v(x))$, $[\boldsymbol{\tau}]_{\mathcal{J}}[v] = D^n([\boldsymbol{\tau}]_{\mathcal{I}}[S \circ v])$.

The proof is by induction on k. For $k = 0$, the result is a consequence of the previous proposition.

Now, let $\boldsymbol{\tau}_i = \theta_i \boldsymbol{x}^{(i)}.\cdots.\theta_1 \boldsymbol{x}^{(1)}.\boldsymbol{t}$, $\boldsymbol{\tau}_{i+1} = \theta_{i+1}\boldsymbol{x}^{(i+1)}.\boldsymbol{\tau}_i$ and let us assume that $[\boldsymbol{\tau}_i]_{\mathcal{J}}[v] = D^n([\boldsymbol{\tau}_i]_{\mathcal{I}}[S \circ v])$. Then $[\boldsymbol{\tau}_{i+1}]_{\mathcal{J}}[v] = \theta_{i+1}\boldsymbol{x}^{(i+1)}.\boldsymbol{h}(\boldsymbol{x}^{(i+1)})$ and $[\boldsymbol{\tau}]_{\mathcal{I}}[S \circ v] = \theta_{i+1}\boldsymbol{x}^{(i+1)}.\boldsymbol{g}(\boldsymbol{x}^{(i+1)})$ where $\boldsymbol{h}(\boldsymbol{x}^{(i+1)}) = [\boldsymbol{\tau}_i]_{\mathcal{J}}[v\{\boldsymbol{x}^{(i+1)}/\boldsymbol{x}^{(i+1)}\}]$ and $\boldsymbol{g}(\boldsymbol{x}^{(i+1)}) = [\boldsymbol{\tau}_i]_{\mathcal{I}}[(S \circ v)\{\boldsymbol{x}^{(i+1)}/\boldsymbol{x}^{(i+1)}\}]$. Since $\{\boldsymbol{x}^{(i+1)}\} \cap ar(\boldsymbol{\tau}) = \emptyset$, we get, by the induction hypothesis, $\boldsymbol{h}(\boldsymbol{x}^{(i+1)}) = D^n(\boldsymbol{g}(S^n(\boldsymbol{x}^{(i+1)})))$. Hence, $\theta_{i+1}\boldsymbol{x}^{(i+1)}.\boldsymbol{h}(\boldsymbol{x}^{(i+1)}) = \theta_{i+1}\boldsymbol{x}^{(i+1)}.D^n(\boldsymbol{g}(S^n(\boldsymbol{x}^{(i+1)})))$.
By Proposition 1.3.12 (page 23), $\theta_{i+1}\boldsymbol{x}^{(i+1)}.D^n(\boldsymbol{g}(S^n(\boldsymbol{x}^{(i+1)}))) =$
$D^n(\theta_{i+1}\boldsymbol{x}^{(i+1)}.\boldsymbol{g}(D^n(S^n(\boldsymbol{x}^{(i+1)})))) = D^n(\theta_{i+1}\boldsymbol{x}^{(i+1)}.\boldsymbol{g}(\boldsymbol{x}^{(i+1)}))$. □

The next remark to do is that if $\mathcal{I}$ is in $\mathcal{C}_\Gamma$ then $\mathcal{J}$ is in $\mathcal{C}_\Gamma$ too. Obviously $\mathcal{P}_{DC}(E_+)$ is completely $\vee$-distributive. Next, we have to show that

$$f^{\mathcal{J}}(E_1, \ldots, E_n) \cap g^{\mathcal{J}}(E'_1, \ldots, E'_{n'}) = [t]_{\mathcal{J}}(E''_1, \ldots, E''_m)$$

with $E''_i = (\bigcap_{j \in I_i} E_j) \cap (\bigcap_{j \in J_i} E'_j)$. But $f^{\mathcal{J}}(E_1, \ldots, E_n) \cap g^{\mathcal{J}}(E'_1, \ldots, E'_{n'})$ $= D(f^{\mathcal{I}}(S(E_1), \ldots, S(E_n))) \cap D(g^{\mathcal{I}}(S(E'_1), \ldots, S(E'_{n'})))$, and, since for any e_1, e_2, $D(e_1) \cap D(e_2) = D(e_1 \wedge e_2)$, this is equal to $D(f^{\mathcal{I}}(S(E_1), \ldots, S(E_n)) \wedge g^{\mathcal{I}}(S(E'_1), \ldots, S(E'_{n'})))$. Since $\mathcal{I} \in \mathcal{C}_\Gamma$,

$$f^{\mathcal{I}}(S(E_1), \ldots, S(E_n)) \wedge g^{\mathcal{I}}(S(E'_1), \ldots, S(E'_{n'})) = [t]_{\mathcal{I}}(e_1, \ldots, e_m)$$

with $e_i = \bigwedge_{j \in I_i} S(E_j) \wedge \bigwedge_{j \in J_i} S(E'_j)$. It is easy to check that $e_i = S(E''_i)$. Since $t \in IFG(\emptyset)$, we get, by Proposition 9.6.7,

$$[t]_{\mathcal{J}}(E''_1, \ldots, E''_m) = D([t]_{\mathcal{I}}(e_1, \ldots, e_m)).$$

Let us come back to the terms $\boldsymbol{\tau}$ and $\boldsymbol{\tau}'$ which we have to compare (see Definition 9.6.3, page 223). The theorem will be proved if the component of index i of $[\boldsymbol{\tau}]_{\mathcal{J}}[D \circ v]$ is equal to some component (R, q) of $[\boldsymbol{\tau}']_{\mathcal{J}}[D \circ v]$ for any interpretation $\mathcal{J} \in \mathcal{C}_\Gamma$ whose domain is $\mathcal{P}(E_+)$ (resp. $\mathcal{P}_{DC}(E_+)$) and for $v : Var \to E$. This is done in the next step.

The second step of the proof. Let $\boldsymbol{\tau}$ and $\boldsymbol{\tau}'$ be the fixed-point terms defined above. Let $\mathcal{J}$ be an interpretation in $\mathcal{C}_\Gamma$ whose domain D is a complete sublattice of $\mathcal{P}(E)$ for inclusion (E need not to be ordered), i.e., if $X_i \in D$ for any $i \in I$ then $\bigcup_{i \in I} X_i$ and $\bigcap_{i \in I} X_i$ are also in D. Let $v : Var \to D$ be any substitution.

Let $\boldsymbol{f}$ be the vector indexed by i such that $\boldsymbol{f}_i = [\boldsymbol{t}_i]_{\mathcal{J}}[id\{v(z)/z\}_{z \in ar(\boldsymbol{\tau})}]$, and $\boldsymbol{g}$ be the vector indexed by the pairs (R, q) such that $\boldsymbol{g}_{R,q} = g_{Im(R),q}$ where

$$g_{I,q} = [\![t_{I,q}]\!]_{\mathcal{J}}[id\{v(z)/z\}_{z\in ar(\tau)}]$$

so that $\boldsymbol{g}_{R,q} = [\![\boldsymbol{t}'_{R,q}]\!]_{\mathcal{J}}[id\{v(z)/z\}_{z\in ar(\tau)}]$.

Let G and G' be the games associated with the closed terms

$$\theta_k \boldsymbol{x}^{(k)}.\cdots.\theta_1 \boldsymbol{x}^{(1)}.\boldsymbol{f} \text{ and } \nu \boldsymbol{v}^{(2m)}.\mu \boldsymbol{v}^{(2m-1)}.\cdots.\nu \boldsymbol{v}^{(2)}.\mu \boldsymbol{v}^{(1)}.\boldsymbol{g},$$

as defined in Definition 4.4.4 (page 97). Let $X = \{\boldsymbol{x}^{(1)}\} \cup \cdots \cup \{\boldsymbol{x}^{(k)}\}$ and $V = \{\boldsymbol{v}^{(1)}\} \cup \cdots \cup \{\boldsymbol{v}^{(2m)}\}$.

(1) Let us consider the winning positional strategy s for Eva which satisfies: if $ind(x) = ind(x') = i \neq 0$ then $s(e,x) = s(e,x')$. We denote this common value by $s(e,i) : Var \to D$. We construct a positional strategy s' for Eva in G' as follows. For any triple (e,I,q) we define $\hat{s}'(e,I,q) : Var \to D$ by

- $\hat{s}'(e,I,q)$ is undefined if there is $i \in I$ such that $s(e,i)$ is undefined (i.e., the position (e,x) is a deadlock and thus loosing for Eva), otherwise
- $\hat{s}'(e,I,q)(v) = \begin{cases} \bigcap_{(i,x)\in R'_{I,y}} s(e,i)(x) & \text{if } v = v(I,y,q) \text{ for some } y, \\ \emptyset & \text{otherwise.} \end{cases}$

Then the strategy s' of Eva at position $p = (e, v^{(j)}_{R,q})$ is to move to $\hat{s}'(e, Im(R), q)$ if it is defined. Otherwise it is undefined.

First let us check that this Eva's move is in the game. If $\hat{s}'(e, Im(R), q)$ is defined then $\forall i \in I, e \in \boldsymbol{f}_i(s(e,i))$, hence, $e \in \bigcap_{i\in I}(\boldsymbol{f}_i(s(e,i)))$. By Proposition 9.6.4, this intersection is equal to

$$g_{I,q}[id\{\bigcap_{(i,x)\in R'_{I,y}} s(e,i)(x)/v(I,y,q)\}_{y\in Y_i}]$$

and, by construction $\bigcap_{(i,x)\in R'_{I,y}} s(e,i)(x) = \hat{s}'(e,I,q)(v(I,y,q))$. It follows that $e \in g_{I,q}[\hat{s}'(e,I,q)]$.

Now let us show that this strategy s' is winning at $(e, v^{j_0}_{R_0,q_0})$ where $R_0 = \{(x_0,x_0)\}$, $q_0 = \delta(q_*, R_0)$, and $j_0 = r(q_0)$, whenever (e_0, x_0) is a winning position in G.

Let $(e_0, v_0), \rho'_0, (e_1, v_1), \rho'_1, \ldots$ be a play consistent with this strategy. To each v_i is associated a relation R_i, a state q_i, and a rank j_i, and we have, by definition of $\hat{s}'$, $\rho'_i(v_{i+1}) = \bigcap_{(x,x')\in R_{i+1}} s(e_i,x)(x')$. If this play is not won by Eva, there are two possibilities

1. the play is finite and its last position is (e_n, v_n) and $\hat{s}'(e_n, Im(R_n), q_n)$ is not defined.
2. the play is infinite and there is $x_0, x_1, \ldots, x_n, \ldots$, with $x_i \in Cod(R_i)$ and $(x_i, x_{i+1}) \in R_{i+1}$, which does not satisfy the parity condition.

In the second case, we have, for all i, $e_{i+1} \in \rho'_i(v_{i+1})$, and since $(x_i, x_{i+1}) \in R_{i+1}$, we have $e_{i+1} \in s(e_i, x_i)(x_{i+1})$.
It follows that $(e_0, x_0), s(e_0, x_0), (e_1, x_1), s(e_1, x_1), \ldots$ is a play in G which is won by Eva, and thus $x_0, x_1, \ldots, x_n, \ldots$ satisfies the parity condition, a contradiction.

In the first case, there exists i_n in $Im(R_n)$, such that $s(e_n, i_n)$ is undefined. Hence, there is $x_n \in Cod(R_n)$ such that $s(e_n, x_n)$ is undefined. But $e_n \in \rho'_{n-1}(v_n) = \bigcap_{(x,x')\in R_n} s(e_{n-1}, x)(x')$.

Since R_n is some $R_{I,y}$ with $I = Im(R_{n-1})$, it follows that there exists $x_{n-1} \in Cod(R_{n-1})$ such that $(x_{n-1}, x_n) \in R_n$. But, since $e_n \in \rho'_{n-1}(v_n)$, we get $e_n \in s(e_{n-1}, x_{n-1})(x_n)$. By the same reasoning, we can find $x_{n-2} \in Cod(R_{n-1})$ such that $e_{n-1} \in s(e_{n-2}, x_{n-2})(x_{n-1})$, and so one, until we arrive at x_0 which is the unique element in $Cod(R_0)$. Thus we have found a play $(e_0, x_0), s(e_0, x_0), (e_1, x_1), s(e_1, x_1), \ldots, s(e_{n-1}, x_{n-1}), (e_n, x_n)$ consistent with the strategy s which is not won by Eva since (e_n, x_n) is not defined, a contradiction.

(2) Let s' be a winning positional strategy for Eva in G' such that $s'(e, v)$ does not depend on the rank of v, so that we may assume that s' is defined on triples (e, R, q), and, of course, $e \in \boldsymbol{g}_{R,q}[s'(e, R, q)] = g_{Im(R),q}[s'(e, R, q)]$, whenever $s'(e, R, q)$ is defined. We define $\hat{s}(e, R, q, i) : Var \to D$ by

$$\hat{s}(e, R, q, x)(x') = \bigcup \{s'(e, R, q)(v(I, y, q)) \mid I = Im(R), (x, x') \in R_{I,y}\}.$$

Obviously, $\hat{s}(e, R, q, x)$ is defined if and only if $s'(e, R, q)$ is defined.

Let us show that the Eva's move from (e, x) to $\hat{s}(e, R, q, x)$ is in G (i.e., $e \in \boldsymbol{f}_i(\hat{s}(e, R, q, x))$ whenever $i = ind(x) \in Im(R)$. By Proposition 9.6.4, for any R such that $Im(R) = I$, $\bigcap_{i\in I} \boldsymbol{f}_i(\hat{s}(e, R, q, x)) = g_{I,q}[id\{E_y/y\}_{y\in Y_I}]$ with $E_y = \bigcap_{(x,x')\in R_{I,y}} \hat{s}(e, R, q, x)(x')$.
But if $(x, x') \in R_{I,y}$, then $s'(e, R, q)(v(I, y, q)) \subseteq \hat{s}(e, R, q, x)(x')$. It follows that $s'(e, R, q)(v(I, y, q)) \subseteq E_y$ and thus,

$$e \in g_{I,q}[s'(e, R, q)] \subseteq g_{I,q}[id\{E_y/y\}_{y\in Y_I}].$$

In case $\hat{s}(e, R, q, x)(x')$ is not empty, there exists y such that $(x, x') \in R_{Im(R),y}$, and $\hat{s}(e, R, q, x)(x') \subseteq s'(e, R, q)(v(I, y, q))$. We choose an arbitrary such y which we denote by by $N(e, R, q, x, x')$.

To define the Eva's strategy, we proceed by induction. With each partial play $p_0, p_1, \ldots, p_n$ in G consistent with the strategy s defined so far, we associate a play $p'_0, p'_1, \ldots, p'_n$ in G' consistent with the strategy s'. In case p_m is an Eva's position, we also define the next Adam's position using p'_m and $\hat{s}$.

At the beginning $p_0 = (e_0, x_0)$ and we set $p'_0 = (e_0, v^{(r(q_0))}_{R_0,q_0})$. Let us assume that p_{2m} is an Eva's position (e_m, x_m) and that $p'_{2m} = (e_m, v^{(r(q_m))}_{R_m,q_m})$.

If $s'(e_m, R_m, q_m)$ is defined (it is always the case if p'_0 is a winning position in G', since s' is a winning strategy), we set $p_{2m+1} = \hat{s}(e_m, R_m, q_m, x_m)$ and $p'_{2m+1} = s'(e_m, R_m, q_m)$. The next position is $p_{2m+2} = (e_{m+1}, x_{m+1})$ with $e_{m+1} \in \hat{s}(e_m, R_m, q_m, x_m)(x_{m+1})$, provided there exists x_{m+1} such that $\hat{s}(e_m, R_m, q_m, x_m)(x_{m+1}) \neq \emptyset$. We take $y = N(e_m, R_m, q_m, x_m, x_{m+1})$ and $p'_{2m+2} = (e_{m+1}, v_{m+1})$ with $v_{m+1} = v(Im(R_m), y, q_m)$. By the choice of y, we have $e_{m+1} \in \hat{s}(e_m, R_m, q_m, x_m)(x_{m+1}) \subseteq s'(e_m, R_m, q_m)(v_{m+1})$, and (p'_{2m+1}, p'_{2m+2}) is a move in G'.

Let $P = (e_0, x_0), \rho_0, (e_1, x_1), \rho_1, \ldots$ be a play consistent with this strategy. If it is finite then it is won by Eva. If it is infinite, let $P' = (e_0, v_0), \rho'_0, (e_1, v_1), \rho'_1 \ldots$ be the corresponding play in G'. If (e_0, v_0) is a winning position then P' is won by Eva. Let us check that P is won by Eva too. By construction, if R_i is the relation associated with v_i, we have $(x_i, x_{i+1}) \in R_{m+1}$. Since $R_0 R_1 \cdots$ is in $L^{\forall}(r)$, $x_0 x_1 \cdots$ satisfies the parity condition and the play P in G is won by Eva.

Since a term in $IFG(\emptyset)$ does not contain the intersection symbol, we have in particular:

Corollary 9.6.9. *For any closed fixed-point term t there exists an intersection-free term t' such that $t =_{\mathcal{C}_{\Gamma}} t'$.*

Proof. By Theorem 2.7.19 (page 68) t is the component of a closed vectorial term $\boldsymbol{\tau}$, with $ar(\boldsymbol{\tau}) = ar(t) = \emptyset$. It follows, by Theorem 9.6.5, that t is equivalent to the component of a closed intersection-free vectorial term $\boldsymbol{\tau}'$, and, by the very definition of a vectorial term (Definition 2.7.2, page 59), is equivalent to a closed intersection-free term. □

9.6.3 The Rabin complementation lemma

The Rabin complementation lemma is a key result in the proof of the decidability of second-order theories [83]. This lemma state that the complement of a recognizable set of trees is recognizable, where, here, recognizable means: accepted by a nondeterministic tree automaton.

The complementation lemma has been proved by Muller and Schupp by considering tree languages recognized by alternating tree automata [71] and showing that they are exactly those recognized by nondeterministic automata [72].

This lemma can also be obtained as a consequence of several previous results: the equivalence between fixed-point terms and automata (Section 7.3, page 180) and the above simulation theorem, that is a generalization of [72].

Let T_{Sig} be the set of all syntactic trees over a signature *Sig* (see Example 6.1.3, page 142). The simulation theorem of Muller and Schupp [72] states that any closed alternating tree automaton is equivalent to a nondeterministic one.

Theorem 9.6.10 (Simulation theorem for tree automata). *Let A be a closed alternating automaton. There exists a nondeterministic automaton B such that $[A]_{\wp\mathcal{T}_{Sig}} = [B]_{\wp\mathcal{T}_{Sig}}$.*

Moreover, if A is a Büchi automaton, B is also a Büchi automaton.

Proof. By Proposition 7.3.5 (page 184), there is a closed term t_A such that $[A]_{\wp\mathcal{T}_{Sig}} = [t_A]_{\wp\mathcal{T}_{Sig}}$. We have seen in Example 9.3.1 (page 212) that the interpretation $\wp\mathcal{T}_{Sig}$ satisfies the conditions $c1 - 3$ of Section 9.3.2 (page 211). It follows, by Corollary 9.6.9 (page 228) that t_A has the same interpretation as an intersection-free term t'.

Therefore the automaton associated with t' is also intersection-free, but, since t' may contain symbols $\tilde{f}$ for $f \in Sig - \{eq\}$, it is not universal (see Definition 8.2.2, page195). However, the same transformation as in the proof of Proposition 8.2.3 (page 195) allows us to transform this automaton into a simply alternating one without adding intersection operators. Thus we get a nondeterministic automaton.

The case for Büchi automata is a consequence of Proposition 9.5.4 (page 219). □

Theorem 9.6.11 (Rabin Complementation Lemma). *If $L \subseteq T_{Sig}$ is the set of trees accepted by a closed automaton A, there is a closed nondeterministic automaton B that accepts all the trees not in L.*

Proof. By Proposition 7.1.10 (page 162), $\overline{L}$ is recognized by the closed automaton $\tilde{A}$ and by the previous theorem $\tilde{A}$ is equivalent to a nondeterministic automaton B. □

9.7 Bibliographic notes and sources

In 1969, Rabin [83] proved the decidability of the monadic second order theory of the full n-ary tree with n successor operations. This result (Rabin Tree Theorem) has been recognized as one of the most powerful decidability results, to which many other decidability questions can be reduced. Rabin based his proof on the same idea that Büchi had successfully used to show the decidability of the monadic second order theory of one successor [21], namely, the reduction of the satisfiability of monadic formulae to the nonemptiness problem of nondeterministic finite automata. However, the closure under complement (Rabin Complementation Lemma), which is an unavoidable step in the proof, already quite hard for Büchi automata on infinite words, turns out to be extremely difficult for automata on infinite trees. Since then, many authors have attempted to simplify the original Rabin's argument. We have already mentioned in Chapter 4 the approach of Gurevich and Harrington [40] based on forgetful determinacy of infinite games (again, the idea can be traced back to Büchi [22]). Another proof with a similar idea was given independently by A. A. Muchnik [68] (see also [69]). In 1985, Muller and Schupp (see [71]) proposed the concept of finite alternating automaton on infinite trees. Although the complementation construction for alternating automata is straightforward, the proof that the complemented automaton behaves as expected is far from obvious, and Muller and Schupp [71] appealed to the Martin Determinacy Theorem for Borel games (see bibliographic notes in Chapter 4). Also, in contrast to the case of nondeterministic automata, the closure of alternating automata under projection (i.e., monadic second order existential quantifier) turned out to be difficult. Apparently, it boils down to the problem of simulation of alternating automata by nondeterministic ones, and Muller and Schupp [71] claimed that it can be best achieved *via* the Gurevich and Harrington Forgetful Determinacy Theorem [40]. However, Emerson and Jutla [33] observed that for alternating automata with *parity* acceptance condition, the simulation by nondeterministic automata is easier, and can be carried out without the appeal to the Gurevich and Harrington's result. This is due to the fact that in the (parity) game associated with an alternating parity automaton, the winner has a positional (or, memoryless) strategy (see Chapter 4). Essentially, the same fact has been independently discovered by Mostowski [66]. Later on, Muller and Schupp [72] gave a general construction for simulating alternating automata by nondeterministic ones; they have considered various acceptance conditions, and provided a detailed complexity analysis.

By the equivalence between the automata and fixed-point terms (Theorem 7.3.6), simulation of an alternating automaton by a nondeterministic one amounts to eliminating intersection ($\wedge$) from the corresponding term. The fact that it can always be done in the powerset algebra of trees (Simulation Theorem), can be inferred as a corollary of the Rabin Tree Theorem, since a fixed-point term (also with $\wedge$) can be easily translated to a monadic formula. (For details of this translation, see, e.g., [10].) However, the advantage of the Simulation Theorem is rather in that it can be used to simplify the proof of the Rabin Tree Theorem (see Section 9.6.3 above). Moreover, in the context of the μ-calculus, we view the Simulation Theorem (especially in the more general setting of Corollary 9.6.9, page 228) worth to be considered for its own.

A related construction in the modal μ-calculus, transforming arbitrary formula into a so-called disjunctive form, was given by Janin and Walukiewicz [45]. This construction is crucial in the Walukiewicz's proof of the Completeness Theorem of the modal μ-calculus [103]. While providing formal proofs is easier for tautologies in disjunctive form than in the general case (as shown already in the primary work of Kozen [55]), the hard part of the Walukiewicz's proof consists essentially in showing that the aforementioned transformation of an arbitrary formula into a disjunctive one (roughly, elimination of intersection) can be carried out within the formal proof system.

A very general version of the Simulation Theorem (called reduction theorem there) for automata over arbitrary complete lattices with monotonic operations was shown by Janin [44].

10. Decision problems

A typical decision question in logic is the *satisfiability* problem: Assuming a class of interpretations $\mathcal{C}$, is there an interpretation in $\mathcal{C}$ satisfying a given sentence φ ? For the μ-calculus, this problem can be phrased as follows. Let a class of μ-interpretations $\mathcal{C}$ be fixed. Given a fixed-point term t, we ask whether there is an interpretation $\mathcal{I}$ in $\mathcal{C}$ such that $[t]_{\mathcal{I}} \neq \bot$. A special instance of this problem is when $\mathcal{C}$ consists of a single interpretation, $\mathcal{C} = \{\mathcal{I}\}$. In this case, we speak of the *nonemptiness* problem for the interpretation $\mathcal{I}$.

In this chapter, we first study the nonemptiness problem for the interpretations satisfying a certain condition which we call *disjunctiveness*; roughly it means that the nonemptiness (i.e. the property of not being $\bot$) of the value of a basic operation is determined by the nonemptiness of its arguments. We show that for a disjunctive interpretation satisfying certain additional requirement of topological flavour, the nonemptiness problem reduces to computing a value of a suitable vectorial Boolean term, and hence is decidable.

Then we show decidability of the satisfiability problem over powerset algebras. The key ingredient here is the Regularity Theorem to which we devote a separate section. This celebrated theorem, essentially due to M. O. Rabin, states that if a finite automaton accepts some tree then it also accepts a "regular" one, i.e., a tree with a finite number of distinct subtrees. Our proof takes advantage of the Selection Property of Section 3.3 (page 75). The use of the Regularity Theorem for the satisfiability problem goes *via* a tree model property for fixed-point terms (over powerset algebras). To the end, we also establish a finite model property: A fixed-point term satisfiable in some powerset algebra has a nonempty denotation in some finite powerset algebra.

Note that we do not give attention to the satisfiability problem over all μ-interpretations. This problem is in fact trivial, since we can interpret any function symbol f by setting $f^{\mathcal{I}}$ to $\top$, for any arguments. We leave to the reader to verify that in such interpretation, the only terms evaluated to $\bot$ are those of the form $\theta y. \cdots \mu x. \cdots \theta' z.x$.

10.1 Disjunctive mappings

Let E be a complete lattice and let $\mathbb{B}$ be its sublattice $\{\bot, \top\}$.

Let $\chi : E \to \mathbb{B}$ be the monotonic mapping defined by $\chi(e) = \bot \Leftrightarrow e = \bot$. χ is extended componentwise into a mapping from E^n into $\mathbb{B}^n$. It is obvious that $\boldsymbol{e} \leq \chi(\boldsymbol{e}) = \chi(\chi(\boldsymbol{e}))$, and that, for $X \subseteq E$, $\chi(\bigvee X) = \bigvee_{x \in X} \chi(x)$.

Definition 10.1.1. If $\boldsymbol{f} : E^n \to E^m$ is a monotonic mapping, we denote by $\boldsymbol{f}^\chi : \mathbb{B}^n \to \mathbb{B}^m$ the monotonic mapping defined by $\boldsymbol{f}^\chi(\boldsymbol{x}) = \chi(\boldsymbol{f}(\boldsymbol{x}))$, for any $\boldsymbol{x} \in \mathbb{B}^n \subseteq E^n$.

Definition 10.1.2. Let C be any subset of E which contains $\mathbb{B}$. A monotonic mapping $\boldsymbol{f} : E^n \to E^m$ is said to be *C-disjunctive* if $\forall \boldsymbol{c} \in C^n$, $\chi(\boldsymbol{f}(\boldsymbol{c})) = \chi(\boldsymbol{f}(\chi(\boldsymbol{c})))$, which is also equal to $\boldsymbol{f}^\chi(\chi(\boldsymbol{c}))$. When C is equal to E, we just say that $\boldsymbol{f}$ is disjunctive (instead of E-disjunctive).

Example 10.1.3. The binary mapping $\vee : E^2 \to E$ is disjunctive: $\chi(e \vee e') = \bot \Leftrightarrow e \vee e' = \bot \Leftrightarrow e = \bot$ and $e' = \bot \Leftrightarrow \chi(e) = \bot$ and $\chi(e') = \bot \Leftrightarrow \chi(e) \vee \chi(e') = \bot \Leftrightarrow \chi(\chi(e) \vee \chi(e')) = \bot$.

The binary mapping $\wedge : E^2 \to E$ need not be disjunctive: Suppose e and e' are such that $e \neq \bot, e' \neq \bot, e \wedge e' = \bot$. Then $\chi(e) = \chi(e') = \top$, and $\chi(e) \wedge \chi(e) = \top$. Hence, $\chi(e \wedge e') = \bot \neq \chi(\chi(e) \wedge \chi(e)) = \top$. □

Proposition 10.1.4. *Let $f^{\wp B}$ and $\tilde{f}^{\wp B}$ from $\mathcal{P}(B)^n$ into $\mathcal{P}(B)$ be such as they are defined in Section 6.1.2 (page 143).*

$f^{\wp B}$ is disjunctive if and only if

(1) if $f^{\wp B}(B, B, \ldots, B) \neq \emptyset$ then $\forall L_1 \neq \emptyset, \ldots, L_n \neq \emptyset$, $f^{\wp B}(L_1, \ldots, L_n) \neq \emptyset$.

$\tilde{f}^{\wp B}$ is disjunctive if and only if

(2) $\forall L_1, \ldots, L_n, \quad f^{\wp B}(L_1, \ldots, L_n) = B \Rightarrow L_1 = \cdots = L_n = B$.

Proof. By definition of $f^{\wp B}$, if there is an i such that $L_i = \emptyset$ then

$$f^{\wp B}(L_1, \ldots, L_n) = \emptyset.$$

It follows that in this case

$$\chi(f^{\wp B}(L_1, \ldots, L_n)) = \chi(f^{\wp B}(\chi(L_1), \ldots, \chi(L_n))) = \emptyset.$$

Thus $f^{\wp B}$ is disjunctive if and only if for $L_1 \neq \emptyset, \ldots, L_n \neq \emptyset$,

$$f^{\wp B}(L_1, \ldots, L_n) \neq \emptyset \Leftrightarrow f^{\wp B}(\chi(L_1), \ldots, \chi(L_n)) = f^{\wp B}(B, \ldots, B)) \neq \emptyset.$$

Since $\tilde{f}^{\wp B}$ is the dual of $f^{\wp B}$, $\tilde{f}^{\wp B}$ is disjunctive if and only if

(3) for any any subsets $L_1, \ldots, L_n$ of B,

$$f^{\wp B}(L_1, \dots, L_n) = B \Leftrightarrow f^{\wp B}(\overline{\chi(\overline{L_1})}, \dots, \overline{\chi(\overline{L_n})}) = B.$$

Since $L = B \Leftrightarrow \overline{\chi(\overline{L})} = B$, it is obvious that (2) implies (3). Now, assume (3) and assume that $f^{\wp B}(L_1, \dots, L_n) = B$.
Then $f^{\wp B}(\overline{\chi(\overline{L_1})}, \dots, \overline{\chi(\overline{L_n})}) = B$, which implies that for any i, $\overline{\chi(\overline{L_i})} \neq \emptyset$, hence, $\overline{\chi(\overline{L_i})} = B = L_i$. □

Example 10.1.5. Let T_{Sig} be the set of syntactic trees over some signature *Sig* of Example 6.1.3, page 142. By definition, $f^{\wp T_{Sig}}(T_1, T_2, \dots, T_m)$ is empty if and only if all the sets $T_1, T_2, \dots, T_m$ are empty. It follows that f is disjunctive and that $f^{\chi} : \mathbb{B}^m \to \mathbb{B}$ satisfies $f^{\chi}(b_1, \dots, b_m) = \bigwedge_{i=1}^{m} b_i$.

The fact that $\tilde{f}^{\wp T_{Sig}}$ is disjunctive can be derived from the previous proposition, or from its definition given on page 196. □

10.2 Decidability of emptiness for disjunctive μ-terms

If under an interpretation $\mathcal{I}$ all symbols f occurring in a closed term τ are interpreted as a disjunctive mapping $f^{\mathcal{I}}$, one can expect to compute the value of $\chi([\tau]_{\mathcal{I}})$ by interpreting this term in $\mathbb{B}$ rather than in the domain of $\mathcal{I}$. This is possible if we have some additional hypothesis on the domain of the interpretation.

10.2.1 Compact lattices

Definition 10.2.1. A *semi-topology* on a complete lattice E is a subset C of E (the elements of C are called *closed* elements) such that

- $\bot \in C$, $\top \in C$,
- $e, e' \in C \Rightarrow e \vee e' \in C$,
- $X \subseteq C \Rightarrow \bigwedge X \in C$.

Definition 10.2.2. An element of E^n is said to be closed if each of its components is closed. A monotonic mapping $\boldsymbol{f} : E^n \to E^m$ is said to be *closed* if $\boldsymbol{f}(\boldsymbol{c})$ is closed whenever $\boldsymbol{c}$ is closed.

Definition 10.2.3. A semi-topology C on E is said to be *compact* if for any $X \subseteq C$ such that $\bigwedge X = \bot$ there exists a finite subset Y of X such that $\bigwedge Y = \bot$.

Example 10.2.4. Let us again consider T_{Sig} equipped with the usual ultrametric distance Δ which is also used in Section 8.2.3 (page 198). This distance induces a topology on T_{Sig}. The family of closed subsets of T_{Sig} is exactly a semi-topology on $\mathcal{P}(T_{Sig})$ in the sense of the previous definition. This semi-topology is compact whenever Sig is finite. Finally, for any $f \in Sig$, if $T_1, \ldots, T_m$ are closed, so is the set $f(T_1, \ldots, T_m)$, hence the mapping $f^{\wp T_{Sig}}$ is closed. □

Lemma 10.2.5. *Assume that C is a compact semi-topology on E. Let β be any ordinal and let $(e_\alpha)_{\alpha<\beta}$ be a nonincreasing sequence of closed elements. If $e_\alpha \neq \bot$ for any $\alpha < \beta$ then $\bigwedge_{\alpha<\beta} e_\alpha \neq \bot$.*

Proof. Assume $\bigwedge_{\alpha<\beta} e_\alpha = \bot$. By compactness, there exist $\alpha_1 < \cdots < \alpha_n < \beta$ such that $\bigwedge_{i=1}^n e_{\alpha_i} = \bot$. But $\bigwedge_{i=1}^n e_{\alpha_i} = e_{\alpha_n} \neq \bot$, a contradiction. □

A straightforward corollary of this lemma is the following:

Corollary 10.2.6. *Assume that C is a compact semi-topology on E. Let β be any ordinal, let $(e_\alpha)_{\alpha<\beta}$ be a nonincreasing sequence of closed elements, and let $d \in \{\bot, \top\}$. If $d \leq \chi(e_\alpha)$ for any $\alpha < \beta$ then $d \leq \chi(\bigwedge_{\alpha<\beta} e_\alpha)$.*

10.2.2 Disjunctiveness of fixed-points

In all which follows, we assume that E has a compact semi-topology defined by its set C of closed elements.

Proposition 10.2.7. *Let $\boldsymbol{f} : E^{n+m} \to E^n$ be monotonic and let $\boldsymbol{g} : E^m \to E^n$ be defined by $\boldsymbol{g}(\boldsymbol{y}) = \theta \boldsymbol{x}.\boldsymbol{f}(\boldsymbol{x}, \boldsymbol{y})$.*

If $\boldsymbol{f}$ is C-disjunctive and closed then

- *$\boldsymbol{g}$ is C-disjunctive and $\boldsymbol{g}^\chi(\boldsymbol{y}) = \theta \boldsymbol{x}.\boldsymbol{f}^\chi(\boldsymbol{x}, \boldsymbol{y})$,*
- *there exists a C-disjunctive and closed monotonic mapping $\boldsymbol{h} : E^m \to E^n$ such that $\boldsymbol{h} \leq \boldsymbol{g}$ and $\boldsymbol{g}^\chi = \boldsymbol{h}^\chi$.*

Proof. Let us remark that $\boldsymbol{g}$ is C-disjunctive if and only if

$$\forall \boldsymbol{c} \in C^m, \quad \chi(\theta \boldsymbol{x}.\boldsymbol{f}(\boldsymbol{x}, \boldsymbol{c})) = \chi(\theta \boldsymbol{x}.\boldsymbol{f}(\boldsymbol{x}, \chi(\boldsymbol{c}))),$$

and that

$$\forall \boldsymbol{b} \in \mathbb{B}^m, \quad \boldsymbol{g}^\chi(\boldsymbol{b}) = \chi(\theta \boldsymbol{x}.\boldsymbol{f}(\boldsymbol{x}, \boldsymbol{b})), \quad \theta \boldsymbol{x}.\boldsymbol{f}^\chi(\boldsymbol{x}, \boldsymbol{b}) = \theta \boldsymbol{x}.\chi(\boldsymbol{f}(\chi(\boldsymbol{x}), \boldsymbol{b})).$$

Moreover, for any $\mathbf{b} \in \mathbb{B}^m$, there exists $\mathbf{c} \in C^m$, such that $\mathbf{b} = \chi(\mathbf{c})$ (it is enough to take $\mathbf{c} = \mathbf{b}$). Therefore, to prove the first point of the result, it is enough to prove

$$\chi(\theta \mathbf{x}.\mathbf{f}(\mathbf{x},\mathbf{c})) = \chi(\theta \mathbf{x}.\mathbf{f}(\mathbf{x},\chi(\mathbf{c}))) = \theta \mathbf{x}.\chi(\mathbf{f}(\chi(\mathbf{x}),\chi(\mathbf{c}))).$$

But, by Proposition 1.3.12 (page 23), the last term can be transformed by

$$\theta \boldsymbol{x}.\chi(\boldsymbol{f}(\chi(\boldsymbol{x}),\chi(\boldsymbol{c}))) = \chi(\theta \boldsymbol{x}.\boldsymbol{f}(\chi(\chi(\boldsymbol{x})),\chi(\boldsymbol{c}))) = \chi(\theta \boldsymbol{x}.\boldsymbol{f}(\chi(\boldsymbol{x}),\chi(\boldsymbol{c}))).$$

Thus, it is enough to show

$$\chi(\theta \boldsymbol{x}.\boldsymbol{f}(\boldsymbol{x},\boldsymbol{c})) = \chi(\theta \boldsymbol{x}.\boldsymbol{f}(\boldsymbol{x},\chi(\boldsymbol{c}))) = \chi(\theta \boldsymbol{x}.\boldsymbol{f}(\chi(\boldsymbol{x}),\chi(\boldsymbol{c}))),$$

i.e., $\chi(\boldsymbol{a}) = \chi(\boldsymbol{b}) = \chi(\boldsymbol{d})$ where

$$\boldsymbol{a} = \theta \boldsymbol{x}.\boldsymbol{f}(\boldsymbol{x},\boldsymbol{c}) = \boldsymbol{g}(\boldsymbol{c}), \boldsymbol{b} = \theta \boldsymbol{x}.\boldsymbol{f}(\boldsymbol{x},\chi(\boldsymbol{c})) = \boldsymbol{g}(\chi(\boldsymbol{c})), \boldsymbol{d} = \theta \boldsymbol{x}.\boldsymbol{f}(\chi(\boldsymbol{x}),\chi(\boldsymbol{c})).$$

Since $\boldsymbol{f}(\boldsymbol{x},\boldsymbol{c}) \leq \boldsymbol{f}(\boldsymbol{x},\chi(\boldsymbol{c})) \leq \boldsymbol{f}(\chi(\boldsymbol{x}),\chi(\boldsymbol{c}))$, we have $\boldsymbol{a} \leq \boldsymbol{b} \leq \boldsymbol{d}$, hence $\chi(\boldsymbol{a}) \leq \chi(\boldsymbol{b}) \leq \chi(\boldsymbol{d})$, and we have only to prove $\chi(\boldsymbol{d}) \leq \chi(\boldsymbol{a})$.

Case $\theta = \nu$. Let $\boldsymbol{g}(\boldsymbol{y}) = \boldsymbol{g}_\gamma(\boldsymbol{y})$ where $\boldsymbol{g}_0(\boldsymbol{y}) = \top^k$, $\boldsymbol{g}_{\alpha+1}(\boldsymbol{y}) = \boldsymbol{f}(\boldsymbol{g}_\alpha(\boldsymbol{y}),\boldsymbol{y})$, $\boldsymbol{g}_\beta(\boldsymbol{y}) = \bigwedge_{\alpha<\beta} \boldsymbol{g}_\alpha(\boldsymbol{y})$. Since $\boldsymbol{f}$ is closed, each $\boldsymbol{g}_\alpha$ is closed. We prove by induction that $\chi(\boldsymbol{d}) \leq \chi(\boldsymbol{g}_\alpha(\boldsymbol{c}))$. The case $\alpha = 0$ is obvious. If $\chi(\boldsymbol{d}) \leq \chi(\boldsymbol{g}_\alpha(\boldsymbol{c}))$, since $\boldsymbol{d} = \boldsymbol{f}(\chi(\boldsymbol{d}),\chi(\boldsymbol{c}))$, we get $\chi(\boldsymbol{d}) \leq \chi(\boldsymbol{f}(\chi(\boldsymbol{g}_\alpha(\boldsymbol{c})),\chi(\boldsymbol{c})))$; since $\boldsymbol{f}$ is C-disjunctive and since $\boldsymbol{c}$ and $\boldsymbol{g}_\alpha(\boldsymbol{c})$ are closed, $\chi(\boldsymbol{f}(\chi(\boldsymbol{g}_\alpha(\boldsymbol{c})),\chi(\boldsymbol{c}))) = \chi(\boldsymbol{f}(\boldsymbol{g}_\alpha(\boldsymbol{c}),\boldsymbol{c})) = \chi(\boldsymbol{g}_{\alpha+1}(\boldsymbol{c}))$. If β is a limit ordinal, the result follows by Corollary 10.2.6.

Since $\boldsymbol{g} = \boldsymbol{g}_\gamma$, $\boldsymbol{g}$ is closed and we can take $\boldsymbol{h} = \boldsymbol{g}$.

Case $\theta = \mu$. Let $\boldsymbol{g}_0(\boldsymbol{y}) = \bot^n$, $\boldsymbol{g}_{\alpha+1}(\boldsymbol{y}) = \boldsymbol{f}(\boldsymbol{g}_\alpha('\boldsymbol{y}),\boldsymbol{y})$, and $\boldsymbol{g}_\beta(\boldsymbol{y}) = \bigvee_{\alpha<\beta} \boldsymbol{g}_\alpha(\boldsymbol{y})$, so that $\boldsymbol{g}(\boldsymbol{y}) = \boldsymbol{g}_\gamma(\boldsymbol{y})$ for some γ. For any $\boldsymbol{y} \in E^n$, $\chi(\boldsymbol{g}_\alpha(\boldsymbol{y}))$ is bounded by $\top^n$ so that $\chi(\boldsymbol{g}_n(\boldsymbol{y})) = \chi(\boldsymbol{g}_{n+1}(\boldsymbol{y})) = \chi(\boldsymbol{g}(\boldsymbol{y}))$. It follows that $\boldsymbol{g}_n \leq \boldsymbol{g}$ and $\boldsymbol{g}_n^\chi = \boldsymbol{g}^\chi$. Then $\chi(\boldsymbol{a}) = \chi(\boldsymbol{g}(\boldsymbol{c})) = \chi(\boldsymbol{g}_n(\boldsymbol{c})) = \chi(\boldsymbol{g}_{n+1}(\boldsymbol{c})) = \chi(\boldsymbol{f}(\boldsymbol{g}_n(\boldsymbol{c}),\boldsymbol{c}))$. Since $\boldsymbol{f}$ and $\boldsymbol{g}_n$ are C-disjunctive, $\chi(\boldsymbol{a}) = \chi(\boldsymbol{g}_n(\chi(\boldsymbol{c}))) = \chi(\boldsymbol{f}(\chi(\boldsymbol{g}_n(\chi(\boldsymbol{c}))),\chi(\boldsymbol{c})))$, and thus, $\chi(\boldsymbol{a}) = \chi(\boldsymbol{f}(\chi(\boldsymbol{a}),\chi(\boldsymbol{c})))$. Hence $\chi(\boldsymbol{a}) \geq \mu \boldsymbol{x}.\chi(\boldsymbol{f}(\boldsymbol{x},\chi(\boldsymbol{c})))$, which is equal, by Proposition 1.3.12, to $\chi(\mu \boldsymbol{x}.\boldsymbol{f}(\chi(\boldsymbol{x}),\chi(\boldsymbol{c}))) = \chi(\boldsymbol{d})$. Thus, we can take $\boldsymbol{h} = \boldsymbol{g}_n$. □

Proposition 10.2.8. *Let $\boldsymbol{h}, \boldsymbol{g} : E^{n+m} \to E^n$ be two monotonic mappings such that*

- $\boldsymbol{h}(\boldsymbol{x},\boldsymbol{y}) \leq \boldsymbol{g}(\boldsymbol{x},\boldsymbol{y})$,
- $\boldsymbol{h}'(\boldsymbol{y}) = \theta \boldsymbol{x}.\boldsymbol{h}(\boldsymbol{x},\boldsymbol{y})$ *is C-disjunctive and* $\boldsymbol{h}'^\chi(\boldsymbol{y}) = \theta \boldsymbol{x}.\boldsymbol{g}^\chi(\boldsymbol{x},\boldsymbol{y})$.

Then $\boldsymbol{g}'(\boldsymbol{y}) = \theta \boldsymbol{x}.\boldsymbol{g}(\boldsymbol{x},\boldsymbol{y})$ is C-disjunctive and $\boldsymbol{g}'^\chi(\boldsymbol{y}) = \theta \boldsymbol{x}.\boldsymbol{g}^\chi(\boldsymbol{x},\boldsymbol{y})$.

Proof. Let $\boldsymbol{c}$ be any closed element of E^m, let

$$\boldsymbol{a} = \theta \boldsymbol{x}.\boldsymbol{g}(\boldsymbol{x}, \boldsymbol{c}), \quad \boldsymbol{b} = \theta \boldsymbol{x}.\boldsymbol{g}(\boldsymbol{x}, \chi(\boldsymbol{c})), \quad \boldsymbol{d} = \theta \boldsymbol{x}.\boldsymbol{g}(\chi(\boldsymbol{x}), \chi(\boldsymbol{c})).$$

We have to prove $\chi(\boldsymbol{a}) = \chi(\boldsymbol{b}) = \chi(\boldsymbol{d})$ where $\chi(\boldsymbol{d})$ is also equal to $\theta \boldsymbol{x}.\boldsymbol{g}^{\chi}(\boldsymbol{x}, \chi(\boldsymbol{c}))$.

Obviously, $\chi(\boldsymbol{a}) \leq \chi(\boldsymbol{b}) \leq \chi(\boldsymbol{d})$. Let

$$\begin{aligned} \boldsymbol{e} &= \boldsymbol{h}'(\boldsymbol{c}) = \theta \boldsymbol{x}.\boldsymbol{h}(\boldsymbol{x}, \boldsymbol{c}), \\ \boldsymbol{e}' &= \boldsymbol{h}'(\chi(\boldsymbol{c})) = \theta \boldsymbol{x}.\boldsymbol{h}(\boldsymbol{x}, \chi(\boldsymbol{c})). \end{aligned}$$

Since $\boldsymbol{h} \leq \boldsymbol{g}$, $\boldsymbol{e} \leq \boldsymbol{a}$. Since $\boldsymbol{h}'$ is C-disjunctive, $\chi(\boldsymbol{e}) = \chi(\boldsymbol{e}')$. Since $\boldsymbol{h}'^{\chi}(\boldsymbol{y}) = \theta \boldsymbol{x}.\boldsymbol{g}^{\chi}(\boldsymbol{x}, \boldsymbol{y})$, $\chi(\boldsymbol{e}') = \chi(\boldsymbol{d})$. □

Proposition 10.2.9. *Let $\boldsymbol{f} : E^{kn+m} \to E^n$ be monotonic and let $\boldsymbol{g} : E^m \to E^n$ be defined by $\boldsymbol{g}(\boldsymbol{y}) = \theta_k \boldsymbol{x}_k. \cdots .\theta_1 \boldsymbol{x}_1.\boldsymbol{f}(\boldsymbol{x}_1, \ldots, \boldsymbol{x}_k, \boldsymbol{y})$.*

If $\boldsymbol{f}$ is C-disjunctive and closed then

- *$\boldsymbol{g}$ is C-disjunctive and $\boldsymbol{g}^{\chi}(\boldsymbol{y}) = \theta_k \boldsymbol{x}_k. \cdots .\theta_1 \boldsymbol{x}_1.\boldsymbol{f}^{\chi}(\boldsymbol{x}_1, \ldots, \boldsymbol{x}_k, \boldsymbol{y})$.*
- *there exists a C-disjunctive and closed monotonic mapping $\boldsymbol{h} : E^m \to E^n$ such that $\boldsymbol{h} \leq \boldsymbol{g}$ and $\boldsymbol{g}^{\chi} = \boldsymbol{h}^{\chi}$.*

Proof. The proof is by induction on k. If $k = 0$ the result trivially holds.

Let $\boldsymbol{g}(\boldsymbol{x}_{k+1}, \boldsymbol{y}) = \theta_k \boldsymbol{x}_k. \cdots .\theta_1 \boldsymbol{x}_1.\boldsymbol{f}(\boldsymbol{x}_1, \ldots, \boldsymbol{x}_k \boldsymbol{x}_{k+1}, \boldsymbol{y})$. We know, by the induction hypothesis, that $\boldsymbol{g}$ is C-disjunctive, that

$$\boldsymbol{g}^{\chi} = \theta_k \boldsymbol{x}_k. \cdots .\theta_1 \boldsymbol{x}_1.\boldsymbol{f}^{\chi}(\boldsymbol{x}_1, \ldots, \boldsymbol{x}_k, \boldsymbol{x}_{k+1}, \boldsymbol{y})$$

and that there is a closed and C-disjunctive $\boldsymbol{h}$ such that $\boldsymbol{h} \leq \boldsymbol{g}$ and $\boldsymbol{h}^{\chi} = \boldsymbol{g}^{\chi}$.

By Proposition 10.2.8, $\theta \boldsymbol{x}_{k+1}.\boldsymbol{h}(\boldsymbol{x}_{k+1}, \boldsymbol{y})$ is C-disjunctive,

$$\begin{aligned} (\theta \boldsymbol{x}_{k+1}.\boldsymbol{h}(\boldsymbol{x}_{k+1}, \boldsymbol{y}))^{\chi} &= \theta \boldsymbol{x}_{k+1}.\boldsymbol{h}^{\chi}(\boldsymbol{x}_{k+1}, \boldsymbol{y}) = \theta \boldsymbol{x}_{k+1}.\boldsymbol{g}^{\chi}(\boldsymbol{x}_{k+1}, \boldsymbol{y}) \\ &= \theta \boldsymbol{x}_{k+1}.\theta_k \boldsymbol{x}_k. \cdots .\theta_1 \boldsymbol{x}_1.\boldsymbol{f}^{\chi}(\boldsymbol{x}_1, \ldots, \boldsymbol{x}_k, \boldsymbol{x}_{k+1}, \boldsymbol{y}), \end{aligned}$$

and there exists a C-disjunctive and closed $\boldsymbol{k}(\boldsymbol{y}) \leq \theta \boldsymbol{x}_{k+1}.\boldsymbol{h}(\boldsymbol{x}_{k+1}, \boldsymbol{y})$ such that $\boldsymbol{k}^{\chi}(\boldsymbol{y}) = \theta \boldsymbol{x}_{k+1}.\boldsymbol{h}^{\chi}(\boldsymbol{x}_{k+1}, \boldsymbol{y})$.

Since $\boldsymbol{h} \leq \boldsymbol{g}$ and $\theta \boldsymbol{x}_{k+1}.\boldsymbol{h}(\boldsymbol{x}_{k+1}, \boldsymbol{y})$ is C-disjunctive, and

$$(\theta \boldsymbol{x}_{k+1}.\boldsymbol{h}(\boldsymbol{x}_{k+1}, \boldsymbol{y}))^{\chi} = \theta \boldsymbol{x}_{k+1}.\boldsymbol{g}^{\chi}(\boldsymbol{x}_{k+1}, \boldsymbol{y}))$$

we get, by Proposition 10.2.9, that $\theta \boldsymbol{x}_{k+1}.\boldsymbol{g}(\boldsymbol{x}_{k+1}, \boldsymbol{y})$ is C-disjunctive and $(\theta \boldsymbol{x}_{k+1}.\boldsymbol{g}(\boldsymbol{x}_{k+1}, \boldsymbol{y}))^{\chi} = \theta \boldsymbol{x}_{k+1}.\boldsymbol{g}^{\chi}(\boldsymbol{x}_{k+1}, \boldsymbol{y}))$.

Finally, since $\boldsymbol{h} \leq \boldsymbol{g}$ we get $\boldsymbol{k}(\boldsymbol{y}) \leq \theta \boldsymbol{x}_{k+1}.\boldsymbol{h}(\boldsymbol{x}_{k+1}, \boldsymbol{y}) \leq \theta \boldsymbol{x}_{k+1}.\boldsymbol{g}(\boldsymbol{x}_{k+1}, \boldsymbol{y})$ with $\boldsymbol{k}^{\chi}(\boldsymbol{y}) = \theta \boldsymbol{x}_{k+1}.\boldsymbol{g}^{\chi}(\boldsymbol{x}_{k+1}, \boldsymbol{y})$ □

In particular, for a C-disjunctive and closed monotonic mapping $f : E^{kn} \to E^n$, we have

$$\chi(\theta_k \boldsymbol{x}_k. \cdots .\theta_1 \boldsymbol{x}_1 . \boldsymbol{f}(\boldsymbol{x}_1, \ldots, \boldsymbol{x}_k)) = \theta_k \boldsymbol{x}_k. \cdots .\theta_1 \boldsymbol{x}_1 . \boldsymbol{f}^{\chi}(\boldsymbol{x}_1, \ldots, \boldsymbol{x}_k).$$

Thus, in a disjunctive interpretation, we can answer the nonemptiness question for a closed term by computing the value of a suitable Boolean term. More precisely, we can state the following.

Corollary 10.2.10. *Let $\mathcal{I}$ be a μ-interpretation of a set of function symbols F. Suppose $\langle D_{\mathcal{I}}, \leq \rangle$ is equipped with a semi-topology, and, for each $f \in F$, the mapping $f^{\mathcal{I}}$ is disjunctive and closed. Suppose furthermore, that for each $f \in F$, the Boolean mapping $(f^{\mathcal{I}})^{\chi}$ is effectively given. Then, given a closed fixed-point term t in fix $\mathcal{T}(F)$, it is decidable whether $[t]_{\mathcal{I}} \neq \bot$.*

Proof. By Corollary 2.7.20 (page 69), t is a component of some vectorial fixed-point term in normal form, say $\theta \boldsymbol{x}_1. \ldots .\theta \boldsymbol{x}_p . \boldsymbol{t}$. By virtue of this form and the assumption, the interpretation $[\boldsymbol{t}]_{\mathcal{I}}$ of the vectorial term $\boldsymbol{t}$ is disjunctive and closed and, moreover, the vectorial Boolean mapping $[\boldsymbol{t}]_{\mathcal{I}}^{\chi}$ can be effectively found. Therefore, by Proposition 10.2.9, the question $[t]_{\mathcal{I}} \neq \bot$ reduces to computing the value of the Boolean vector $\theta \boldsymbol{x}_1. \ldots .\theta \boldsymbol{x}_p . [\boldsymbol{t}]_{\mathcal{I}}^{\chi}$. This, of course, can be done effectively (see Chapter 11 for discussion of the algorithms for this problem). □

10.2.3 Emptiness of nondeterministic tree automata

Let A be a closed nondeterministic tree automaton over *Sig* (see Section 7.1.5, page 162). Let B be the result of converting A into an abstract form according to the transformation described on page 162. Each rule of its abstract form B has one of the two following forms

- $x = eq(x_1, x_2)$ with $qual(x) = \exists$,
- $x = f(x_1, \ldots, x_{\rho(f)})$ with $f \in Sig$ and $qual(x) = \forall$.

Note that the rules of the second kind are the rules of the original nondeterministic automaton and that applying rules of the first kind amounts to selecting the rule of the second kind that will be applied.

By Proposition 7.3.5, page 184, there is a closed fixed-point term t_B such that $[A]_{\wp T_{Sig}} = [B]_{\wp T_{Sig}} = [t_B]_{\wp T_{Sig}} \subseteq T_{Sig}$.

By construction, t_B is a component of a vectorial term $\theta_1 \boldsymbol{x}_1. \cdots .\theta_k \boldsymbol{x}_k . \boldsymbol{t}$ and each component of $\boldsymbol{t}$ has the form $\tilde{eq}(x_1, x_2)$ or $f(x_1, \ldots, x_{\rho(f)})$, according to whether this component is associated with a rule of the first or second kind. For notational simplicity, let us assume that t_B is the first component of this vectorial term.

Now, using Example 10.1.3 and Proposition 10.1.4 (page 234), it is easy to see that the powerset algebra $\wp\mathcal{T}_{Sig}$ restricted to the signature $Sig \cup \{\tilde{eq}\}$ satisfies the assumptions of Corollary 10.2.10. (The semi-topology is that of Example 10.2.4, page 236.)

More precisely, let $\boldsymbol{s}$ be the vectorial Boolean term obtained by substituting in $\boldsymbol{t}$, $x_1 \vee x_2$ for $\tilde{eq}(x_1, x_2)$ and $x_1 \wedge \cdots \wedge x_{\rho(f)}$ for $f(x_1, \ldots, x_{\rho(f)})$. Then we have

Proposition 10.2.11. *The set* $\lfloor A \rfloor_{\wp\mathcal{T}_{Sig}} = \lfloor B \rfloor_{\wp\mathcal{T}_{Sig}}$ *is not empty if and only if the first component of* $\theta_1\boldsymbol{x}_1.\cdots.\theta_k\boldsymbol{x}_k.\boldsymbol{s}$ *is equal to 1.*

Now let us recall that the rules $x = \tilde{eq}(x_1, x_2)$ of B have been obtained by decomposing the rules $x = \tau_1 \vee \cdots \vee \tau_{n_x}$ of A. Using the results of Section 1.4, in particular, Propositions 1.4.4 (page 28) and 1.4.14 (page 36), we can further transform $\theta_1\boldsymbol{x}_1.\cdots.\theta_k\boldsymbol{x}_k.\boldsymbol{s}$ into a vectorial Boolean fixed-point term indexed by the states of A, satisfying the following proposition.

Proposition 10.2.12. *Let A be a closed nondeterministic automaton of initial state x_I. Then we can construct a closed Boolean vectorial fixed-point term*

$$\boldsymbol{b} = \theta_1\boldsymbol{x}_1.\cdots.\theta_k\boldsymbol{x}_k.\boldsymbol{s}$$

indexed by the states of A, such that

- *the set recognized by A is not empty if and only if the component of index x_I of $\boldsymbol{b}$ is equal to 1,*
- *for any rule $x = \bigvee_{i \in I(x)} f_i(x_{i,1}, \ldots x_{i,\rho(f_i)})$ of A, the component of index x of $\boldsymbol{s}$ is $\sum_{i \in I(x)} \prod_{j=1}^{\rho(f_i)} x_{i,j}$.*

10.3 The regularity theorem

10.3.1 Quotients of trees

Let t be any tree over the signature Sig (see Example 6.1.3, page 142). We define the equivalence relation $\sim$ on dom t by $w \sim w'$ if and only if for any $u \in \omega^*$, $wu \in$ dom $t \Leftrightarrow w'u \in$ dom t and if $wu \in$ dom t then $t(wu) = t(w'u)$. In other words, two nodes are equivalent if the two subtrees of t induced by these nodes are equal (see Example 6.3.3, page 148).

The quotient dom $t/_\sim$ can be made a semi-algebra, denoted by $\mathbf{t}/_\sim$, by $w/_\sim \doteq f^{\mathbf{t}/\sim}(v_1/_\sim, \ldots, v_n/_\sim)$ if and only if $w \doteq f^{\mathbf{t}}(v_1, \ldots, v_n)$.

Proposition 10.3.1. *The mapping that sends w on its equivalence class $w/_\sim$ is a surjective reflective homomorphism.*

Proof. By definition of the quotient semi-algebra $\mathbf{t}/_\sim$, the mapping h that maps w to $w/_\sim$ is a surjective homomorphism. Let us show that it is reflective (see Definition 6.3.2, page 148), i.e., if $t(w) = f$ and if if $w \sim w'$, then $t(w') = f$ and $wi \sim w'i$ for $i = 1, \ldots, n$.

Since $w \sim w'$ we have $t(w) = t(w') = f$ and for any $u \in \omega^*$, we have $t(wiu) = t(w'iu)$, hence $wi \sim w'i$. □

It follows, by Proposition 6.3.7 (page 150), that the interpretation of a closed fixed-point τ in $\wp\mathbf{t}$ is empty if and only if it is empty in the quotient semi-algebra.

Definition 10.3.2. A syntactic tree t over the signature *Sig* is said to be *regular* if its equivalence $\sim$ is of finite index.

Proposition 10.3.3. *A tree t over Sig is regular if and only if there is a surjective reflective homomorphism h from* $\mathbf{t}$ *into a finite semi-algebra B over Sig.*

Proof. The necessity of the condition is a direct consequence of the definition of a regular tree and of Proposition 10.3.1. To prove that the condition is sufficient, it is enough to prove that $h(w) = h(w') \Rightarrow w \sim w'$.

First, since $w \doteq f^{\mathbf{t}}(w.1, \ldots, w.n)$ with $f = t(w)$, we have $h(w) \doteq f^B(h(w.1), \ldots, h(w.n))$, and since $h(w) = h(w')$, there exists $v_1, \ldots v_n$, with $h(v_i) = h(w.i)$ and $w' = f^{\mathbf{t}}(v_1, \ldots, v_n)$, which implies $t(w') = t(w)$. We have also, for any $i = 1, \ldots, n$, $v_i = w'.i$, and thus $h(w.i) = h(w'.i)$.

By induction on the length of $u \in \omega^*$, we get $t(w.u) = t(w'.u)$. □

10.3.2 The regularity theorem

In this section we prove Rabin's Regularity Theorem [84].

Theorem 10.3.4. *Every nonempty set of trees recognizable by a finite automaton contains at least one regular tree.*

Proof. Let A be a finite nondeterministic automaton of initial state x_I recognizing the set L. Let us assume that L is not empty.

By Proposition 10.2.12 (page 240) the component of index x_I of

$$\boldsymbol{b} = \theta_1 \boldsymbol{x}_1 . \cdots . \theta_k \boldsymbol{x}_k . \boldsymbol{s}$$

is equal to 1, where the component of index x of $\boldsymbol{s}$ is $\sum_{i \in I(x)} \prod_{j=1}^{\rho(f_i)} x_{i,j}$.

By the selection property (Theorem 3.3.5, page 76), for each state x, there exists $i(x) \in I(x)$ such that $\boldsymbol{b} = \theta_1 \boldsymbol{x}_1 . \cdots . \theta_k \boldsymbol{x}_k . \boldsymbol{s}'$ where the component of index x of $\boldsymbol{s}'$ is $\prod_{j=1}^{\rho(f_{i(x)})} x_{i(x),j}$.

Now, let us consider the automaton A' that has the same states as A and where the rule for a state x is $x = f_{i(x)}(x_{i(x),1}, \ldots x_{i(x),\rho(f_{i(x)})})$. The language L' recognized by A' with initial state x_I is obviously included in L. Moreover, again by Proposition 10.2.12, it is easy to see that L' is not empty if and only if the component of index x_I of $\theta_1\boldsymbol{x}_1.\cdots.\theta_k\boldsymbol{x}_k.\boldsymbol{s}'$. It follows that L' is not empty.

But, clearly, from any state, A' accepts at most one tree, and that if this tree exists, it is regular. Indeed, it is the tree whose quotient is isomorphic to the following finite semi-algebra: Its elements are the states of A' reachable from x and $x \doteq f(x_1, \ldots, x_n)$ if and only if $x = f(x_1, \ldots, x_n)$ is a rule of A'.

□

10.4 The satisfiability over powerset algebras

We call a semi-algebra $\mathcal{B}$ a *model* of a closed fixed-point term t if $[\![t]\!]_{\wp\mathcal{B}} \neq \emptyset$. In this case we also say that $\mathcal{B}$ *satisfies* t. The *satisfiability problem* mentioned in the title is the question whether a closed fixed-point term has a model. In the case of the modal μ-calculus presented in Section 6.2, where the fixed-point terms can be identified with the modal formulas, this terminology is the usual one.

We show that the satisfiability problem for fixed-point terms can be reduced to the nonemptiness problem of a nondeterministic tree automaton, and hence, by the results presented in Section 10.2.3, is decidable. We also show that whenever a fixed-point term t has a model, it also has a finite model whose size is bounded by a recursive function of t, which gives another argument for decidability.

The restriction to closed terms is not essential. We could consider an apparently more general question for an arbitrary fixed-point term t, whether there exists a semi-algebra $\mathcal{B}$ and a valuation $val : ar(t) \to \mathcal{P}(B)$ such that $[\![t]\!]_{\wp\mathcal{B}}(val) \neq \emptyset$. However, this property is easily reducible to the satisfiability of a closed fixed-point term t' obtained from t by replacing each free variable of t by a term whose interpretation is always maximal (e.g., by $\nu z.z$).

10.4.1 Bounded decomposition

Definition 10.4.1. Let Sig be a signature. We call any mapping K from $Sig - \{eq\}$ to $\mathbb{N}$ a *type*. A semi-algebra $\mathcal{B}$ of universe B over Sig is said to be of *decomposition types bounded by* K (or K*-bounded* for short) if, for any $b \in B$ and any $f \in Sig - \{eq\}$, there are at most $K(f)$ distinct tuples $(b_1, \ldots, b_{\rho(f)})$ such that $b \doteq f((b_1, \ldots, b_{\rho(f)})$. It is said to be *codeterministic* if for any element $b \in B$, there is one and only one tuple $(f, b_1, \ldots, b_{\rho(f)}) \in (Sig - \{eq\}) \times B^*$ such that $b \doteq f((b_1, \ldots, b_{\rho(f)})$.

Example 10.4.2. A semi-algebra **t** associated with a syntactic tree $t \in T_{Sig}$ is codeterministic, and so is the algebra $\mathcal{T}_{Sig}$ of all syntactic trees over *Sig* (Example 6.1.3).

Clearly, not every infinite semi-algebra is K-bounded for any K. We will show, however, that if a closed fixed-point term has a model, it also has model which is K-bounded. To this end, it will be convenient to use the correspondence between the fixed-point terms and automata.

Let $A = \langle Sig, Q, \emptyset, x_I, Tr, qual, rank \rangle$ be an automaton with an empty set of variables. We define the mapping $K : Sig - \{eq\} \to \mathbb{N}$ by letting $K(f)$ be the cardinality of $Q_f \subseteq Q$, where Q_f is the set of all states x such that $qual(x) = \forall$ and $Tr(x) = f(y_1, \ldots, y_{\rho(f)})$ for some states $y_1, \ldots, y_{\rho(f)}$.

Proposition 10.4.3. *Let $\mathcal{B}$ be a semi-algebra over Sig on the universe B. There exists a K-bounded semi-algebra $\mathcal{B}'$ over Sig of the same universe B such that $[A]_{\wp\mathcal{B}} \subseteq [A]_{\wp\mathcal{B}'}$.*

Proof. Let $G = G(A, \mathcal{B})$ be the game used to define $[A]_{\wp\mathcal{B}}$ (see Definition 7.1.2, page 157) and let s be a globally winning positional strategy for Eva (see Theorem 4.3.8, page 92).

We define the semi-algebra $\mathcal{B}'$ on the universe B, by letting, for $f \in Sig - \{eq\}$, $b \doteq f^{\mathcal{B}'}(d_1, \ldots, d_{\rho(f)})$ if and only if there is a state $x \in Q_f$ such that $s : (b, x) \mapsto (x = Tr(x), \langle d_1, \ldots, d_{\rho(f)}, b \rangle)$. (Note that, for any $x \in Q_f$ the position (b, x) belongs to Eva.) The symbol *eq* is interpreted as usual.

By definition, $\mathcal{B}'$ is K-bounded. Moreover, $b \doteq f^{\mathcal{B}'}(d_1, \ldots, d_{\rho(f)})$ clearly implies $b \doteq f^{\mathcal{B}}(d_1, \ldots, d_{\rho(f)})$. (In algebraic terms, we could say that $\mathcal{B}'$ is a *sub*–semi-algebra of $\mathcal{B}$.) It follows that the positions of the game $G' = G(A, \mathcal{B}')$ form a subset of the positions of the game G, and any move in G' is also a move in G. Let s' be the restriction of the strategy s to the set of the positions of Eva in G'. It easy to see that s' is a strategy for Eva in G'; we will show that it is winning at any position (b, x_I^A), such that $b \in [A]_{\wp\mathcal{B}}$.

For, let P be a play in G' starting from such a position and consistent with s'. It follows easily by the definitions, that any initial segment of P coincides with an initial segment of some play in G consistent with s. Hence, if P is infinite, it amounts itself to a play in G starting from a winning position (b, x_I^A) and consistent with s; thus P is won by Eva.

If P is finite, it ends with a position p such that no move is possible from p; we have to show that p is a position of Adam. By the remark above, p occurs in some play in G starting from (b, x_I^A) and consistent with s, hence p is a position winning for Eva in G. Actually, there are two cases where p is a deadlock in G':

- p is a transition position $(x = c, c)$ with $\rho(c) = 0$. Then it is also a deadlock in G, and thus it does not belongs to Eva.

– p is a state position (b, x) with $x = f(x_1, \ldots, x_n)$ and there is no tuple $(d_1, \ldots, d_n)$ such that $b \doteq f^{\mathcal{B}'}(d_1, \ldots, d_n)$ (which implies $f \neq eq$!). If it is an Eva's position, it is not a deadlock in G, and hence there must be $d_1, \ldots, d_n$ such that $b \doteq f^{\mathcal{B}}(d_1, \ldots, d_n)$. Therefore The Eva's strategy s is defined at the position p. If $(d'_1, \ldots, d'_n)$ is the tuple selected by s, then $b \doteq f^{\mathcal{B}'}(d'_1, \ldots, d'_n)$, a contradiction.

Hence, in any case P is won by Eva, as desired. □

By the equivalence of the automata and fixed-point terms (Theorem 7.3.6), we immediately get the following

Corollary 10.4.4. *For any closed fixed-point term t over $Sig_\sim$, one can compute a type K, such that if t has any model, it has also a model which is K-bounded.*

10.4.2 Tree models

As mentioned in the introduction to the section, we wish to resolve the satisfiability problem by reducing it to the question of nonemptiness of tree automata. An intermediate step will be the satisfiability over tree models. We have shown in the previous subsection that any model can be turned into a K-bounded one, but, in general, it is not possible to turn a model into a tree in the sense of Example 10.4.2, over the same signature. For example, a term $eq(c, d)$ (in other writing $c \wedge d$) is not satisfied in a tree algebra. Then we introduce an auxiliary signature, and show that any K-bounded semi-algebra can be in a sense interpreted in a codeterministic semi-algebra over the same universe. The latter can be in turn unwinded into a tree.

Given two types $K_1, K_2 : Sig - \{eq\} \to \mathbb{N}$, we write $K_1 \leq K_2$ to mean that, for all $f \in Sig - \{eq\}$, $K_1(f) \leq K_2(f)$.

Definition 10.4.5. For any type K, we define the signature Sig^K as the set $\{F_k \mid k \leq K\} \cup \{eq\}$, where the arity of F_k is

$$\rho(F_k) = \sum_{f \in Sig - \{eq\}} k(f) \cdot \rho(f).$$

The $\rho(F_k)$ arguments of F_k will be indexed by the triples (f, i, j) such that $k(f) \cdot \rho(f) > 0, 1 \leq i \leq k(f), 1 \leq j \leq \rho(f)$.

Definition 10.4.6. If $\mathcal{C}$ is a semi-algebra of universe B over the signature Sig^K, we define the *derived semi-algebra* $\delta(\mathcal{C})$ on the same universe B, but over the signature Sig, by $b \doteq f^{\delta(\mathcal{C})}(b_1, \ldots, b_{\rho(f)})$ if and only if there is a symbol F_k, a number i such that $1 \leq i \leq k(f)$, and a decomposition

$b \doteq F_k^{\mathcal{C}}(c_1, \dots, c_{\rho(F_k)})$ such that its argument of index (f, i, j) $(1 \le j \le \rho(f))$ is b_j.

The interpretation of eq is defined as usual.

Clearly, if $\mathcal{C}$ is K-bounded for some type K, so is $\delta(\mathcal{C})$, for a suitable type K'. On the other hand, the following property comes easily by the definitions.

Proposition 10.4.7. *For any K-bounded semi-algebra $\mathcal{B}$ over Sig, there is a codeterministic semi-algebra $\mathcal{C}$ over Sig^K such that $\mathcal{B} = \delta(\mathcal{C})$.*

Proof. Let $k_b(f) \le K(f)$ be the number of tuples $(b_1, \dots, b_n)$ such that $b \doteq f^{\mathcal{B}}(b_1, \dots, b_n)$. Then we define $\mathcal{C}$ by setting $b \doteq F_{k_b}^{\mathcal{C}}(c_1, \dots, c_n)$ if and only if the argument c_ℓ whose index is (f, i, j) is the j-th argument of the i-th decomposition $b \doteq f^{\mathcal{B}}(b_1, \dots, b_n)$. □

Given a semi–algebra $\mathcal{C}$ over signature Sig^K, we can, for both $\mathcal{C}$ and $\delta(\mathcal{C})$, consider the powerset algebras over the corresponding signatures (see Section 6.1.2). Let, for $f \in Sig$ and $F_k \in Sig^K$, $f^{\wp\delta(\mathcal{C})}$ and $F_k^{\wp\mathcal{C}}$ be the mappings of Definition 6.1.4 (page 143).

For any $f \in Sig$, for any F_k and any i such that $1 \le i \le k(f)$, for any subsets $B_1, \dots, B_{\rho(f)}$, and C of B, we denote by

$$F_k^{\wp\mathcal{C}}((f,i) \looparrowright (B_1, \dots, B_{\rho(f)}), \star \looparrowright C)$$

the value of $F_k^{\wp\mathcal{C}}(C_1, \dots, C_{\rho(F_k)})$ where the argument C_{ind} of index ind is B_j if $ind = (f, i, j)$, and C otherwise.

The value of $\tilde{F}_k^{\wp\mathcal{C}}((f,i) \looparrowright (B_1, \dots, B_{\rho(f)}), \star \looparrowright C)$ is defined in the same way. It follows that it is equal to the complement of $F_k^{\wp\mathcal{C}}((f,i) \looparrowright (\overline{B_1}, \dots, \overline{B_{\rho(f)}}), \star \looparrowright \overline{C})$.

The following proposition is an immediate consequence of the definition of the semi-algebra $\delta(\mathcal{C})$.

Proposition 10.4.8. *For any f in $Sig - \{eq\}$, and any $B_1, \dots, B_n \subseteq B$, where $n = \rho(f)$,*

$$f^{\wp\delta(\mathcal{C})}(B_1, \dots, B_n) = \bigcup_{k \le K} \bigcup_{1 \le i \le k(f)} F_k^{\wp\mathcal{C}}((f,i) \looparrowright (B_1, \dots, B_n), \star \looparrowright B))$$

$$\tilde{f}^{\wp\delta(\mathcal{C})}(B_1, \dots, B_n) = \bigcap_{k \le K} \bigcap_{1 \le i \le k(f)} \tilde{F}_k^{\wp\mathcal{C}}((f,i) \looparrowright (B_1, \dots, B_n), \star \looparrowright \emptyset))$$

We will use the above property to transfer the interpretation of fixed-point terms between the two powerset algebras.

For clarity, we will write $t_1 \vee t_2$ instead of $\tilde{eq}(t_1, t_2)$ and $t_1 \wedge t_2$ instead of $eq(t_1, t_2)$. We will also use the abbreviations $\bigvee_{i \in I} t_i$ and $\bigwedge_{i \in I} t_i$ with the obvious meaning. We fix two closed terms, τ_{tt} and τ_{ff}, such that, for any semi-algebra $\mathcal{D}$ over Sig^K, $[\tau_{tt}]_{\wp\delta(\mathcal{D})} = D$ and $[\tau_{ff}]_{\wp\delta(\mathcal{D})} = \emptyset$. (We can take, for example, $\tau_{tt} = \nu z.z$ and $\tau_{ff} = \mu z.z$.)

Now, let $f \in Sig - \{eq\}$ be of arity n, let $k \leq K$, and $1 \leq i \leq k(f)$. Let $t_1, \ldots, t_n$, s, be some terms over Sig^K. We define $F_k((f,i) \looparrowright (t_1, \ldots, t_n), \star \looparrowright s)$ as the term obtained from $F_k(y_1, \ldots, y_{\rho(F_k)})$ by replacing the argument y_{ind} of index ind by t_j if $ind = (f,i,j)$, and by s otherwise.

We then define the terms $f^K(t_1, \ldots, t_n)$ and $\tilde{f}^K(t_1, \ldots, t_n)$ by

$$f^K(t_1, \ldots, t_n) =_{def} \bigvee_{k \leq K} \bigvee_{1 \leq i \leq k(f)} F_k((f,i) \looparrowright (t_1, \ldots, t_n), \star \looparrowright \tau_{tt})$$

$$\tilde{f}^K(t_1, \ldots, t_n) =_{def} \bigwedge_{k \leq K} \bigwedge_{1 \leq i \leq k(f)} \tilde{F}_k((f,i) \looparrowright (t_1, \ldots, t_n), \star \looparrowright \tau_{ff})$$

Definition 10.4.9. Let K be a type. For each fixed-point term t over Sig, we define the term t^K over Sig^K inductively as follows.

- $x^K = x$,
- $(eq(t_1,t_2))^K = eq(t_1^K, t_2^K)$, and $(\tilde{eq}(t_1,t_2))^K = \tilde{eq}(t_1^K, t_2^K)$,
- for $f \in Sig - \{eq\}$,
  - $(f(t_1, \ldots, t_n))^K = f^K(t_1^K, \ldots, t_n^K)$,
  - $(\tilde{f}(t_1, \ldots, t_n))^K = \tilde{f}^K(t_1^K, \ldots, t_n^K)$,
- $(\theta x.t)^K = \theta x.t^K$.

We will show that the translation $t \mapsto t^K$ preserves the semantics.

Proposition 10.4.10. *Let $\mathcal{C}$ be a semi-algebra on the universe B over the signature Sig^K, and $\delta(\mathcal{C})$ be the derived algebra. Let t be a closed term over $Sig_{\sim}$. Then*

$$[t]_{\wp\delta(\mathcal{C})} = [t^K]_{\wp\mathcal{C}}$$

Proof. By induction on the term t (not necessarily closed), we show that, for any valuation $val : ar(t) \to \mathcal{P}(B)$, $[t]_{\wp\delta(\mathcal{C})}(val) = [t^K]_{\wp\mathcal{C}}(val)$. It follows easily by the definition of the semantics of fixed-point terms, and from Proposition 10.4.8. □

From Corollary 10.4.4, and Propositions 10.4.7 and 10.4.10, we immediately get that if a fixed-point term t has a model then t^K has a model which is a codeterministic semi-algebra. We close the section by showing that the latter can be unwinded into a tree.

Let, in general, $\mathcal{C}$ be a codeterministic semi-algebra over a signature Sig^K of a universe B, and let $b_0 \in B$. We construct a semi-algebra $\mathbf{T}_{\mathcal{C},b_0}$ by unfolding the graph of $\mathcal{C}$ issued from b_0 into a syntactic tree $T_{\mathcal{C},b_0}$ over $Sig^K - \{eq\}$.

More formally, we define, by induction on the length of $w \in \mathbb{N}^*$, the mappings $T_{\mathcal{C},b_0} : \mathbb{N}^* \to Sig^K \cup \{\bot\}$, and $\alpha : \mathbb{N}^* \to B \cup \{\bot\}$. (Here we assume that $\bot$ is an auxiliary element not belonging neither to Sig^K nor to B.) The following invariant will be maintained during the construction: if $\alpha(w) = b \in B$ then $T_{\mathcal{C},b_0}(w) = F$, where F is the unique symbol in $Sig^K - \{eq\}$ such that the operation $F^{\mathcal{C}}$ decomposes b in $\mathcal{C}$.

We let $\alpha(\varepsilon) = b_0$, and $T_{\mathcal{C},b_0}(\varepsilon) = F$, where F is the unique symbol in $Sig^K - \{eq\}$, such that $b_0 \doteq F^{\mathcal{C}}(b_1, \ldots, b_{\rho(F)})$, for some $b_1, \ldots, b_{\rho(F)} \in B$. Now suppose that $T_{\mathcal{C},b_0}$ and α are already defined for all words of some length n, and the invariant is satisfied. For each $w \in \mathbb{N}^*$ with $|w| = n$ such that $\alpha(w) = b \in B$, we have $T_{\mathcal{C},b_0}(w) = F$, for some F, and if $\rho(F) > 0$ then b can be uniquely decomposed in $\mathcal{C}$ by $b \doteq F^{\mathcal{C}}(b_1, \ldots, b_{\rho(F)})$, for some $b_1, \ldots, b_{\rho(F)} \in B$. We let, for $i = 1, \ldots, \rho(F)$, $\alpha(wi) = b_i$, and $T_{\mathcal{C},b_0}(wi) = F_i$, where F_i is the unique symbol in $Sig^K - \{eq\}$, such that $b_i \doteq F_i^{\mathcal{C}}(d_1, \ldots, d_{\rho(F_i)})$. For all words v of length $n+1$ that are not successors of such w's, we let $\alpha(v) = T_{\mathcal{C},b_0}(v) = \bot$. Clearly, the invariant is maintained.

Finally, we let dom $T_{\mathcal{C},b_0} = \{w : T_{\mathcal{C},b_0}(w) \neq \bot\}$.

Recall that, in general, with any syntactic tree T over $Sig_K - \{eq\}$ we associate a semi-algebra $\mathbf{T}$ over Sig^K, with *eq* interpreted as always, and the remaining symbols interpreted as in Example 6.1.3 (see also Section 7.1.5).

The following property comes easily from the above construction.

Lemma 10.4.11. *$T_{\mathcal{C},b_0} \restriction dom\ T_{\mathcal{C},b_0}$ is a syntactic tree over $Sig^K - \{eq\}$ and $\alpha \restriction dom\ T_{\mathcal{C},b_0}$ is a reflective homomorphism from the semi-algebra $\mathbf{T}_{\mathcal{C},b_0}$ to $\mathcal{C}$, such that $\alpha(\varepsilon) = b_0$.*

In the sequel, by abuse of notation, we will denote $T_{\mathcal{C},b_0} \restriction$ dom $T_{\mathcal{C},b_0}$ and $\alpha \restriction$ dom $T_{\mathcal{C},b_0}$ simply by $T_{\mathcal{C},b_0}$ and α, respectively.

We are ready to state the following tree–model property.

Proposition 10.4.12. *For any closed fixed-point term t over $Sig_\sim$, one can compute a type K, such that the following conditions are equivalent.*

1. *t has a model;*
2. *t^K (given by Definition 10.4.9) has a model which is a semi-algebra $\mathbf{T}$ associated with a syntactic tree over Sig^K;*
3. *t^K has a model $\mathbf{T}$ as above, and moreover $\varepsilon \in [t^K]_{\wp\mathbf{T}}$.*

Proof. Let K be the type computed in Corollary 10.4.4.

(1 $\Rightarrow$ 3). From Corollary 10.4.4 and Propositions 10.4.7 and 10.4.10, we know that if t has any model then t^K is satisfied in some codeterministic semi-algebra $\mathcal{C}$. Let $b_0 \in [t^K]_{\wp\mathcal{C}}$. We construct the tree semi–algebra $\mathbf{T}_{\mathcal{C},b_0}$ and the homomorphism $\alpha : \mathbf{T}_{\mathcal{C},b_0} \to \mathcal{C}$ as above. Since α is reflective, we

have, by Corollary 6.3.8 (page 151), $w \in [t^K]_{\wp \mathbf{T}_{C,b_0}} \Leftrightarrow \alpha(w) \in [t^K]_{\wp \mathbf{T}_{C,b_0}}$; in particular, $\varepsilon \in [t^K]_{\wp \mathbf{T}_{C,b_0}} \Leftrightarrow b_0 \in [t^K]_{\wp \mathbf{T}_{C,b_0}}$.

(3 $\Rightarrow$ 2) is trivial.

(2 $\Rightarrow$ 1). If $\mathbf{T}$ is a model of t^K then, again by Proposition 10.4.10, the derived semi-algebra $\delta(\mathbf{T})$ is a model of t. □

10.4.3 Finite models and decidability

Recall that the satisfiability problem for fixed-point terms is the question whether, for a given closed term t over $Sig_{\sim}$, there exists a model, i.e. a semi-algebra $\mathcal{B}$ such that $[t]_{\wp \mathcal{B}}$ is nonempty.

We are ready to state the decidability result.

Theorem 10.4.13. *The satisfiability problem for closed fixed-point terms is decidable.*

Proof. From Proposition 10.4.12, we know that the problem whether a closed fixed-point term t over $Sig_{\sim}$ has a model can be effectively reduced to the question whether, for some syntactic tree T over Sig^K, $\varepsilon \in [t^K]_{\wp \mathbf{T}}$. By Proposition 6.3.9, (page 151), this last condition is equivalent to $T \in [t]_{\wp \mathcal{T}_{Sig^K}}$ i.e., to the membership of T in the interpretation of t^K over the powerset algebra of all syntactic trees over Sig^K. Now, by the equivalence of automata and fixed-point terms (Theorem 7.3.6, page 186), we can construct an automaton A_{t^K} over Sig^K such that $T \in [t^K]_{\wp \mathcal{T}_{Sig^K}}$ if and only if $T \in [A_{t^K}]_{\wp \mathcal{T}_{Sig^K}}$. Therefore, the original question can be effectively reduced to the problem whether $[A_{t^K}]_{\wp \mathcal{T}_{Sig^K}} \neq \emptyset$; in other words, to the nonemptiness problem of an automaton interpreted in the powerset tree algebra. By the simulation theorem for tree automata (Theorem 9.6.10, page 229), we can effectively construct a *nondeterministic automaton* A'_{t^K} such that $[A'_{t^K}]_{\wp \mathcal{T}_{Sig^K}} = [A_{t^K}]_{\wp \mathcal{T}_{Sig^K}}$. The decidability of the nonemptiness problem for the nondeterministic tree automata is shown in Section 10.2.3, page 239.

This remark completes the proof. □

From the equivalence of automata and fixed-point terms (Theorem 7.3.6, page 186), we immediately get the following.

Corollary 10.4.14. *It decidable whether, for a given automaton A (without variables), there exists a semi-algebra $\mathcal{B}$ such that $[A]_{\wp \mathcal{B}} \neq \emptyset$.*

Another consequence is the decidability of the equivalence problem.

Corollary 10.4.15. *It is decidable whether two fixed-point terms have the same interpretation in all powerset algebras.*

Proof. Assume first that fixed-point terms t_1 and t_2 are closed. By Proposition 6.1.6, page 144, for each closed term t we can construct its dual $\tilde{t}$ such that, for any semi-algebra $\mathcal{B}$, $[\tilde{t}]_{\wp\mathcal{B}} = \overline{[t]_{\wp\mathcal{B}}}$. Now, the inequality $[t_1]_{\wp\mathcal{B}} \neq [t_2]_{\wp\mathcal{B}}$ amounts to $([t_1]_{\wp\mathcal{B}} \cap [\tilde{t_2}]_{\wp\mathcal{B}}) \cup ([\tilde{t_1}]_{\wp\mathcal{B}} \cap [t_2]_{\wp\mathcal{B}}) \neq \emptyset$, i.e., to

$$[\tilde{eq}(eq(t_1, \tilde{t_2}), eq(\tilde{t_1}, t_2))]_{\wp\mathcal{B}} \quad \neq \quad \emptyset$$

If t_1, t_2 are not closed, we have $ar(t_1) \cup ar(t_2) = \{z_1, \ldots, z_m\}$. Then we enrich the signature by some fresh constants $d_1, \ldots, d_m$. Now, it is easy to see that the terms t_1 and t_2 do not have the same interpretation in a powerset algebra $\wp\mathcal{B}$ if and only if there exists an extension $\mathcal{B}'$ of the semi-algebra $\mathcal{B}$ by interpretation of the constants $d_1, \ldots, d_m$, such that

$$[t_1\{d_1/z_1, \ldots, d_m/z_m\}]_{\wp\mathcal{B}'} \quad \neq \quad [t_2\{d_1/z_1, \ldots, d_m/z_m\}]_{\wp\mathcal{B}'}$$

Thus, in any case, the non-equivalence of t_1 and t_2 reduces to the non-emptiness of a closed fixed-point term. □

An analogous claim can be, of course, stated for automata.

The second main result of this section is the following.

Theorem 10.4.16 (Finite Model Property). *If a fixed-point term has a model, it has also a finite model.*

Proof. Like in the proof of the previous theorem, we will use the existence of an automaton A_{t^K}, and a nondeterministic automaton A'_{t^K} such that t has any model if and only if $[A'_{t^K}]_{\wp\mathcal{T}_{Sig^K}} \neq \emptyset$. Then, if t has a model then A'_{t^K} accepts some syntactic tree over Sig^K. The latter, by Regularity Theorem (page 241), implies that A'_{t^K} accepts some regular tree, T say. Since the automaton A'_{t^K} is equivalent to A_{t^K} over the powerset tree algebra, and A_{t^K} is semantically equivalent to t^K, we have $T \in [t^K]_{\wp\mathcal{T}_{Sig^K}}$. By Proposition 6.3.9, this implies $\varepsilon \in [t^K]_{\wp\mathbf{T}}$.

Now, by Proposition 10.3.3 (page 241), the quotient of $\mathbf{T}$ under the equivalence $\sim$ is a finite semi-algebra $\mathbf{T}/_\sim$ over Sig^K, and the canonical homomorphism $h : w \mapsto [w]_\sim$ is reflective (where $[w]_\sim$ denotes the equivalence class of w under $\sim$). Again, by Corollary 6.3.8 (page 151), $\varepsilon \in [t^K]_{\wp\mathbf{T}}$ implies $[\varepsilon]_\sim \in [t^K]_{\wp\mathbf{T}/_\sim}$. By Proposition 10.4.10 we conclude that $[\varepsilon]_\sim \in [t]_{\wp\delta(\mathbf{T}/_\sim)}$, and hence the derived semi-algebra $\delta(\mathbf{T}/_\sim)$ is the desired finite model of t. □

Remark on complexity. Let $n = |t|$. It follows by the definition of the homomorphism *auto* translating fixed-point terms to automata (Proposition 7.3.3, page 181), that the automaton A_t equivalent to a fixed-point term t can have a size linear in the size of t. Thus the size of the signature Sig^K constructed in Corollary 10.4.4 can be bounded by $n^{|Sig|}$. If we consider the signature Sig as fixed, we can estimate the size of Sig^K as $n^{\mathcal{O}(1)}$, and the same can be said about the size of the term t^K constructed in Definition 10.4.9. Now, again, the size of the automaton A_{t^K} equivalent to t^K (see the proof of Theorem 10.4.13) is linear in the size of t^K, and hence can be estimated by $n^{\mathcal{O}(1)}$. Now, the complexity of the whole procedure relies mainly on the complexity of the translation of the automaton A_{t^K} into the nondeterministic automaton A'_{t^K}. This is probably best accomplished in the work by Muller and Schupp [72], who translate an alternating automaton with the Rabin chain condition (equivalent to the parity condition) of m states into a nondeterministic tree automaton with $2^{m^{\mathcal{O}(1)}}$ states and of Mostowski index $m^{\mathcal{O}(1)}$. (The way presented in this book, via the Simulation Theorem, is less direct.) This gives us a nondeterministic tree automaton of the size $2^{n^{\mathcal{O}(1)}}$ and of Mostowski index $n^{\mathcal{O}(1)}$ whose nonemptiness is equivalent to the satisfiability of t (with $n = |t|$). Now the size of a minimal regular tree accepted by a nondeterministic automaton (if any) is polynomial in the number of states of this automaton (in fact, linear as remarked by Emerson [31]). The size of a minimal finite model of t amounts to the size of a minimal regular tree accepted by A_{t^K}. The complexity of the nonemptiness problem of a nondeterministic tree automaton with m states and a Mostowski index k is best estimated by Emerson and Jutla [32] as $m^{\mathcal{O}(k)}$. Thus we can test the nonemptiness of A_{t^K} in time $2^{n^{\mathcal{O}(1)} \cdot n^{\mathcal{O}(1)}} = 2^{n^{\mathcal{O}(1)}}$.

This finally allows us to estimate the complexity of the satisfiability problem of fixed-point terms, as well as the size of a finite model of a satisfiable term t, by a single exponential function in the size of t.

It should also be noticed that, since all the constructions used in the proof of the above theorem are effective, if t is satisfiable, one can effectively construct a finite model of t.

10.5 Bibliographic notes and sources

Let us review a number of related decidability results.

The decidability of the nonemptiness problem for Rabin automata on infinite trees was shown by Rabin [83]; it constituted one of the steps in the proof of the Rabin Tree Theorem (see notes after the preceding chapter). The Regularity Theorem was slightly later discovered also by Rabin [84]. Emerson [31] used it to show that the aforementioned nonemptiness problem is in class *NP* (in fact, *NP*-complete, as shown later by Emerson and Jutla [32]).

The decidability of the satisfiability problem for the modal μ-calculus was first shown by Kozen and Parikh [56], by a reduction to the monadic second order theory of binary tree, and application of the Rabin Tree Theorem. Streett and Emerson [91] gave an elementary decision procedure by a direct reduction to the nonemptiness problem of tree automata; they also established the small model theorem for the modal μ-calculus. A single–exponential deterministic–time upper bound was shown later by Emerson and Jutla [32], building on the result of Safra [86] on the complexity of determinization of automata on infinite words. The exponential lower bound follows by the work of Fischer and Ladner [37] on the propositional dynamic logic. This lower bound applies, of course, also to the satisfiability problem for our μ-calculus over arbitrary powerset algebras.

A simple, polynomial time procedure for the nonemptiness problem in the powerset algebra of trees without intersection, as well as for the satisfiability problem over all powerset algebras without intersection, was given in [78]. A similar result for disjunctive formulas of the modal μ-calculus (see bibliographic notes in Chapter 9) was independently shown by Janin and Walukiewicz [45]. Both procedures use a reduction to (scalar) terms of the Boolean μ-calculus.

The decidability (in exponential time) of the satisfiability problem for the μ-calculus over powerset algebras (with intersection and dualities), shown in Section 10.4 can be also derived as a corollary to the results of McAllester, Givan, Witty and Kozen on Tarskian set constraints [60].

11. Algorithms

We have seen in the previous chapter, that some important decision questions for the μ-calculus amounts to computing the value of a finite vectorial Boolean fixed-point term. We can also note that computing the value of $[t]_{\mathcal{I}}$ in any *finite* μ-interpretation $\mathcal{I}$ can be reduced to this problem, by first constructing the induced powerset interpretation (c.f. Section 2.5, page 53), and then exploiting the correspondence between powerset interpretations and Boolean terms (explained in Section 3.2, page 72). A similar reduction applies to the problem of determining the winner at a given position of a parity game. (By the results of Chapter 4, the two problems are, in fact equivalent.)

The problem of evaluating finite vectorial Boolean fixed-point terms is, by the time we are writing, a fascinating challenge. No algorithm polynomial in the size of the term has been discovered. On the other hand, we know that the problem (more precisely, the decision version of it, namely, whether the first component is 1) is in the complexity class *NP* $\cap$ *co* $-$ *NP* [34] (even in *UP* $\cap$ *co* $-$ *UP* [49]). Thus we may believe that the question is tantalizingly close to feasible. We devote this last chapter to presenting some known algorithmic solutions to the problem.

11.1 Evaluation of vectorial fixed-point terms

We already know that if h is a monotonic mapping from $\mathbb{B}$ to $\mathbb{B}$, $\nu x.h(x) = h(1)$ and $\mu x.h = h(0)$. It follows that the evaluation of a Boolean fixed point term is in time linear in the size of this term.

Very often, however, fixed points to be evaluated are given in the vectorial form

$$\theta_1 \boldsymbol{x}_1.\theta_2 \boldsymbol{x}_2.\cdots.\theta_k \boldsymbol{x}_k.t(\boldsymbol{x}_1, \boldsymbol{x}_2, \ldots, \boldsymbol{x}_k).$$

Of course, this form can be transformed into scalar ones (indeed, it is its very definition: See Section 2.7, page 58) but it costs an exponential blow-up in the size of the term to be evaluated. Fortunately, more efficient algorithms exist, which directly deal with vectorial expressions.

First of all, for any monotonic function $\boldsymbol{t} : \mathbb{B}^{\boldsymbol{x}} \to \mathbb{B}^{\boldsymbol{x}}$, where $\boldsymbol{x}$ is a vector of variables of length n (for sake of notational simplicity, here we identify a vector of variables with the set of its components), let us consider the following sequences of elements of $\mathbb{B}^{\boldsymbol{x}}$:

$$\begin{aligned} \boldsymbol{a}_0 &= \boldsymbol{0}, \\ \boldsymbol{a}_1 &= \boldsymbol{t}(\boldsymbol{0}), \\ \boldsymbol{a}_2 &= \boldsymbol{t}(\boldsymbol{a}_1), \\ &\dots \\ \boldsymbol{a}_{i+1} &= \boldsymbol{t}(\boldsymbol{a}_i), \\ &\dots \end{aligned}$$

and

$$\begin{aligned} \boldsymbol{b}_0 &= \boldsymbol{1}, \\ \boldsymbol{b}_1 &= \boldsymbol{t}(\boldsymbol{1}), \\ \boldsymbol{b}_2 &= \boldsymbol{t}(\boldsymbol{b}_1), \\ &\dots \\ \boldsymbol{b}_{i+1} &= \boldsymbol{t}(\boldsymbol{b}_i), \\ &\dots \end{aligned}$$

Lemma 11.1.1.

$$\begin{aligned} \mu \boldsymbol{x}.\boldsymbol{t}(\boldsymbol{x}) &= \boldsymbol{a}_n, \\ \nu \boldsymbol{x}.\boldsymbol{t}(\boldsymbol{x}) &= \boldsymbol{b}_n. \end{aligned}$$

Proof. Obviously, $\boldsymbol{a}_n \leq \mu \boldsymbol{x}.\boldsymbol{t}(\boldsymbol{x})$. Since any increasing sequence of elements of $\mathbb{B}^n$ has at most $n+1$ elements, the increasing sequence $\boldsymbol{a}_0, \boldsymbol{a}_1, \dots, \boldsymbol{a}_{n+1}$ contains at least two equal elements $\boldsymbol{a}_i$ and $\boldsymbol{a}_j$ with $i < j \leq n+1$, hence, $\boldsymbol{a}_i = \boldsymbol{a}_{i+1}$. It follows that $\boldsymbol{a}_i = \boldsymbol{a}_p$ for $i \leq p$. Hence, $\boldsymbol{a}_n = \boldsymbol{a}_{n+1} = \boldsymbol{t}(\boldsymbol{a}_n)$, and the result follows.

The proof is similar for the greatest fixed point. □

Lemma 11.1.2. *Let $\boldsymbol{t} : \mathbf{B}^{\boldsymbol{x}} \times \mathbf{B}^{\boldsymbol{y}} \to \mathbf{B}^{\boldsymbol{x}}$ be monotonic, where $\boldsymbol{x}$ is a vector of variables of length n. Let $\boldsymbol{g}(\boldsymbol{y}) = \theta \boldsymbol{x}.\boldsymbol{t}(\boldsymbol{x}, \boldsymbol{y})$, and let $\boldsymbol{b}$ be any element of $\mathbf{B}^{\boldsymbol{y}}$. Let us consider the following sequence of elements of $\mathbf{B}^{\boldsymbol{x}}$.*

$$\begin{aligned} \boldsymbol{a}_1 &= \boldsymbol{t}(\boldsymbol{v}_\theta, \boldsymbol{b}), \\ \boldsymbol{a}_2 &= \boldsymbol{t}(\boldsymbol{a}_1, \boldsymbol{b}), \\ &\dots \\ \boldsymbol{a}_{i+1} &= \boldsymbol{t}(\boldsymbol{a}_i, \boldsymbol{b}) \\ &\dots \end{aligned}$$

where $\boldsymbol{v}_\theta$ *is equal to* $\mathbf{0}$ *if* $\theta = \mu$ *and to* $\mathbf{1}$ *if* $\theta = \nu$.
Then, $\boldsymbol{g}(\boldsymbol{b}) = \boldsymbol{a}_n$.

Proof. Since $\boldsymbol{g}(\boldsymbol{b}) = \theta\boldsymbol{x}.\boldsymbol{t}(\boldsymbol{x}, \boldsymbol{b})$, the results follows from the previous lemma. □

An immediate corollary of this lemma is the following proposition.

Proposition 11.1.3. *For any monotonic function* $\boldsymbol{t} : \mathbb{B}^{\boldsymbol{x}} \times \mathbb{B}^{\boldsymbol{y}} \to \mathbb{B}^{\boldsymbol{x}}$, *where* $\boldsymbol{x}$ *is a vector of variables of length* n, *let us consider the following sequences of monotonic functions from* $\mathbb{B}^{\boldsymbol{y}}$ *to* $\mathbb{B}^{\boldsymbol{x}}$*:*

$$\begin{aligned}
\boldsymbol{g}_0(\boldsymbol{y}) &= \mathbf{0}, \\
\boldsymbol{g}_1(\boldsymbol{y}) &= \boldsymbol{t}(\mathbf{0}, \boldsymbol{y}), \\
\boldsymbol{g}_2(\boldsymbol{y}) &= \boldsymbol{t}(\boldsymbol{g}_1(\boldsymbol{y}), \boldsymbol{y}), \\
&\dots \\
\boldsymbol{g}_{i+1}(\boldsymbol{y}) &= \boldsymbol{t}(\boldsymbol{g}_i(\boldsymbol{y}), \boldsymbol{y}), \\
&\dots
\end{aligned}$$

and

$$\begin{aligned}
\boldsymbol{h}_0(\boldsymbol{y}) &= \mathbf{1}, \\
\boldsymbol{h}_1(\boldsymbol{y}) &= \boldsymbol{t}(\mathbf{1}, \boldsymbol{y}), \\
\boldsymbol{h}_2(\boldsymbol{y}) &= \boldsymbol{t}(\boldsymbol{h}_1(\boldsymbol{y}), \boldsymbol{y}), \\
&\dots \\
\boldsymbol{h}_{i+1}(\boldsymbol{y}) &= \boldsymbol{t}(\boldsymbol{h}_i(\boldsymbol{y}), \boldsymbol{y}), \\
&\dots
\end{aligned}$$

Then $\boldsymbol{g}_n(\boldsymbol{y}) = \mu\boldsymbol{x}.\boldsymbol{t}(\boldsymbol{x}, \boldsymbol{y})$ *and* $\boldsymbol{h}_n(\boldsymbol{y}) = \nu\boldsymbol{x}.\boldsymbol{t}(\boldsymbol{x}, \boldsymbol{y})$.

Proof. Let $\boldsymbol{g}(\boldsymbol{y}) = \mu\boldsymbol{x}.\boldsymbol{t}(\boldsymbol{x}, \boldsymbol{y})$. We have to show that $\boldsymbol{g}(\boldsymbol{b}) = \boldsymbol{g}_n(\boldsymbol{b})$ for any $\boldsymbol{b} \in \mathbb{B}^{\boldsymbol{y}}$. It is easy to see that the sequence $(\boldsymbol{g}_i(\boldsymbol{b}))_{i \geq 1}$ is exactly the sequence $(\boldsymbol{a}_i)_{i \geq 1}$ of the previous lemma. The proof is similar for the greatest fixed point. □

Example 11.1.4. Let $\boldsymbol{x} = \langle x_1, x_2, x_3 \rangle$, $\boldsymbol{y} = \langle y_1, y_2, y_3, y_4 \rangle$, and

$$\boldsymbol{t}(\boldsymbol{x}, \boldsymbol{y}) = \langle x_2 y_3 + x_3 y_2, x_1 + x_3 + y_1, x_2 y_4 \rangle.$$

Then

$$\begin{aligned}
\boldsymbol{g}_1(\boldsymbol{y}) &= \boldsymbol{t}(\mathbf{0}, \boldsymbol{y}) \\
&= \langle 0, y_1, 0\rangle, \\
\boldsymbol{g}_2(\boldsymbol{y}) &= \boldsymbol{t}(\langle 0, y_1, 0\rangle, \boldsymbol{y}) \\
&= \langle y_1y_3, y_1, y_1y_4\rangle, \\
\boldsymbol{g}_3(\boldsymbol{y}) &= \boldsymbol{t}(\langle y_1y_3, y_1, y_1y_4\rangle, \boldsymbol{y}) \\
&= \langle y_1y_3 + y_1y_2y_4, y_1y_3 + y_1y_4 + y_1, y_1y_4\rangle, \\
&= \langle y_1y_3 + y_1y_2y_4, y_1, y_1y_4\rangle, \\
\boldsymbol{g}_4(\boldsymbol{y}) &= \boldsymbol{t}(\langle y_1y_3 + y_1y_2y_4, y_1, y_1y_4\rangle, \boldsymbol{y}) \\
&= \langle y_1y_3 + y_1y_2y_4, y_1y_3 + y_1y_2y_4 + y_1y_4 + y_1, y_1y_4\rangle \\
&= \langle y_1y_3 + y_1y_2y_4, y_1, y_1y_4\rangle, \\
&= \boldsymbol{g}_3(\boldsymbol{y}).
\end{aligned}$$

Thus $\mu\boldsymbol{x}.\boldsymbol{t}(\boldsymbol{x}, \langle y_1, y_2, y_3, y_4\rangle) = \langle y_1y_3 + y_1y_2y_4, y_1, y_1y_4\rangle$.

To get the greatest fixed point, we compute

$$\begin{aligned}
\boldsymbol{h}_1(\boldsymbol{y}) &= \boldsymbol{t}(\mathbf{1}, \boldsymbol{y}) \\
&= \langle y_2 + y_3, 1, y_4\rangle, \\
\boldsymbol{h}_2(\boldsymbol{y}) &= \boldsymbol{t}(\langle y_2 + y_3, 1, y_4\rangle, \boldsymbol{y}) \\
&= \langle y_3 + y_2y_4, y_1 + y_2 + y_3 + y_4, y_4\rangle, \\
\boldsymbol{h}_3(\boldsymbol{y}) &= \boldsymbol{t}(\langle y_3 + y_2y_4, y_1 + y_2 + y_3 + y_4, y_4\rangle, \boldsymbol{y}) \\
&= \langle (y_1 + y_2 + y_3 + y_4)y_3 + y_2y_4, y_3 + y_2y_4 + +y_4 + y_1, \\
&\quad (y_1 + y_2 + y_3 + y_4)y_4\rangle, \\
&= \langle y_3 + y_2y_4, y_3 + y_4 + y_1, y_4\rangle, \\
\boldsymbol{h}_4(\boldsymbol{y}) &= \boldsymbol{t}(\langle y_3 + y_2y_4, y_1 + y_3 + y_4, y_4\rangle, \boldsymbol{y}) \\
&= \langle (y_1 + y_3 + y_4)y_3 + y_2y_4, y_3 + y_2y_4 + y_4 + y_1, (y_1 + y_3 + y_4)y_4\rangle \\
&= \langle y_3 + y_2y_4, y_3 + y_4 + y_1, y_4\rangle, \\
&= \boldsymbol{h}_3(\boldsymbol{y}).
\end{aligned}$$

Thus $\nu\boldsymbol{x}.\boldsymbol{t}(\boldsymbol{x}, \langle y_1, y_2, y_3, y_4\rangle) = \langle y_3 + y_2y_4, y_3 + y_4 + y_1, y_4\rangle$.

□

Using several times the method in the above proposition one can evaluate

$$\theta_1\boldsymbol{x}_1\theta_2\boldsymbol{x}_2\cdots\theta_k\boldsymbol{x}_k\boldsymbol{t}(\boldsymbol{x}_1, \boldsymbol{x}_2, \ldots, \boldsymbol{x}_k)$$

in the following way:
first compute $\boldsymbol{f}_k(\boldsymbol{x}_1, \boldsymbol{x}_2, \ldots, \boldsymbol{x}_{k-1}) = \theta_k\boldsymbol{x}_k\boldsymbol{t}(\boldsymbol{x}_1, \boldsymbol{x}_2, \ldots, \boldsymbol{x}_k)$
then $\boldsymbol{f}_{k-1}(\boldsymbol{x}_1, \boldsymbol{x}_2, \ldots, \boldsymbol{x}_{k-2}) = \theta_{k-1}\boldsymbol{x}_{k-1}.\boldsymbol{f}_k(\boldsymbol{x}_1, \boldsymbol{x}_2, \ldots, \boldsymbol{x}_{k-1})$,
and so on, until $\boldsymbol{f}_1 = \theta_1\boldsymbol{x}_1.\boldsymbol{f}_2(\boldsymbol{x}_1)$.

In the above, we assume that $\boldsymbol{t}(\boldsymbol{x}_1, \boldsymbol{x}_2, \ldots, \boldsymbol{x}_k)$ is presented by a vector of functional Boolean terms (written with $\wedge$, $\vee$, $\mathbf{0}$ and $\mathbf{1}$, but without μ and ν), whose variables are in $\{\boldsymbol{x}_1, \ldots, \boldsymbol{x}_k\}$. Thus the size $|\boldsymbol{t}|$ of $\boldsymbol{t}$ is well defined as the joint lenght of these terms. In the sequel, we often do not make notational distinction between a vectorial functional Boolean term and the vectorial Boolean function it denotes.

However the above is not necessarily an efficient way of computing: In the above proposition, the time needed to compute $\boldsymbol{g}_{i+1}(\boldsymbol{y})$ from $\boldsymbol{g}_i(\boldsymbol{y})$ is obviously linear in the size $|\boldsymbol{t}|$ of $\boldsymbol{t}$, (not taking into account possible simplifications using the algebraic laws of a Boolean algebra), so that the time needed to compute $\mu\boldsymbol{x}.\boldsymbol{t}(\boldsymbol{x}, \boldsymbol{y})$ and $\nu\boldsymbol{x}.\boldsymbol{t}(\boldsymbol{x}, \boldsymbol{y})$ is linear in $n|\boldsymbol{t}|$ where n is the length of $\boldsymbol{x}$. Thus the time needed to compute $\theta_1\boldsymbol{x}_1\theta_2\boldsymbol{x}_2\cdots\theta_k\boldsymbol{x}_k\boldsymbol{t}(\boldsymbol{x}_1, \boldsymbol{x}_2, \ldots, \boldsymbol{x}_k)$ is equal to $n(|\boldsymbol{t}| + |\boldsymbol{f}_k| + |\boldsymbol{f}_{k-1}| + \cdots + |\boldsymbol{f}_2|)$ plus the time for simplifications, as one can see on the previous example.

We have noticed that the time needed to evaluate $\boldsymbol{t}$ is proportional to its size $|\boldsymbol{t}|$. Therefore, in the sequel, we assume for simplicity, that this time is equal to $|\mathsf{t}|$.

11.1.1 A naive algorithm

From an algorithmic point of view, it is better to proceed as follows: to compute

$$\theta_1\boldsymbol{x}_1.\theta_2\boldsymbol{x}_2.\cdots.\theta_k\boldsymbol{x}_k.\boldsymbol{t}(\boldsymbol{x}_1, \boldsymbol{x}_2, \ldots, \boldsymbol{x}_k) = \boldsymbol{f}_1 = \theta_1\boldsymbol{x}_1.\boldsymbol{f}_2(\boldsymbol{x}_1)$$

we need only to compute $\boldsymbol{a}_1 = \boldsymbol{f}_2(\boldsymbol{v}_1)$, where $\boldsymbol{v}_1$ is $\mathbf{0}$ or $\mathbf{1}$ according to the value of θ_1, $\boldsymbol{a}_2 = \boldsymbol{f}_2(\boldsymbol{a}_1), \ldots, \boldsymbol{a}_n = \boldsymbol{f}_2(\boldsymbol{a}_{n-1})$. Now to compute

$$\boldsymbol{f}_2(\boldsymbol{a}_j) = (\theta_2\boldsymbol{x}_2.\boldsymbol{f}_3(\boldsymbol{x}_1, \boldsymbol{x}_2))(\boldsymbol{a}_j) = \theta_2\boldsymbol{x}_2.\boldsymbol{f}_3(\boldsymbol{a}_j, \boldsymbol{x}_2),$$

we need to compute $\boldsymbol{b}_1 = \boldsymbol{f}_3(\boldsymbol{a}_j, \boldsymbol{v}_2), \boldsymbol{b}_2 = \boldsymbol{f}_3(\boldsymbol{a}_j, \boldsymbol{b}_1), \ldots, \boldsymbol{b}_n = \boldsymbol{f}_3(\boldsymbol{a}_j, \boldsymbol{b}_{n-1})$. We use the same procedure for computing $\boldsymbol{f}_3(\boldsymbol{a}_j, \boldsymbol{b}_i)$ and so on.

Example 11.1.5. Let $\boldsymbol{x} = \langle x_1, x_2, x_3\rangle$, $\boldsymbol{y} = \langle y_1, y_2, y_3\rangle$, and

$$\boldsymbol{t}(\boldsymbol{x}, \boldsymbol{y}) = \langle x_2y_3 + x_3y_2, x_1 + x_3 + y_1, 0\rangle.$$

We want to compute $\nu\boldsymbol{y}.\mu\boldsymbol{x}.\boldsymbol{t}(\boldsymbol{x}, \boldsymbol{y})$. Let $\boldsymbol{f}(\boldsymbol{y}) = \mu\boldsymbol{x}.\boldsymbol{t}(\boldsymbol{x}, \boldsymbol{y})$. By substituting 0 for y_4 in Example 11.1.4, we know that $\boldsymbol{f}(y_1, y_2, y_3) = \langle y_1y_3, y_1, 0\rangle$. Then, $\nu\boldsymbol{y}.\mu\boldsymbol{x}.\boldsymbol{t}(\boldsymbol{x}, \boldsymbol{y}) = \boldsymbol{a}_3$ where $\boldsymbol{a}_1 = \boldsymbol{f}(\mathbf{1})$, $\boldsymbol{a}_2 = \boldsymbol{f}(\boldsymbol{a}_1)$, $\boldsymbol{a}_3 = \boldsymbol{f}(\boldsymbol{a}_2)$.

We have $\boldsymbol{a}_1 = \boldsymbol{f}(\mathbf{1}) = \mu\boldsymbol{x}.\boldsymbol{t}(\boldsymbol{x}, \mathbf{1}) = \boldsymbol{b}_{1,3}$ where

$$\begin{aligned}
\boldsymbol{b}_{1,1} &= \boldsymbol{t}(\mathbf{0}, \mathbf{1}) = \langle 0, 1, 0\rangle, \\
\boldsymbol{b}_{1,2} &= \boldsymbol{t}(\boldsymbol{b}_{1,1}, \mathbf{1}) = \boldsymbol{t}(\langle 0, 1, 0\rangle, \langle 1, 1, 1\rangle) = \langle 1, 1, 0\rangle, \\
\boldsymbol{b}_{1,3} &= \boldsymbol{t}(\boldsymbol{b}_{1,2}, \mathbf{1}) = \boldsymbol{t}(\langle 1, 1, 0\rangle, \langle 1, 1, 1\rangle) = \langle 1, 1, 0\rangle.
\end{aligned}$$

Thus, $\boldsymbol{a}_1 = \langle 1,1,0\rangle$ and we compute $\boldsymbol{a}_2 = \boldsymbol{b}_{2,3}$ where

$$\begin{aligned}
\boldsymbol{b}_{2,1} &= \boldsymbol{t}(\mathbf{0},\boldsymbol{a}_1) = \boldsymbol{t}(\langle 0,0,0\rangle,\langle 1,1,0\rangle) = \langle 0,1,0\rangle,\\
\boldsymbol{b}_{2,2} &= \boldsymbol{t}(\boldsymbol{b}_{2,1},\boldsymbol{a}_1) = \boldsymbol{t}(\langle 0,1,0\rangle,\langle 1,1,0\rangle) = \langle 0,1,0\rangle,\\
\boldsymbol{b}_{2,3} &= \boldsymbol{t}(\boldsymbol{b}_{2,2},\boldsymbol{a}_1) = \boldsymbol{t}(\langle 0,1,0\rangle,\langle 1,1,0\rangle) = \langle 0,1,0\rangle.
\end{aligned}$$

Finally, $\boldsymbol{a}_3 = \boldsymbol{b}_{3,3}$ where

$$\begin{aligned}
\boldsymbol{b}_{3,1} &= \boldsymbol{t}(\mathbf{0},\boldsymbol{a}_2) = \boldsymbol{t}(\langle 0,0,0\rangle,\langle 0,1,0\rangle) = \langle 0,0,0\rangle,\\
\boldsymbol{b}_{3,2} &= \boldsymbol{t}(\boldsymbol{b}_{3,1},\boldsymbol{a}_2) = \boldsymbol{t}(\langle 0,0,0\rangle,\langle 0,1,0\rangle) = \langle 0,0,0\rangle,\\
\boldsymbol{b}_{3,3} &= \boldsymbol{t}(\boldsymbol{b}_{3,2},\boldsymbol{a}_2) = \boldsymbol{t}(\langle 0,0,0\rangle,\langle 0,1,0\rangle) = \langle 0,0,0\rangle,
\end{aligned}$$

□

The time complexity of this procedure relies on the following lemma, which is an immediate corollary of Lemma 11.1.2 (page 254).

Lemma 11.1.6. *Let $\boldsymbol{t} : \mathbb{B}^{\boldsymbol{x}} \times \mathbb{B}^{\boldsymbol{y}} \to \mathbb{B}^{\boldsymbol{x}}$ be monotonic, and $\boldsymbol{g}(\boldsymbol{y}) = \theta\boldsymbol{x}.\boldsymbol{t}(\boldsymbol{x},\boldsymbol{y})$. If for any $\boldsymbol{a} \in \mathbb{B}^{\boldsymbol{x}}$ and any $\boldsymbol{b} \in \mathbb{B}^{\boldsymbol{y}}$, the time needed for computing $\boldsymbol{t}(\boldsymbol{a},\boldsymbol{b})$ is bounded from above by T, then the time needed for computing $\boldsymbol{g}(\boldsymbol{b})$ is bounded from above by nT, where n is the length of the vector $\boldsymbol{x}$.*

By iteratively applying this lemma, we get the following result.

Proposition 11.1.7. *The time needed for computing*

$$\theta_1\boldsymbol{x}_1.\theta_2\boldsymbol{x}_2.\cdots.\theta_k\boldsymbol{x}_k.\boldsymbol{t}(\boldsymbol{x}_1,\boldsymbol{x}_2,\ldots,\boldsymbol{x}_k) = \boldsymbol{f}_1 = \theta_1\boldsymbol{x}_1.\boldsymbol{f}_2(\boldsymbol{x}_1)$$

is at most $n^k|\boldsymbol{t}|$.

11.1.2 Improved algorithms

The naive algorithm can be improved in several ways.

Avoiding useless recomputations In the naive algorithm for computing $\theta_1\boldsymbol{x}_1.\theta_2\boldsymbol{x}_2.\cdots.\theta_k\boldsymbol{x}_k.\boldsymbol{t}(\boldsymbol{x}_1,\boldsymbol{x}_2,\ldots,\boldsymbol{x}_k)$, each fixed point

$$\theta_i\boldsymbol{x}_i.\theta_{i+1}\boldsymbol{x}_{i+1}.\cdots.\theta_k\boldsymbol{x}_k.\boldsymbol{t}(\boldsymbol{b}_1,\boldsymbol{b}_2,\ldots,\boldsymbol{b}_{i-1},\boldsymbol{x}_i,\boldsymbol{x}_{i+1}\ldots\boldsymbol{x}_k),$$

is computed a number of times, for different values $\boldsymbol{b}_1, \ldots, \boldsymbol{b}_{i-1}$. Each time, the iterative computation of this fixed point starts with the initial value $\mathbf{0}$ or $\mathbf{1}$.

The algorithm proposed in [20] is based on the observation that it is not always necessary to start the computation with **0** or **1** (see Lemma 11.1.9 below) and thus avoids a lot of useless computations, that makes the algorithm run in time $n^{k/2}|\boldsymbol{t}|$, instead of $n^k|\boldsymbol{t}|$ for the naive algorithm.

Let us state this result more precisely. Without loss of generality we may restrict ourselves to vectorial Boolean terms

$$\theta_1\boldsymbol{x}_1.\theta_2\boldsymbol{x}_2.\cdots.\theta_k\boldsymbol{x}_k.\boldsymbol{t}(\boldsymbol{x}_1,\boldsymbol{x}_2,\ldots,\boldsymbol{x}_k),$$

where θ_i is equal to μ if i is odd and ν if i is even. (By the golden lemma: Propsition 1.3.2, page 19, any vectorial term can be reduced to such a form without increasing its size, see also Section 2.7.4, page 63.) We may also assume that k is odd, since $\boldsymbol{t}(\boldsymbol{x}_1,\boldsymbol{x}_2,\ldots,\boldsymbol{x}_k) = \theta\boldsymbol{z}.\boldsymbol{t}(\boldsymbol{x}_1,\boldsymbol{x}_2,\ldots,\boldsymbol{x}_k)$ when any variable in $\boldsymbol{z}$ does not occur in any $\boldsymbol{x}_i$.

Proposition 11.1.8. *There is an algorithm which computes*

$$\mu\boldsymbol{x}_k.\nu\boldsymbol{y}_k.\mu\boldsymbol{x}_{k-1}.\nu\boldsymbol{y}_{k-1}\cdots\mu\boldsymbol{x}_1.\nu\boldsymbol{y}_1.\mu\boldsymbol{x}_0.\boldsymbol{t}(\boldsymbol{x}_k,\boldsymbol{y}_k,\boldsymbol{x}_{k-1},\boldsymbol{y}_{k-1},\ldots,\boldsymbol{x}_1,\boldsymbol{y}_1,\boldsymbol{x}_0)$$

in time $(1+(k+1)n)n^k|\boldsymbol{t}| \leq (k+2)n^{k+1}|\boldsymbol{t}|$.

The proof is given below, as a corollary of Proposition 11.1.10. The basic lemma which shows that some computations can be avoided is the following. It is a direct consequence of Proposition 1.2.13 (page 12).

Lemma 11.1.9. *Let* $\boldsymbol{f}$ *be a monotonic mapping from* $\mathbb{B}^n$ *into* $\mathbb{B}^n$. *Let* $\boldsymbol{b}$ *in* $\mathbb{B}^n$ *be such that (i)* $\boldsymbol{b} \leq \boldsymbol{f}(\boldsymbol{b})$ *and (ii)* $\boldsymbol{b} \leq \mu\boldsymbol{x}.\boldsymbol{f}(\boldsymbol{x})$.

Then $\mu\boldsymbol{x}.\boldsymbol{f}(\boldsymbol{x})$ *can be computed by the following algorithm.*

```
i := 0; a0 := b;
repeat i := i+1;
       ai := f(ai-1)
until ai := ai-1;
a := ai;
```

Proposition 11.1.10. *Let* $\boldsymbol{f} : \mathbb{B}^{pn} \to \mathbb{B}^n$ *and* $\boldsymbol{g} : \mathbb{B}^{(p+1)n} \to \mathbb{B}^n$, *for some* $p > 0$, *be defined by*

$$\boldsymbol{f}(\boldsymbol{u}) = \mu\boldsymbol{x}_k.\nu\boldsymbol{y}_k.\mu\boldsymbol{x}_{k-1}.\nu\boldsymbol{y}_{k-1}.\cdots.\mu\boldsymbol{x}_1.\nu\boldsymbol{y}_1.\mu\boldsymbol{x}_0.$$
$$\boldsymbol{t}(\boldsymbol{u},\boldsymbol{x}_k,\boldsymbol{y}_k,\boldsymbol{x}_{k-1},\boldsymbol{y}_{k-1},\ldots,\boldsymbol{x}_1,\boldsymbol{y}_1,\boldsymbol{x}_0)$$

and $\boldsymbol{g}(\boldsymbol{v}) =$

$$\nu\boldsymbol{y}_k.\mu\boldsymbol{x}_{k-1}.\nu\boldsymbol{y}_{k-1}.\cdots.\mu\boldsymbol{x}_1.\nu\boldsymbol{y}_1.\mu\boldsymbol{x}_0.\boldsymbol{t}'(\boldsymbol{v},\boldsymbol{y}_k,\boldsymbol{x}_{k-1},\boldsymbol{y}_{k-1},\ldots,\boldsymbol{x}_1,\boldsymbol{y}_1,\boldsymbol{x}_0).$$

Let $\boldsymbol{a}_1,\ldots,\boldsymbol{a}_m$ *be an increasing sequence of elements of* $\mathbb{B}^{pn}$ *and* $\boldsymbol{a}'_1,\ldots,\boldsymbol{a}'_m$ *be an increasing sequence of elements of* $\mathbb{B}^{(p+1)n}$.

Then one can compute (i) all the elements $\boldsymbol{f}(\boldsymbol{a}_1),\ldots,\boldsymbol{f}(\boldsymbol{a}_m)$, *in this order, in global time at most* $(m+n+kn)n^k|\boldsymbol{t}|$, *and (ii) all the elements* $\boldsymbol{g}(\boldsymbol{a}'_1),\ldots,\boldsymbol{g}(\boldsymbol{a}'_m)$, *in this order, in global time at most* $(m+kn)n^k|\boldsymbol{t}|$.

Proof. The proof is by induction on k.

Let us consider the case (i) and let $\boldsymbol{f}(\boldsymbol{u}) = \mu \boldsymbol{x}_k.\boldsymbol{h}(\boldsymbol{u}, \boldsymbol{x}_k)$ with $\boldsymbol{h}(\boldsymbol{u}, \boldsymbol{x}_k) =$

$\nu \boldsymbol{y}_k.\mu \boldsymbol{x}_{k-1}.\nu \boldsymbol{y}_{k-1}.\cdots.\mu \boldsymbol{x}_1.\nu \boldsymbol{y}_1.\mu \boldsymbol{x}_0.\boldsymbol{t}(\boldsymbol{u}, \boldsymbol{x}_k, \boldsymbol{y}_k, \boldsymbol{x}_{k-1}, \boldsymbol{y}_{k-1}, \ldots, \boldsymbol{x}_1, \boldsymbol{y}_1, \boldsymbol{x}_0)$

if $k > 0$, or $\boldsymbol{h}(\boldsymbol{u}, \boldsymbol{x}_k) = \boldsymbol{t}(\boldsymbol{u}, \boldsymbol{x}_k)$ if $k = 0$.

Let $\boldsymbol{a}'_i = \boldsymbol{f}(\boldsymbol{a}_i) = \mu \boldsymbol{x}_k.\boldsymbol{h}(\boldsymbol{a}_i, \boldsymbol{x}_k)$. Since $\boldsymbol{a}_{i-1} \leq \boldsymbol{a}_i$ we have

$$\boldsymbol{a}'_{i-1} = \boldsymbol{f}(\boldsymbol{a}_{i-1}) \leq \boldsymbol{f}(\boldsymbol{a}_i) = \mu \boldsymbol{x}_k.\boldsymbol{h}(\boldsymbol{a}_i, \boldsymbol{x}_k).$$

Moreover,

$$\boldsymbol{a}'_{i-1} = \mu \boldsymbol{x}_k.\boldsymbol{h}(\boldsymbol{a}_{i-1}, \boldsymbol{x}_k) = \boldsymbol{h}(\boldsymbol{a}_{i-1}, \boldsymbol{a}'_{i-1}) \leq \boldsymbol{h}(\boldsymbol{a}_i, \boldsymbol{a}'_{i-1}).$$

Thus we can apply the algorithm of Lemma 11.1.9, to compute $\mu \boldsymbol{x}_k.\boldsymbol{h}(\boldsymbol{a}_i, \boldsymbol{x}_k)$, starting with $\boldsymbol{a}'_{i-1}$ so that the sequence $\boldsymbol{a}'_1, \ldots, \boldsymbol{a}'_m$ is computed by

```
j := 0; b1,0 := 0;
repeat j:= j+1; b1,j := h(a1, b1,j-1)
until b1,j = b1,j-1;
a'1 := b1,j;

for i := 2 to m do
   j := 0; bi,0 := a'i-1;
  repeat j:= j+1; bi,j := h(ai, bi,j-1)
  until bi,j = bi,j-1;
  a'i := bi,j
  end
```

For each $i = 1, \ldots, m$, let us denote by j_i the value of j such that $\boldsymbol{a}'_i = \boldsymbol{b}_{i,j_i} = \boldsymbol{b}_{i,j_i+1}$. Then the intermediate values used to compute $\boldsymbol{a}'_i$ are

$$\boldsymbol{b}_{i,0} < \boldsymbol{b}_{i,1} < \cdots < \boldsymbol{b}_{i,j_i} = \boldsymbol{b}_{i,j_i+1}.$$

This sequence contains j_i strictly increasing steps and $\boldsymbol{h}$ has been computed $j_i + 1$ times. Because $\boldsymbol{b}_{i,j_i+1} = \boldsymbol{b}_{i+1,0}$, we concatenate all these sequences, and we obtain an increasing sequence of elements of $\mathbb{B}^n$ so that the number $\sum_{i=1}^{m} j_i$ of strictly increasing steps in this sequence is less than or equal to n. Thus, the total number of computations of $\boldsymbol{h}$ is $\sum_{i=1}^{m}(j_i+1) = m+\sum_{i=1}^{m} j_i \leq m + n$.

In the case $k = 0$, $\boldsymbol{h}(\boldsymbol{u}, \boldsymbol{x}_1) = \boldsymbol{t}(\boldsymbol{u}, \boldsymbol{x}_1)$, the time needed for computing $\boldsymbol{h}$ is $|\boldsymbol{t}|$, thus the global time for computing the sequence is less than $(m+n)|\boldsymbol{t}|$. In the case $k > 0$, $\boldsymbol{h}$ has to be computed on the values of the increasing sequence

$$\langle \boldsymbol{a}_1, \boldsymbol{b}_{1,0}\rangle, \langle \boldsymbol{a}_1, \boldsymbol{b}_{1,1}\rangle, \ldots, \langle \boldsymbol{a}_1, \boldsymbol{b}_{1,j_1}\rangle,$$
$$\langle \boldsymbol{a}_2, \boldsymbol{b}_{1,j_1+1}\rangle, \langle \boldsymbol{a}_2, \boldsymbol{b}_{2,1}\rangle, \ldots, \langle \boldsymbol{a}_2, \boldsymbol{b}_{2,j_2}\rangle,$$
$$\ldots$$
$$\langle \boldsymbol{a}_m, \boldsymbol{b}_{m-1,j_{m-1}+1}\rangle, \langle \boldsymbol{a}_m, \boldsymbol{b}_{m,1}\rangle, \ldots, \langle \boldsymbol{a}_m, \boldsymbol{b}_{m,j_m}\rangle.$$

By case (ii) of the induction hypothesis this can be done in time $(m+n+kn)n^k|\boldsymbol{t}|$.

Now let us consider the case (ii). In this case $\boldsymbol{g}(\boldsymbol{v}) = \nu \boldsymbol{y}_k.\boldsymbol{f}'(\boldsymbol{v}, \boldsymbol{y}_k)$, and the computation of $\boldsymbol{g}(\boldsymbol{a}'_1), \ldots, \boldsymbol{g}(\boldsymbol{a}'_m)$ is done by

for i := 1 to m do $\boldsymbol{b}_{i,0}$:= $\mathbf{1}$; for j := 1 to n do $\boldsymbol{b}_{i,j}$:= $\boldsymbol{f}'(\boldsymbol{a}'_i, \boldsymbol{b}_{i,j-1})$ end

First, let us show

$$\forall i : 1 \leq i \leq m-1, \forall j : 0 \leq j \leq n,\ \boldsymbol{b}_{i,j} \leq \boldsymbol{b}_{i+1,j}$$

by induction on j. Obviously, $\boldsymbol{b}_{i,0} = \boldsymbol{b}_{i+1,0} = \mathbf{1}$. If $\boldsymbol{b}_{i,j-1} \leq \boldsymbol{b}_{i+1,j-1}$, then $\boldsymbol{b}_{i,j} = \boldsymbol{f}'(\boldsymbol{a}'_i, \boldsymbol{b}_{i,j-1}) \leq \boldsymbol{f}'(\boldsymbol{a}'_{i+1}, \boldsymbol{b}_{i+1,j-1}) = \boldsymbol{b}_{i+1,j}$.

For each $j, (1 \leq j \leq n)$, the values of the sequence

$$\begin{aligned} \boldsymbol{b}_{1,j} &= \boldsymbol{f}'(\boldsymbol{a}'_1, \boldsymbol{b}_{1,j-1}), \\ \boldsymbol{b}_{2,j} &= \boldsymbol{f}'(\boldsymbol{a}'_2, \boldsymbol{b}_{2,j-1}), \\ &\ldots \\ \boldsymbol{b}_{m,j} &= \boldsymbol{f}'(\boldsymbol{a}'_m, \boldsymbol{b}_{m,j-1}) \end{aligned}$$

are computed in this order (interleaved with computations of values of analogous sequences for other j). Therefore, since $\boldsymbol{f}'$ is computed on an increasing sequence of arguments, the computation of $\boldsymbol{f}'(\boldsymbol{a}'_{i+1}, \boldsymbol{b}_{i+1,j-1})$ can start with $\boldsymbol{b}_{i,j}$ which has been previously computed. By case (i) of the induction hypothesis, the global time needed to compute this sequence is at most $(m+kn)n^{k-1}|\boldsymbol{t}|$. Since such a sequence is computed n times, (once for each value of j), the global time needed to compute $\boldsymbol{g}(\boldsymbol{a}'_1), \ldots, \boldsymbol{g}(\boldsymbol{a}'_m)$ is $n(m+kn)n^{k-1}|\boldsymbol{t}| = (m+kn)n^k|\boldsymbol{t}|$.

□

Proof of Proposition 11.1.8 We may consider that

$$\mu \boldsymbol{x}_k.\nu \boldsymbol{y}_k.\mu \boldsymbol{x}_{k-1}.\nu \boldsymbol{y}_{k-1}.\cdots.\mu \boldsymbol{x}_1.\nu \boldsymbol{y}_1.\mu \boldsymbol{x}_0.\boldsymbol{t}(\boldsymbol{x}_k, \boldsymbol{y}_k, \boldsymbol{x}_{k-1}, \boldsymbol{y}_{k-1}, \ldots, \boldsymbol{x}_1, \boldsymbol{y}_1, \boldsymbol{x}_0)$$

is a constant function defined on $\mathbb{B}^0$, so that we can apply case (i) of the previous proposition with $p = 0$ and $m = 1$.

□

A linear algorithm for alternation-depth 1 fixed-point terms Another improvement consists of an algorithm for computing

$$\theta \boldsymbol{x}.\boldsymbol{t}(\boldsymbol{b}_1, \dots, \boldsymbol{b}_{n-1}, \boldsymbol{x})$$

in time proportional to $|\boldsymbol{t}|$ instead of $n|\boldsymbol{t}|$, which can be used when $\boldsymbol{t} = \langle t_1, t_2, \dots, t_n \rangle$ has the following form: each t_i is either a disjunction or a conjunction of a set of variables.

This algorithm is originated in a model-checking algorithm for alternation-depth 1 terms of the modal μ-calculus (see Section 6.2, page 145). Arnold and Crubillé [6] found a model-checking algorithm linear in the size of the model and quadratic in the size of the term. Cleaveland and Steffen [25] slightly improved this algorithm to make it linear also in the size of the formula. Simultaneously and independently, Andersen [1] and Vergauwen and Lewi [99] showed that evaluating a term of the modal μ-calculus on a finite graph amounts to evaluating a vectorial Boolean μ-term, of the same alternation-depth, of size equal to the size of the graph times the size of the term, and which is in the form mentioned above (see Section 3.2, page 72).

Let $X = \{x_1, \dots, x_n\}$ and let $t_i = C_i X_i$ where X_i is a subset of X and C_i is either $\bigvee$ or $\bigwedge$ (Remember that $\bigvee \emptyset = 0$ and $\bigwedge \emptyset = 1$).

We say that a variable x_i is a *successor* of x_j, or that x_j is a *predecessor* of x_i, if $x_j \in X_i$. An *edge* is a pair $\langle x_j, x_i \rangle$ such that x_i is a successor of x_j. The number of edges is $e = \sum_{i=1}^{n} |X_i|$.

Let us compute $\boldsymbol{b} = \mu \langle x_1, \dots, x_n \rangle . \langle t_1, \dots, t_n \rangle$.

With each variable x_i we associate a counter n_i and a Boolean variable z_i. Intuitively, the counter n_i records how many predecessors of x_i have still to take the value 1 for x_i getting value 1. Initially, if C_i is $\bigvee$, it is enough that one element of X_i gets the value 1 for x_i getting the value 1, thus n_i is initialized to 1; if C_i is $\bigwedge$, all elements of X_i have to get value 1, and n_i is initialized to $|X_i|$. The Boolean variables are initialized to 0. The time needed to initialize the counters and the Boolean variables is $\alpha_0 n + \beta_0 e$.

The algorithm consists in executing the main loop:

```
for j := 1 to n do if n_j = 0 and z_j = 0 then visit(j) end;
```

where the recursive procedure `visit(j)` is defined by:

```
z_j := 1;
for all successors x_i of x_j do
      { traversal of the edge ⟨x_j, x_i⟩:}
      if n_i > 0 then n_i := n_i − 1;
      if n_i = 0 and z_i = 0 then visit(i);
      end
```

The procedure `visit` is executed at most once for each i, since `visit(i)` is executed when $z_i = 0$ and sets this variable to 1. It follows that each edge is traversed at most once. Hence, the complexity of the main loop is $\alpha n + \beta e$.

During the execution of the algorithm, the different values successively taken by the vector $z = \langle z_1, \ldots, z_n \rangle$ form an increasing sequence (since a variable can only be set from 0 to 1) of length $p \leq n+1$. Let $z(n) = \langle z_1(n), \ldots, z_n(n) \rangle$, for $1 \leq n \leq p$, be the nth element of this sequence. We claim that $z(p) = b$.

Let us show that $z(p) \leq b$. Since $z(1) = \mathbf{0} \leq b$, it is sufficient to prove that for for any $n, 1 < n \leq p$, $\forall i, z_i(n) \leq t_i(z(n-1))$, from which one can derive $z(n) \leq t(z(n-1))$ and, by induction on n, $\forall n : 1 \leq n \leq p, z(n) \leq b$. Let us assume that $z_i(n) = 1$. If $z_i(n-1)$ is also equal to 1, and thus, $n-1 > 1$, then $z_i(n) = z_i(n-1) \leq t_i(z(n-2)) \leq t_i(z(n-1))$. If $z_i(n-1) = 0$, z_i has just been set to 1 by executing `visit(i)`, that implies that n_i was equal to 0 just before. Either n_i was initially 0, then $t_i = \bigwedge \emptyset = 1$ and $t_i(z(n-1)) = 1$, or either 1 or $|X_i|$ edges of target x_i (according to whether C_i is $\bigvee$ or $\bigwedge$) have been previously traversed in order to set n_i to 0. Since such an edge is traversed only after visiting its source, and since visiting j sets z_j to 1, either one or all of the z_j such that $x_j \in X_i$ have been previously set to 1, and $z_j(n-1) = 1$ for such j's. It follows that $t_i(z(n-1)) = 1$.

To prove that $b \leq z(p)$, it is enough to show that $t(z(p)) \leq z(p)$. Let x_i be a variable such that $t_i(z(p)) = C_i\{z_j(p) \mid x_j \in X_i\} = 1$, and let us show that $z_i(p) = 1$. If $C_i = \bigvee$, then X_i is not empty and for at least one $x_j \in X_i$ (i.e., x_i is a successor of x_j), $z_j(p) = 1$; but when the variable z_j has been set to 1 (by `visit(j)`), the edge $\langle x_j, x_i \rangle$ was to be traversed; if at this time, the variable z_i was not already 1, then the counter n_i (initialized to 1) was still equal to 1 (otherwise, z_i should have been set to 1 just after n_i was set to 0), and in this case traversing the edge causes n_i be set to 0, then `visit(i)` is executed, which sets z_i to 1. Therefore, $z_i(p) = 1$. If $C_i = \bigwedge$, we consider two cases: $X_i = \emptyset$ and $X_i \neq \emptyset$. In the first case n_i is initialized to 0 and `visit(i)` is surely executed (at least at the level of the main loop). In the second case, for each $x_j \in X_i$, $z_j(p) = 1$, which implies that `visit(j)` has been executed and the edge $\langle x_j, x_i \rangle$ has been traversed. But each time such an edge is traversed, the counter n_i, initialized to $|X_i|$, is decremented so that when traversing for the last time an edge of target x_i, n_i is set to 0, and if z_i was not already 1, it is set to 1 by `visit(i)`. Hence, $z_i(p) = 1$.

The unfolding method Combining the above linear algorithm together with Proposition 11.1.3 (page 255), H. Seidl [89] has given another way of obtaining an algorithm to compute

$$\mu \boldsymbol{x}_k.\nu \boldsymbol{y}_k.\mu \boldsymbol{x}_{k-1}.\nu \boldsymbol{y}_{k-1} \cdots \mu \boldsymbol{x}_1.\nu \boldsymbol{y}_1.\mu \boldsymbol{x}_0.\boldsymbol{t}(\boldsymbol{x}_k, \boldsymbol{y}_k, \boldsymbol{x}_{k-1}, \boldsymbol{y}_{k-1}, \ldots, \boldsymbol{x}_1, \boldsymbol{y}_1, \boldsymbol{x}_0)$$

in time $n^k|\boldsymbol{t}|$ which is a result similar to Proposition 11.1.8 (page 259).

This result is based on the following property (Proposition 11.1.11 below). Before stating this property we need some definition. Let $I = \{i_1, i_2, \ldots, i_q\}$ be a subset of $\{1, \ldots, p\}$ with $i_1 < i_2 < \cdots < i_q$. For any vector of variables $\boldsymbol{x} = \langle x_1, x_2, \ldots, x_p \rangle$ of length p, we denote by $\pi_I \boldsymbol{x}$ the vector $\langle x_{i_1}, x_{i_2}, \ldots, x_{i_q} \rangle$ of length q. Similarly, if $\boldsymbol{g}(\boldsymbol{y}) = \langle g_1(\boldsymbol{y}), g_2(\boldsymbol{y}), \ldots, g_p(\boldsymbol{y}) \rangle : \mathbb{B}^m \to \mathbb{B}^p$ then we denote by $\pi_I \boldsymbol{g}(\boldsymbol{y})$ the mapping $\langle g_{i_1}(\boldsymbol{y}), g_{i_2}(\boldsymbol{y}), \ldots, g_{i_q}(\boldsymbol{y}) \rangle : \mathbb{B}^m \to \mathbb{B}^q$.

Proposition 11.1.11. *Let $\boldsymbol{f}(\boldsymbol{x}, \boldsymbol{y}, \boldsymbol{u}) : \mathbb{B}^{pq} \times \mathbb{B}^q \times \mathbb{B}^r \to \mathbb{B}^{pq}$ be a monotonic vectorial Boolean function presented by a vectorial functional Boolean term, with $p > 0$, $q > 0$. Let I be a subset of $\{1, \ldots, pq\}$ of cardinality q. Then there exist a vectorial functional Boolean term $\boldsymbol{h}(\boldsymbol{z}, \boldsymbol{u}) : \mathbb{B}^{pq^2} \times \mathbb{B}^r \to \mathbb{B}^{pq^2}$ and a subset J of $\{1, \ldots, pq^2\}$ of cardinality q such that $\nu \boldsymbol{y}.\pi_I \mu \boldsymbol{x}.\boldsymbol{f}(\boldsymbol{x}, \boldsymbol{y}, \boldsymbol{u}) = \pi_J \mu \boldsymbol{z}.\boldsymbol{h}(\boldsymbol{z}, \boldsymbol{u})$.*

Moreover, the size of $\boldsymbol{h}$ is equal to $q \times |\boldsymbol{f}|$ and the construction of $\boldsymbol{h}$ is done in time $q \times |\boldsymbol{f}|$.

Proof. Let $\mathbf{1}$ be $\langle 1, \ldots, 1 \rangle \in \mathbb{B}^q$, let $\boldsymbol{z} = \langle \boldsymbol{z}_1, \ldots, \boldsymbol{z}_q \rangle$ where the $\boldsymbol{z}_i$ are distinct vectors of variables of length pq, and let $\boldsymbol{h}(\boldsymbol{z}, \boldsymbol{u}) = \boldsymbol{h}(\boldsymbol{z}_1, \ldots, \boldsymbol{z}_q, \boldsymbol{u}) =$

$$\langle \boldsymbol{f}(\boldsymbol{z}_1, \mathbf{1}, \boldsymbol{u}), \boldsymbol{f}(\boldsymbol{z}_2, \pi_I \boldsymbol{z}_1, \boldsymbol{u}), \ldots, \boldsymbol{f}(\boldsymbol{z}_i, \pi_I \boldsymbol{z}_{i-1}, \boldsymbol{u}), \ldots, \boldsymbol{f}(\boldsymbol{z}_q, \pi_I \boldsymbol{z}_{q-1}, \boldsymbol{u}) \rangle.$$

Let $\mu \boldsymbol{z}.\boldsymbol{h}(\boldsymbol{z}, \boldsymbol{u})$ be $\langle \boldsymbol{h}_1(\boldsymbol{u}), \boldsymbol{h}_2(\boldsymbol{u}), \ldots, \boldsymbol{h}_q(\boldsymbol{u}) \rangle$. We claim that for $i = 1, \ldots, q$, $\boldsymbol{h}_i(\boldsymbol{u}) = \mu \boldsymbol{x}.\boldsymbol{f}(\boldsymbol{x}, \pi_I \boldsymbol{h}_{i-1}(\boldsymbol{u}), \boldsymbol{u})$ where, by convention, $\pi_I \boldsymbol{h}_0(\boldsymbol{u}) = \mathbf{1}$.

By Proposition 1.3.2 (page 19) we have $\mu \boldsymbol{z}.\boldsymbol{h}(\boldsymbol{z}, \boldsymbol{u}) =$
$\mu \langle \boldsymbol{z}_1, \ldots, \boldsymbol{z}_q \rangle.\langle \boldsymbol{f}(\boldsymbol{z}_1, \mathbf{1}, \boldsymbol{u}), \boldsymbol{f}(\boldsymbol{z}_2, \pi_I \boldsymbol{z}_1, \boldsymbol{u}), \ldots, \boldsymbol{f}(\boldsymbol{z}_q, \pi_I \boldsymbol{z}_{q-1}, \boldsymbol{u}) \rangle =$
$\mu \langle \boldsymbol{x}_1, \ldots, \boldsymbol{x}_q \rangle.\mu \langle \boldsymbol{y}_1, \ldots, \boldsymbol{y}_q \rangle.\langle \boldsymbol{f}(\boldsymbol{x}_1, \mathbf{1}, \boldsymbol{u}), \boldsymbol{f}(\boldsymbol{x}_2, \pi_I \boldsymbol{y}_1, \boldsymbol{u}), \ldots, \boldsymbol{f}(\boldsymbol{x}_q, \pi_I \boldsymbol{y}_{q-1}, \boldsymbol{u}) \rangle$
which is also equal, by Proposition 1.3.3 (page 19), to

$$\mu \langle \boldsymbol{x}_1, \ldots, \boldsymbol{x}_q \rangle.\langle \boldsymbol{f}(\boldsymbol{x}_1, \pi_I \boldsymbol{h}_0(\boldsymbol{u}), \boldsymbol{u}), \boldsymbol{f}(\boldsymbol{x}_2, \pi_I \boldsymbol{h}_1(\boldsymbol{u}), \boldsymbol{u}), \ldots, \boldsymbol{f}(\boldsymbol{x}_q, \pi_I \boldsymbol{h}_{q-1}(\boldsymbol{u}), \boldsymbol{u}) \rangle.$$

It follows obviously that $\boldsymbol{h}_i(\boldsymbol{u}) = \boldsymbol{f}(\boldsymbol{h}_i(\boldsymbol{u}), \pi_I \boldsymbol{h}_{i-1}(\boldsymbol{u}), \boldsymbol{u})$, hence

$$\mu \boldsymbol{x}.\boldsymbol{f}(\boldsymbol{x}, \pi_I \boldsymbol{h}_{i-1}(\boldsymbol{u}), \boldsymbol{u}) \leq \boldsymbol{h}_i(\boldsymbol{u}).$$

Conversely, let $\boldsymbol{g}_i(\boldsymbol{u}) = \mu \boldsymbol{x}.\boldsymbol{f}(\boldsymbol{x}, \pi_I \boldsymbol{h}_{i-1}(\boldsymbol{u}), \boldsymbol{u})$.
We have $\boldsymbol{g}_i(\boldsymbol{u}) = \boldsymbol{f}(\boldsymbol{g}_i(\boldsymbol{u}), \pi_I \boldsymbol{h}_{i-1}(\boldsymbol{u}), \boldsymbol{u})$, hence $\langle \boldsymbol{g}_1(\boldsymbol{u}), \ldots, \boldsymbol{g}_q(\boldsymbol{u}) \rangle = \langle \boldsymbol{f}(\boldsymbol{g}_1(\boldsymbol{u}), \pi_I \boldsymbol{h}_0(\boldsymbol{u}), \boldsymbol{u}), \ldots \boldsymbol{f}(\boldsymbol{g}_q(\boldsymbol{u}), \pi_I \boldsymbol{h}_{q-1}(\boldsymbol{u}), \boldsymbol{u}) \rangle$. It follows that

$$\begin{aligned} \mu \boldsymbol{z}.\boldsymbol{h}(\boldsymbol{z}, \boldsymbol{u}) &= \mu \langle \boldsymbol{x}_1, \ldots, \boldsymbol{x}_q \rangle.\langle \boldsymbol{f}(\boldsymbol{x}_1, \pi_I \boldsymbol{h}_0(\boldsymbol{u}), \boldsymbol{u}), , \ldots, \boldsymbol{f}(\boldsymbol{x}_q, \pi_I \boldsymbol{h}_{q-1}(\boldsymbol{u}), \boldsymbol{u}) \rangle \\ &\leq \langle \boldsymbol{g}_1(\boldsymbol{u}), \ldots, \boldsymbol{g}_q(\boldsymbol{u}) \rangle. \end{aligned}$$

Now, by Proposition 11.1.3 (page 255), $\nu \boldsymbol{y}.\pi_I \mu \boldsymbol{x}.\boldsymbol{f}(\boldsymbol{x}, \boldsymbol{y}, \boldsymbol{u})$ is the qth element of the following sequence;

$$\begin{aligned} \pi_I \mu \boldsymbol{x}.\boldsymbol{f}(\boldsymbol{x}, \mathbf{1}, \boldsymbol{u}) &= \pi_I \boldsymbol{h}_1(\boldsymbol{u}) \\ \pi_I \mu \boldsymbol{x}.\boldsymbol{f}(\boldsymbol{x}, \pi_I \boldsymbol{h}_1(\boldsymbol{u}), \boldsymbol{u}) &= \pi_I \boldsymbol{h}_2(\boldsymbol{u}) \\ &\cdots \\ \pi_I \mu \boldsymbol{x}.\boldsymbol{f}(\boldsymbol{x}, \pi_I \boldsymbol{h}_{i-1}(\boldsymbol{u}), \boldsymbol{u}) &= \pi_I \boldsymbol{h}_i(\boldsymbol{u}) \\ &\cdots \end{aligned}$$

But $\boldsymbol{h}_q(\boldsymbol{u}) = \pi_{I'}\mu\boldsymbol{z}.\boldsymbol{h}(\boldsymbol{z},\boldsymbol{u})$ with $I' = \{(q-1)p+1,(q-1)p+2,\ldots,(q-1)p+p\}$, hence $\nu\boldsymbol{y}.\pi_I\mu\boldsymbol{x}.\boldsymbol{f}(\boldsymbol{x},\boldsymbol{y},\boldsymbol{u}) = \pi_I\boldsymbol{h}_q(\boldsymbol{u}) = \pi_J\mu\boldsymbol{z}.\boldsymbol{h}(\boldsymbol{z},\boldsymbol{u})$ where $J = \pi_I I'$ is equal to $\{(q-1)p+i_1,(q-1)p+i_2,\ldots,(q-1)p+i_q\}$. □

Since, by Proposition 1.3.12 (page 23), $\mu\boldsymbol{x}.\pi_I\boldsymbol{f}(\boldsymbol{x},\boldsymbol{u}) = \pi_I\mu\boldsymbol{y}.\boldsymbol{f}(\pi_I\boldsymbol{y},\boldsymbol{u})$, we can repeatedly apply the previous proposition. Each suppression of a ν operator causes a multiplicative factor of n, and we get that

$$\mu\boldsymbol{x}_k.\nu\boldsymbol{y}_k.\mu\boldsymbol{x}_{k-1}.\nu\boldsymbol{y}_{k-1}\cdots\mu\boldsymbol{x}_1.\nu\boldsymbol{y}_1.\mu\boldsymbol{x}_0.\boldsymbol{t}(\boldsymbol{x}_k,\boldsymbol{y}_k,\boldsymbol{x}_{k-1},\boldsymbol{y}_{k-1},\ldots,\boldsymbol{x}_1,\boldsymbol{y}_1,\boldsymbol{x}_0)$$

is equal to

$$\pi_I\mu\boldsymbol{z}_k.\mu\boldsymbol{z}_{k-1}.\cdots.\mu\boldsymbol{z}_1.\mu\boldsymbol{z}_0.\boldsymbol{t}'(\boldsymbol{z}_k,\boldsymbol{z}_{k-1},\ldots,\boldsymbol{z}_1,\boldsymbol{z}_0)$$

which is equal, by Proposition 1.3.2, (page 19), to

$$\pi_I\mu\boldsymbol{z}.\boldsymbol{t}'(\boldsymbol{z},\boldsymbol{z},\ldots,\boldsymbol{z},\boldsymbol{z})$$

where $\boldsymbol{z}$ is of length n^{k+1}, $\boldsymbol{t}'$ is of size $n^k|\boldsymbol{t}|$ and constructed in time $n^k|\boldsymbol{t}|$. By applying the linear algorithm to $\mu\boldsymbol{x}.\boldsymbol{t}'$ we get that

$$\mu\boldsymbol{x}_k.\nu\boldsymbol{y}_k.\mu\boldsymbol{x}_{k-1}.\nu\boldsymbol{y}_{k-1}\cdots\mu\boldsymbol{x}_1.\nu\boldsymbol{y}_1.\mu\boldsymbol{x}_0.\boldsymbol{t}(\boldsymbol{x}_k,\boldsymbol{y}_k,\boldsymbol{x}_{k-1},\boldsymbol{y}_{k-1},\ldots,\boldsymbol{x}_1,\boldsymbol{y}_1,\boldsymbol{x}_0)$$

can be computed in time $n^k|\boldsymbol{t}|$.

11.2 Winning positions and winning strategies

11.2.1 Computations of winning positions

We have seen in Section 4.3.1 (page 88) that computing the winning position in a finite game G amounts to computing the value of a Boolean vectorial fixed-point term $\nu\boldsymbol{x}^{(2k)}.\mu\boldsymbol{x}^{(2k-1)}.\cdots.\nu\boldsymbol{x}^{(2)}.\mu\boldsymbol{x}^{(1)}.\boldsymbol{f}(\boldsymbol{x}^{(1)},\ldots,\boldsymbol{x}^{(2k)})$ for some $\boldsymbol{f}:\mathbb{B}^{2kn}\to\mathbb{B}^n$ where n is the number of positions of G, k is the maximal rank of a position, and the size of $\boldsymbol{f}$ is equal to the size of G. Therefore the set of winning positions can be computed in time $n^k|G|$.

Using the converse relationship between games and the Boolean μ-calculus, we can show the following result.

Proposition 11.2.1. *Computing*

$$\boldsymbol{b} = \nu\boldsymbol{x}^{(2k)}.\mu\boldsymbol{x}^{(2k-1)}.\cdots.\nu\boldsymbol{x}^{(2)}.\mu\boldsymbol{x}^{(1)}.\boldsymbol{f}(\boldsymbol{x}^{(1)},\ldots,\boldsymbol{x}^{(2k)})$$

for some $\boldsymbol{f}:\mathbb{B}^{2kn}\to\mathbb{B}^n$ *is in NP (and also in co-NP).*

Proof. Let G be the associated game, that is of size $|G| = |\boldsymbol{f}|$ and that can be constructed in linear time. A position p of G is winning if and only if there exists a positional strategy s such that p is a winning position in the one-player game $G(s)$ (see Lemma 4.2.1, page 87). Our claim is thus a consequence of the following property (Lemma 11.2.2). □

Lemma 11.2.2. *For any positional strategy s, the set of winning positions of $G(s)$ can be computed in time $k|G|$.*

Proof. For each $r \in \{1, \dots, k\}$ consider the subgraph $G_r(s)$ of $G(s)$ induced by the variables of rank less than or equal to $2r - 1$. Let $\mathcal{C}(r)$ be the set of strongly connected components of $G_r(s)$ that contain at least one variable of rank $2r-1$. If z is a variable belonging to some strongly connected component $C \in \mathcal{C}(r)$ then z is the origin of an infinite 0-path. Thus, if i is the index of a variable that is the origin of a finite path (in $G(s)$), ending in $V(r) = \{z \mid \exists C \in \mathcal{C}(r) : z \in C\}$ then $\boldsymbol{b}_i = 0$. Conversely, if $\boldsymbol{b}_i = 0$, there exists a finite path $z = z_0, \dots, z_p$ with $\text{rank}(z_p) = r$ and a finite path from z_p to z_p which is a path in G_r. All the variables in this second path belong to a strongly connected component $C \in \mathcal{C}(r)$. Thus $z_p \in V(r)$. Now let V' be the set of variables from which there is a finite path ending in $V(r)$ for some r. We have shown that $\boldsymbol{b}_i = 0 \Leftrightarrow \boldsymbol{x}_i^{(1)} \in V'$.

Now, by slightly modifying the famous Tarjan's algorithm for computing strongly connected components of a graph (see [3]), it is easy to compute each $V(r)$ in time $O(|G(s)|)$ and then V' also in time $O(|G(s)|) \leq O(|G|)$. □

11.2.2 Computations of winning strategies

Computing the strategy that wins in all winning positions (see Theorem 4.3.8, page 92) is the same problem as computing a selection function for the Boolean μ-calculus (see Theorem 3.3.7, page 77).

Assume that T is the time needed to compute the value of a Boolean vectorial fixed-point term $\nu\boldsymbol{x}^{(2k)}.\mu\boldsymbol{x}^{(2k-1)}.\cdots.\nu\boldsymbol{x}^{(2)}.\mu\boldsymbol{x}^{(1)}.\boldsymbol{f}(\boldsymbol{x}^{(1)}, \dots, \boldsymbol{x}^{(2k)})$ of length n.

Let us assume that this term depends on $2m$ additional variables, two for each occurrence of a disjunction in $\boldsymbol{f}$ (see Theorem 3.3.7, page 77), i.e., $\nu\boldsymbol{x}^{(2)}.\mu\boldsymbol{x}^{(1)}.\boldsymbol{f}(\boldsymbol{x}^{(1)}, \dots, \boldsymbol{x}^{(2k)}) = \boldsymbol{g}(z_1, \dots, z_i, \dots, z_m, z'_1, \dots, z'_i, \dots, z'_m)$ so that $\boldsymbol{b} = \boldsymbol{g}(1, \dots, 1, \dots, 1, 1, \dots, 1, \dots, 1)$. We have to find $b_1, \dots, b_m \in \mathbb{B}$ such that $\mathbf{b} = \boldsymbol{g}(b_1, \dots, b_i, \dots, b_m, \overline{b_1}, \dots, \overline{b_i}, \dots, \overline{b_m})$. Note that for any $b_1, \dots, b_m, b'_1, \dots, b'_m \in \mathbb{B}$, $\boldsymbol{g}(b_1, \dots, b_i, \dots, b_m, b'_1, \dots, b'_i, \dots, b'_m)$ is still computable in time T. Then the required $b_1, \dots, b_m$ are computed in time $T + m(n + T)$ as follows.

- Compute $\boldsymbol{b} = \boldsymbol{g}(1, \dots, 1, \dots, 1, 1, \dots, 1, \dots, 1)$ in time T.

– Assume we know $b_1, \ldots, b_{i-1}$. Compute

$$\boldsymbol{b}_i = \boldsymbol{g}(b_1, \ldots, b_{i-1}, 1, 1, \ldots, 1, \overline{b_1}, \ldots, \overline{b_{i-1}}, 0, 1, \ldots, 1)$$

in time T and compare it to $\boldsymbol{b}$ in time $O(n)$. Set b_i to 1 if they are equal, to 0 otherwise.

Since $m \leq \boldsymbol{f}$, we have, for $T = n^k|\boldsymbol{f}|$, an upper bound of $O(n^k|\boldsymbol{f}|^2)$. An interesting question is whether this bound can be lowered to $n^k|\boldsymbol{f}|$.

11.3 Bibliographic notes and sources

The credits to the work by Browne, Clarke, Jha, Long and Marrero [20], Cleaveland and Steffen [25], Andersen [1], Vergauwen and Lewi [99], Seidl [89], and Arnold and Crubillé [6], were given in the text.

The fact that computing a component of a Boolean vectorial fixed-point term can be reduced to determining the winner in a parity game, and hence (by the positional determinacy) is in *NP*, and by symmetry also in *co-NP* (Proposition 11.2.1), was observed by Emerson, Jutla, and Sistla [34], in a slightly different setting, namely, for the model–checking problem for the modal μ-calculus. Recently, Jurdziński [49] showed that the problem is actually in the class *UP* ∩ *co-UP* (where *UP* stands for *unambiguous* nondeterministic polynomial time, see, e.g. Papadimitriou [79]). In [50], Jurdziński presented another algorithm, which is not worse in time complexity than the algorithms by Browne *et al.* [20] and Seidl [89] but which, in contrast to those algorithms, uses only a polynomial amount of space.

Bibliography

1. H. R. Andersen. Model checking and boolean graphs. *Theoretical Computer Science*, 126:3–30, 1994.
2. D. N. Arden. Delayed logic and finite state machines. In *Theory of Computing and Machine Design*, pages 1–35. Univ. of Michigan Press, 1960.
3. A. Arnold. *Finite transition systems.* Prentice Hall Int., 1994.
4. A. Arnold. A selection property of the boolean μ-calculus and some of its applications. *RAIRO–Theoretical Informatics and Applications*, 31:371–384, 1997.
5. A. Arnold. The μ-calculus alternation-depth hierarchy is strict on binary trees. *RAIRO-Theoretical Informatics and Applications*, 33:329–339, 1999.
6. A. Arnold and P. Crubillé. A linear algorithm to solve fixed-point equations on transition systems. *Information Processing Letters*, 29:57–66, 1988.
7. A. Arnold and M. Nivat. Non deterministic recursive programs. In *FCT'77*, pages 12–21. Lect. Notes Comput. Sci. 56, 1977.
8. A. Arnold and M. Nivat. The metric space of infinite trees. Algebraic and topological properties. *Fundamenta Informaticae*, 4:445–476, 1980.
9. A. Arnold and D. Niwiński. Fixed point characterization of Büchi automata on infinite trees. *J. Inf. Process. Cybern. EIK*, 26:453–461, 1990.
10. A. Arnold and D. Niwiński. Fixed point characterization of weak monadic logic definable sets of trees. In M. Nivat and A. Podelski, editors, *Tree automata and Languages*, pages 159–188. Elsevier, 1992.
11. H. P. Barendregt. The type free lambda calculus. In J. Barwise, editor, *Handbook of Mathematical Logic*, pages 1091–1132. North Holland, 1977.
12. S. K. Basu and R. T. Yeh. Strong verification of programs. *IEEE Trans. on Software Engineering*, 1:339–345, 1975.
13. H. Bekič. Definable operations in general algebras, and the theory of automata and flowcharts. Technical report, IBM Labaoratory, Wien, 1967.
14. L. Bernátsky and Z. Esik. Semantics of flowchart programs and the free Conway theories. *RAIRO–Theoretical Informatics and Applications*, 32:35–78, 1998.
15. G. Birkhoff. On the combination of subalgebras. *Proc. Camb. Phil. Soc.*, 29, 1933.
16. G. Birkhoff. *Lattice theory.* AMS (3rd edition), 1979.
17. J. C. Bradfield. The modal mu-calculus alternation hierarchy is strict. *Theoretical Computer Science*, 195:133–153, 1997.
18. J. C. Bradfield. Simplifying the modal mu-calculus alternation hierarchy. In M. Morvan, C. Meinel, and D. Krob, editors, *Proc. STACS '98*, pages 39–49. Lect. Notes Comput. Sci. 1373, 1998.
19. J. C. Bradfield. Fixpoint alternation: arithmetic, transition systems, and the binary tree. *RAIRO-Theoretical Informatics and Applications*, 33:341–356, 1999.

20. A. Browne, E. M. Clarke, S. Jha, D. E. Long, and W. Marrero. An improved algorithm for the evaluation of fixpoint expressions. *Theoretical Computer Science*, 178:237–255, 1997.
21. J. R. Büchi. On a decision method in restricted second order arithmetic. In E. Nagl, editor, *Logic, Methodology, and Philosophy of Science*, pages 1–11. Stanford Univ. Press, 1960.
22. J. R. Büchi. Using determinacy to eliminate quantifiers. In M. Karpinski, editor, *Fundamentals of Computation Theory*, volume 56, pages 367–378. Lect. Notes Comput. Sci., 1977.
23. J. R. Büchi and L. H. Landweber. Solving sequential conditions by finite-state strategies. *Trans. Am. Math. Soc.*, 138:295–311, 1969. (reprinted in: S. Mac Lane and D. Siefkes, editors, *The collected works of J. Richard Büchi.* Springer-Verlag, 1990.).
24. A. K. Chandra, D. Kozen, and L. Stockmeyer. Alternation. *J. ACM*, 28:114–133, 1981.
25. R. Cleaveland and B. Steffen. Computing behavioural relations, logically. In *18th Int. Coll. on Automata, Languages and Programming*, pages 127–138. Lect. Notes Comput. Sci. 510, 1991.
26. M. Davis. Infinite games of perfect information. *Ann. Math. Studies (Advances in game theory)*, 52:85–101, 1964.
27. E. W. Dijkstra. *A Discipline of Programming.* Prentice Hall, 1976.
28. J. Doner. Tree acceptors and some of their applications. *J. Comput. System Sci.*, 4:406–451, 1970.
29. H.-D. Ebbinghaus and J. Flum. *Finite model theory.* Springer Verlag, 1995.
30. A. Ehrenfeucht and J. Mycielski. Positional strategies for mean payoff games. *Int. J. of Game Theory*, 8:109–113, 1979.
31. E. A. Emerson. Automata, tableaux, and temporal logic. In R. Parikh, editor, *Logics of Programs*, pages 79–88. Lect. Notes Comput. Sci. 193, 1985.
32. E. A. Emerson and C. S. Jutla. The complexity of tree automata and logics of programs. In *Proceedings 29th Annual IEEE Symp. on Foundations of Comput. Sci.*, 1988.
33. E. A. Emerson and C. S. Jutla. Tree automata, mu-calculus and determinacy. In *Proceedings 32th Annual IEEE Symp. on Foundations of Comput. Sci.*, pages 368–377. IEEE Computer Society Press, 1991.
34. E. A. Emerson, C. S. Jutla, and A. P. Sistla. On model-checking for fragments of the μ-calculus. In C. Courcoubetis, editor, *Computer Aided Verification*, pages 385–396. Lect. Notes Comput. Sci. 697, 1993.
35. E. A. Emerson and C-L. Lei. Efficient model checking in fragments of the propositional μ-calculus. In *Symp. on Logic in Comput. Sci.*, pages 267–278, 1986.
36. E.A. Emerson and E.M. Clarke. Characterizing correctness properties of parallel programs using fixpoints. In *ICALP'80*, pages 169–181. Lect. Notes Comput. Sci. 85, 1980.
37. M. J. Fischer and R. E. Ladner. Propositional dynamic logic of regular programs. *J. Computer System Sci.*, 18:194–211, 1979.
38. L. Flon and N. Suzuki. Consistent and complete proof rules for the total correctness of parallel programs. In *Proc. 19th IEEE Symp. on Foundations of Computer Science*, 1978.
39. G. Grätzer. *Universal algebra.* Van Nostrand, 1968.
40. Y. Gurevich and L. Harrington. Trees, automata and games. In *Proc. 14th ACM Symp. on the Theory of Computing*, pages 60–65, 1982.

41. D. Harel, D. Kozen, and J. Tiuryn. *Dynamic Logic*. MIT Press, Cambridge, MA, 2000.
42. N. Immerman. Relational queries computable in polynomial time. In *Proc. 14th ACM Symp. on Theory of Computation*, pages 147–152, 1982.
43. N. Immerman. *Descriptive complexity*. Springer Verlag, 1998.
44. D. Janin. Automata, tableaus and a reduction theorem for fixpoint calculi in arbitrary complete lattices. In *Proc. 12th IEEE Symp. on Logic in Comput. Sci.*, pages 172–182, 1997.
45. D. Janin and I. Walukiewicz. Automata for the μ–calculus and related results. In *Proc. MFCS'95*, volume 969, pages 552–562. Lect. Notes Comput. Sci., 1995.
46. J.Hopcroft and J. Ullman. *Introduction to Automata Theory, Languages, and Computation*. Addison–Wesley, 1979.
47. B. Jónnson and A. Tarski. Boolean algebras with operators. part i. *Amer. J. Math.*, 73:891–939, 1951.
48. B. Jónnson and A. Tarski. Boolean algebras with operators. part ii. *Amer. J. Math.*, 74:127–167, 1952.
49. M. Jurdziński. Deciding the winner in parity games is in UP $\cap$ co-UP. *Information Processing Letters*, 68:119–124, 1998.
50. M. Jurdziński. Small progress measures for solving parity games. In *Proc. 17th Symp. Theoretical Aspects of Computer Science*, volume 1770, pages 290–301. Lect. Notes Comput. Sci., 2000.
51. R. Kaivola. On modal mu-calculus and Büchi tree automata. *Information Processing Letters*, 54:17–22, 1995.
52. N. Klarlund. Progress measures, immediate determinacy, and a subset construction of tree automata. *Ann. Pure Appl. Logic*, 69:243–268, 1994.
53. S. C. Kleene. Representation of events in nerve nets and finite automata. In *Automata Studies*, volume 3-4. Princeton Univ. Press, 1956.
54. B. Knaster. Un théorème sur les fonctions des ensembles. *Ann. Soc. Polon. Math.*, 6:133–134, 1928.
55. D. Kozen. Results on the propositional μ-calculus. *Theoretical Computer Science*, 27:333–354, 1983.
56. D. Kozen and R. J. Parikh. A decision procedure for the propositional mu–calculus. In *2nd Workshop on Logics of Programs*, volume 164, pages 313–325. Lect. Notes Comput. Sci., 1983.
57. G. Lenzi. A hierarchy theorem for the mu-calculus. In F. Meyer auf der Heide and B. Monien, editors, *Proc. ICALP '96*, pages 87–109. Lect. Notes Comput. Sci. 1099, 1996.
58. R. S. Lubarsky. μ–definable sets of integers. *J. Symbolic Logic*, 58:291–313, 1993.
59. A. Mader. *Verification of modal properties using boolean equation systems*. PhD thesis, Fakultät Informatik, Technische Universität München, 1997.
60. D. McAllester, R. Givan, C. Witty, and D. Kozen. Tarskian set constraints. In *Proc. 11th IEEE Symp. on Logic in Comput. Sci.*, pages 138–147, 1996.
61. R. McNaughton. Testing and generating infinite sequences by a finite automaton. *Information and Control*, 9:521–530, 1966.
62. R. McNaughton. Infinite games played on finite graphs. *Annals of Pure and Applied Logic*, 65:149–184, 1993.
63. R. Milner. *Communication and concurrency*. Prentice-Hall, 1989.
64. Y. Moschovakis. *Elementary induction on abstract structures*. North Holland, 1974.

65. A. W. Mostowski. Regular expressions for infinite trees and a standard form of automata. In *Computation Theory*, pages 157–168. Lect. Notes Comput. Sci. 208, 1985.
66. A. W. Mostowski. Games with forbidden positions. Technical Report Technical Report 78, Instytut Matematyki, University of Gdansk, 1991.
67. A. W. Mostowski. Hierarchies of weak automata and weak monadic formulas. *Theoretical Computer Science*, 83:323–335, 1991.
68. A. A. Muchnik. Games on infinite trees and automata with dead-ends: a new proof of the decidability of the monadic theory of two successors (in russian). *Semiotics and Information*, 24:17–40, 1984.
69. A. A. Muchnik. Games on infinite trees and automata with dead-ends: a new proof for the decidability of the monadic second order theory of two successors. *Bull. EATCS*, 48, 1992.
70. D. E. Muller. Infinite sequences and finite machines. In *Proc. 4th Annual IEEE Symposium on Switching Circuit Theory and Logical Design*, pages 3–16, 1963.
71. D. E. Muller and P. E. Schupp. Alternating automata on infinite trees. *Theoretical Computer Science*, 54:267–276, 1987.
72. D. E. Muller and P. E. Schupp. Simulating alternating tree automata by nondeterministic automata: New results and new proofs of the theorems of Rabin, McNaughton and Safra. *Theoretical Computer Science*, 141:69–107, 1995.
73. J. Mycielski. Games with perfect information. In R. J. Aumann and S. Hart, editors, *Handbook of Game Theory with Economic Applications*, volume 1, pages 41–70. North-Holland, 1992.
74. D. Niwiński. Equational μ-calculus. In A. Skowron, editor, *Computaion theory*, pages 169–176. Lect. Notes Comput. Sci. 208, 1985.
75. D. Niwiński. On fixed point clones. In L. Kott, editor, *Proc. 13th ICALP*, pages 464–473. Lect. Notes Comput. Sci. 226, 1986.
76. D. Niwiński. *Hierarchy of objects definable in the fixed point calculus.* PhD thesis, University of Warsaw, 1987.
77. D. Niwiński. Fixed points vs. infinite generation. In *Proc. 3rd IEEE Symp. on Logic in Comput. Sci.*, pages 402–409, 1988.
78. D. Niwiński. Fixed points characterization of infinite behaviour of finite state systems. *Theoretical Computer Science*, 189:1–69, 1997.
79. Ch. Papadimitriou. *Computational complexity.* Addison–Wesley Publishing Company, 1994.
80. D. M. R. Park. On the semantics of fair parallelism. In *Abstract Software Specification*, pages 504–526. Lect. Notes Comput. Sci. 86, 1980.
81. D. M. R. Park. Concurrency and automata on infinite sequences. In *5th G.I. Conf. on Theoret. Comput. Sci.*, pages 167–183. Lect. Notes Comput. Sci. 104, 1981.
82. V. Pratt. A decidable μ-calculus. In *Proc. 22nd IEEE Symp. on Foundations of Computer Science*, pages 421–427, 1981.
83. M. O. Rabin. Decidability of second-order theories and automata on infinite trees. *Trans. Amer. Soc*, 141:1–35, 1969.
84. M. O. Rabin. Automata on infinite objects and Church's problem. *American Math. Soc., CBMS Lecture Series*, 13:1–22, 1972.
85. M. O. Rabin. Weakly definable relations and special automata. In Y. Bar-Hillel, editor, *Mathematical Logic and Foundation of Set Theory*, pages 1–23. North–Holland, Amsterdam, 1990.
86. S. Safra. On the complexity of ω-automata. In *Proc.29th Symp. Foundations of Computer Science*, pages 319–327, 1988.

87. D. Scott. Lattice of flow diagrams. In *Symposium on Semantics of Algorithmic Languages*, volume 188 of *Lecture Notes in Mathematics*, 1971.
88. D. S. Scott and J. W. de Bakker. *A theory of programs.* IBM Vienna, 1969.
89. H. Seidl. Fast and simple nested fixpoints. *Information Processing Letters*, 59:303–308, 1996.
90. H. Seidl and D. Niwiński. On distributive fixed-point expressions. *RAIRO–Theoretical Informatics and Applications*, 33:427–446, 1999.
91. R. S. Streett and E. A. Emerson. An automaton theoretic decision procedure for the propositional μ-calculus. *Information and Computation*, 81:249–264, 1989.
92. M. Takahashi. The greatest fixed points and rational omega–tree languages. *Theoretical Computer Science*, 44:259–274, 1986.
93. A. Tarski. A lattice theoretical fixpoint theorem and its applications. *Ann. Soc. Polon. Math.*, 5:285–309, 1955.
94. J. W. Thatcher and J. B. Wright. Generalized finite automata with an application to a decision problem of second-order logic. *Math. Systems Theory*, 2:57–82, 1968.
95. W. Thomas. Automata on infinite objects. In J. Van Leeuwen, editor, *Handbook of Theoretical Computer Science (vol. B)*, volume B, pages 133–191. Elsevier Science Pub., 1990.
96. W. Thomas. On the synthesis of strategies in infinite games. In E. W. Mayr and C. Puech, editors, *STACS'95*, pages 1–13. Lect. Notes Comput. Sci. 900, 1995.
97. W. Thomas. Languages, automata, and logic. In G. Rozenberg and A. Salomaa, editors, *Handbook of Formal Languages*, volume 3, pages 389–455. Springer-Verlag, 1997.
98. M. Y. Vardi. Complexity of relational query languages. In *Proc. 14th ACM Symp. on Theory of Computation*, pages 137–146, 1982.
99. B. Vergauwen and J. Lewi. A linear algorithm for solving fixed-point equations on transition systems. In J.-C. Raoult, editor, *CAAP'92*, pages 321–341. Lect. Notes Comput. Sci. 581, 1992.
100. K. Wagner. Eine topologische Charakterisierung einiger Klassen regulärer Folgenmengen. *J. Inf. Process. Cybern. EIK*, 13:473–487, 1977.
101. K. Wagner. On ω–regular sets. *Inform. and Contr.*, 43:123–177, 1979.
102. I. Walukiewicz. Monadic second-order logic on tree-like structures. In C. Puech and R. Reischuk, editors, *Proc. STACS '96*, pages 401–414. Lect. Notes Comput. Sci. 1046, 1996.
103. I. Walukiewicz. Completeness of Kozen's axiomatisation of the propositional μ–calculus. *Information and Computation*, 157:142–182, 2000.
104. J. Wei. Correctness of fixpoint transformations. *Theoretical Computer Science*, 129:123–142, 1994.
105. W. Zielonka. Infinite games on finitely coloured graphs with applications to automata on infinite trees. *Theoretical Computer Science*, 200:135–183, 1998.

Index

www.ingramcontent.com/pod-product-compliance
Lightning Source LLC
Chambersburg PA
CBHW050128240326
41458CB00124B/1646
9780444506207